D0161521

Groundbreaking Scientific Experiments, Inventions, and Discoveries of the Ancient World

Recent Titles in
Groundbreaking Scientific Experiments, Inventions and Discoveries through the Ages

Groundbreaking Scientific Experiments, Inventions and Discoveries of the 17th Century
Michael Windelspecht

Groundbreaking Scientific Experiments; Inventions and Discoveries of the 18th Century
Jonathan Schectman

Groundbreaking Scientific Experiments, Inventions and Discoveries of the 19th Century
Michael Windelspecht

GROUNDBREAKING SCIENTIFIC EXPERIMENTS, INVENTIONS, AND DISCOVERIES OF THE ANCIENT WORLD

ROBERT E. KREBS AND CAROLYN A. KREBS

Groundbreaking Scientific Experiments, Inventions and Discoveries through the Ages
Robert E. Krebs, Series Adviser

GREENWOOD PRESS
Westport, Connecticut • London

Library of Congress Cataloging-in-Publication Data

Krebs, Robert E., 1922-
Groundbreaking scientific experiments, inventions and discoveries of the ancient world / Robert E. Krebs and Carolyn A. Krebs.
p. cm.—(Groundbreaking scientific experiments, inventions and discoveries through the ages)
Includes bibliographical references and index.
ISBN 0–313–31342–3 (alk. paper)
1. Science, Ancient. 2. Technology–History. I. Krebs, Carolyn A. II. Title. III. Series.
Q124.95.K74 2003
509.3—dc21 2003045530

British Library Cataloguing in Publication Data is available.

Library of Congress Catalog Card Number: 2003045530
ISBN: 0–313–31342–3

First published in 2003

Greenwood Press, 88 Post Road West, Westport, CT 06881
An imprint of Greenwood Publishing Group, Inc.
www.greenwood.com

Printed in the United States of America

The paper used in this book complies with the Permanent Paper Standard issued by the National Information Standards Organization (Z39.48–1984).

10 9 8 7 6 5 4 3 2 1

For Evan, Kira, and Heather, who are the future...

Contents

ILLUSTRATIONS

Figures

Table

SERIES FOREWORD

The material contained in five volumes in this series of historical groundbreaking experiments, discoveries, and inventions encompasses many centuries from the pre-historic period up to the twentieth century. Topics are explored from the time of pre-historic humans, the age of classical Greek and Roman science, the Christian era, the Middle Ages, the Renaissance period from the years 1350 to 1600, the beginnings of modern science of the 17th century, and great inventions, discoveries, and experiments of the 18th and 19th centuries. This historical approach to science by Greenwood Press is intended to provide students with the materials needed to examine science as a specialized discipline. The authors present the topics for each historical period alphabetically and include information about the women and men responsible for specific experiments, discoveries, and inventions.

All volumes concentrate on the physical and life sciences and follow the same historical format that describes the scientific developments of that period. In addition to the science of each historical period, the authors explore the implications of how historical groundbreaking experiments, discoveries, and inventions influenced the thoughts and theories of future scientists, and how these developments affected people's lives.

As readers progress through the volumes, it will become obvious that the nature of science is cumulative. In other words, scientists of one historical period draw upon and add to the ideas and theories of earlier periods. This is evident in contrast to the recent irrationalist philosophy of the history and sociology of science that views science, not as a unique, self-correcting human empirical inductive activity, but as just another social or cultural activity where scientific knowledge is conjectural, scientific laws are contrived, scientific theories are all false, scientific facts are fickle, and scientific truths are relative. These volumes belie postmodern deconstructionist assertions that no scientific idea has greater validity than any other idea, and that all "truths" are a matter of opinion.

For example, in 1992 the plurality opinion by three jurists of the U.S. Supreme Court in *Planned Parenthood v. Case* restated the "right" to abortion by stating: *"at the heart of liberty is the right to define one's own concept of existence, of meaning of the universe, and of the mystery of human life."* This is a remarkable deconstructionist statement, not because it supports the right to abortion, but because the Court supports the relativistic premise that anyone's concept of the universe is whatever that person wants it to be, and not what the universe actually is based on: what science has determined by experimentation, the use of statistical probabilities, and empirical inductive logic.

When scientists develop factual knowledge as to the nature of nature they understand that "rational assurance is not the same thing as perfect certainty." By applying statistical probability to new factual data this knowledge provides the basis for building scientific hypotheses, theories, and laws over time. Thus, scientific knowledge becomes self-correcting as well as cumulative.

In addition, this series refutes the claim that each historical theory is based on a false paradigm (a methodological framework) that is discarded and later is just superseded by a new more recent theory also based on a false paradigm. Scientific knowledge is of a sequential nature that revises, adds to, and builds upon old ideas and theories as new theories are developed based on new knowledge.

Astronomy is a prime example of how science progressed over the centuries. Lives of people who lived in the pre-historical period were geared to the movement of the sun, moon, and stars. Cultures in all countries developed many rituals based on observations of how nature affected the flow of life, including the female menstrual cycle, their migrations to follow food supplies, or adaptations to survive harsh winters. Later, after the discovery of agriculture at about 8000 or 9000 B.C.E., people learned to relate climate and weather, the phases of the moon, and the periodicity of the sun's apparent motion to the Earth as these astronomical phenomena seemed to determine the fate of their crops.

The invention of bronze by alloying first arsenic and later tin with copper occurred in about 3000 B.C.E. Much later, after discovering how to use the iron found in celestial meteorites and still later, in 1000 B.C.E. when people learned how to smelt iron from its ore, civilization entered the Iron Age. The people in the Tigris-Euphrates region invented the first calendar based on the phases of the moon and seasons in about 2800 B.C.E. During the ancient and classical Greek and Roman periods (about 700 B.C.E. to A.D. 100) mythical gods were devised to explain what was viewed in the heavens or to justify their behavior. Myths based on astronomy, such as the sun and planet gods as well as Gaia the Earth mother, were part of their religions affecting their way of life. This period was the beginning of the philosophical thoughts of Aristotle and others concerning astronomy and nature in general that predated modern science. In about 235 B.C.E. the Greeks first proposed a heliocentric relationship of the sun and planets. Ancient people in Asia, Egypt, and India invented fantastic structures to assist the unaided eye in viewing the positions and motions of the moon, stars, and sun. These instruments were the forerunners of the invention of modern telescopes and other devices that made modern astronomical discoveries possible. Ancient astrology was based on the belief that the positions of bodies in the heavens controlled one's life. Astrology is still confused with the science of astronomy, and it still is not based on any reliable astronomical data.

The ancients knew that a dewdrop of water on a leaf seemed to magnify the leaf's surface. This led to invention of a glass bead that could be used as a magnifying glass. In 1590 Zacharias Janssen, a spectacle-maker, discovered that two convex lenses, one at each end of a tube, increased the magnification. In 1608 Hans Lippershey's assistant turned the instrument around and discovered that distant objects appeared closer, thus the telescope was discovered. The telescope has been used for both navigation and astronomical observations from the 17th century up to the present time. The inventions of new instruments, such as the microscope and telescope, led to new discoveries such as the cell by Robert Hooke and the four moons of Jupiter by Galileo, who made this important discovery that revolutionized astronomy with a telescope of his own design and construction. These inventions and discoveries enabled the expansion of astronomy from an ancient "eyeball" science to an ever-expanding series of experiments and discoveries leading to many new theories about the universe. Others invented and improved astronomical instruments, such as the reflecting telescope combined with photography, the spectroscope, and Earth-orbiting astronomical instruments resulting in the discovery of new planets, galaxies, and new theories related to astronomy and the universe in the 20th century. The age of "enlightenment" through the 18th and 19th centuries culminated in an explosion of new knowledge of the universe that continued through the 20th and into the 21st centuries. Scientific laws, theories, and facts we now know about astronomy and the universe are grounded in the experiments, discoveries, and inventions of the past centuries, just as they are in all areas of science.

The books in the series Groundbreaking Scientific Experiments, Inventions and Discoveries through the Ages are written in easy to understand language with a minimum of scientific jargon. They are appropriate references for middle and senior high school audiences, as well as for the college level nonscience major, and for the general public interested in the development and progression of science over the ages.

Robert E. Krebs
University of Illinois at Chicago

INTRODUCTION

According to the big-bang theory, the universe was formed about 14 to 15 billion years ago. The solar system and planets, including Earth, were formed about 4.5 billion years ago. And the first primitive organic matter and archaebacteria appeared on Earth about 3.5 to 4 billion years ago. During the Mesozoic Era, about 150 million to 250 million years ago, small primitive mammals existed along with the more dominant dinosaurs. Once the dinosaurs became extinct, around the beginning of the Cenozoic Era 65 million years ago, mammals, particularly large mammals, flourished. It is estimated that about 200,000 species of mammals existed in the distant past, and about 6,500 species of protoprimates are presumed to have existed at one time. Today, biologists recognize only about 200 species of living primates. Thus, only about 3.8 percent of all the primate species that lived throughout the Cenozoic Era (65 million years ago to the present) still exist.

Our early ancestors' entry onto the scene came as late as 4 to 5 million years ago when a small African hominid emerged, now referred to as *Australopithecus*. It was not until about 2 million years ago that the *Homo* genus evolved as the more advanced *Homo habilis*. Evolution then produced the even more advanced *Homo erectus* about 1.8 million years ago. *Homo sapiens*, the branch of hominids to which humans belong, emerged about 250,000 to 400,000 years ago. However, the subspecies *Homo sapiens sapiens*, the hominids that most closely resemble modern humans, appeared only about 120,000 years ago. The *Cro-Magnon* culture that closely resembled modern-day humans is only about 40,000 years old. *Homo sapiens* (modern man) is the only species of the *Homo* genus still in existence.

This raises the question, "What is ancient?" Ancient means "old" or "archaic." These are subjective terms indicating that *ancient* can actually mean anytime from the beginning of the universe to the appearance of mammals on the Earth, to the comparatively recent days of the horse and buggy, or even to last year's fashions. Some children may think of their parents as "ancient" and/or their grandparents as very "ancient." Anthropologists may consider events prior to recorded history in Mesopotamia or Egypt as ancient. For our purposes, "ancient" means when humans first used fire, developed tools and weapons, or discovered how to control and alter their environment for their own purposes and needs (e.g., through agriculture). Early humans no doubt "invented" useful objects by applying learned technologies. They discovered many things about their environment, and conducted early experiments by trial-and-error methods. Modern humans, by the way, are still discovering things about their universe, inventing objects for their use to control their environment, and experimenting by using some of the same methods employed by our ancient ancestors.

Ancient humans also further developed the ability to categorize and list things that could be useful and/or harmful. This required a degree of abstracting, generalizing, stereotyping, predicting, and relating cause with effect beyond that which other animals were capable of. Also involved was the development of a larger brain, an advanced level of intelligence, and consciousness beyond their primate ancestors, which most likely developed in conjunction with the improved flexible forearm and hand with opposable thumb. There is evidence that primates other than ancient humans also developed some simple tools and weapons for securing food and protection from predators. But early humans, with their larger brains, advanced intelligence, and opposable thumbs, developed the technology for producing improved stone tools and weapons. The "ancient" period addressed in this book includes the span of history of early discoveries, experiments, and inventions from the Stone Age, the Bronze Age, and the Iron Age, through early civilizations in Africa, Asia, the Middle East, Europe, and the Americas to the ancient and classical Greek and Roman civilizations. The "ancient" period for this book also includes the early Christian Era up to about 476 C.E., which is the accepted date for the fall of the Roman Empire. (Note: C.E. represents the "Common Era," which replaces A.D.; B.C.E. stands for "Before the Common Era" and replaces B.C.; and B.P. refers to "Before the Present.")

Many ancient discoveries provided the background for additional discoveries, inventions, and experiments during the later Middle Ages and Renaissance. These in turn provided the background for the advancement of science and technology during the Age of Enlightenment, and even now ancient concepts affect the nature of our current social, economic, and political lives. For instance, prehistoric domestication of plants and animals and the ancient practice of selecting the "best" seeds to store and plant, or the "best" animals to breed, led to the nature of inherited characteristics discovered by Gregor Mendel, the concept of organic evolution by Darwin, the nature of the **DNA** double helix by Watson and Crick, and, more recently, the nature of the entire human genome, which has great promise in potential cures for many human illnesses. (Words in bold type can be found in the Glossary.) Thus, the ancient knowledge of breeding plants and animals is the forerunner of modern genetic engineering. Another example is the ancient Greek concept of the indestructible and indivisible "atom" devised by Leucippus and Democritus, which has been contemplated, revised, and reconfirmed by scientists through experiments that led to discoveries concerning the nature of matter. New experiments are providing information regarding the structure of the atom and its ultimate parts. By inventing new instruments and conducting new experiments, theoretical particle physicists expect to discover the "ultimate" particle of matter and sooner or later arrive at the "Grand Unification Theory" (GUT) or the "Theory of Everything" (TOE) for the universe. Archaeologists have traced the progressive inventions of ancient humans in developing stone, and later metal, tools and weapons from simple handheld sharpened stones, to attaching handles to stones and metal blades, to using pointed spears and arrows as missiles. These early tool and weapon inventions not only improved over time, but also affected and changed our lives over many centuries. Ultimately, the ancient concept of the atom and ancient missiles (spears) were combined (nuclear weapons carried by ICBMs), altering the nature of society and providing the possibility of ending civilization as we know it. These and many other ancient discoveries, inventions, and experiments are presented in this book.

This volume on the Ancient World for the series Groundbreaking Scientific Experiments, Inventions and Discoveries through the Ages gives credence to the concept that science is not always clean and precise, especially in its formative stages. This reference book is designed for high school and undergraduate students, as well as the general public, who are interested in learning more about the foundations of science and technology that our early ancestors built and that succeeding generations added to and improved.

1

Agriculture and Animal Domestication

Background and History (Agriculture)

Given the timeline for the evolution of our species that is described in this book's introduction, it is extraordinary that all of the agricultural, technological, and cultural advancements that have so enriched our lives have taken place since the Neolithic Period that began around 10,000 B.C.E., following the retreat of the last ice age. Fossils and artifacts from periods prior to the Neolithic Period indicate that archaic humans, even those who more closely resembled modern-day humans, fashioned stones to use as knives and scrapers. This was a hunting and gathering society that used animals for nourishment, clothing, and in some cases shelter, and relied on the flora of the period, such as berries and nuts, as a supplement to their diet and for medicinal uses. Scholars believe that the development of modern humans progressed in four stages: (1) hunting/gathering; (2) nomadic, with some domestication of animals; (3) farming; and finally, (4) civilization.

Agriculture, sometimes referred to as settled farming, is defined as the deliberate clearing and cultivating of arable soil, planting and covering seeds, irrigating the fields, weeding unwanted or harmful vegetation, harvesting the resultant crops, and finally storing those crops for future use. Its inception has had more of an impact on our environment and culture than any other institution. Historians have not determined an exact date for the beginning of agriculture or farming. Most agree that farming began in several places and in different forms. For example, hunters and gatherers during the Upper Paleolithic Period of 40,000 years ago engaged in hobby or part-time farming. It was not a full-time endeavor for them. Early humans followed herds of game, primarily mammoths and mastodons, across the plains that opened up when glaciers receded. They then settled into temporary communities during the hunt. Using a wooden stick or perhaps an antler from a slain elk or deer, early humans scratched at the soil and "planted" the seeds from plants native to that area and then ate the harvest. It may have been more serendipitous than planned, but it provided them with a source of food in addition to what they could gather or kill. (Humans are one of the few omnivorous mammals. Both plant and animal proteins are staples of our diets, yet humans are able to survive on one without the other.) These communities were short-lived as early humans, who were basically nomads, traveled on to the next hunting site.

The same was true for early farmers. As the soil was exhausted from overcultivation, humans simply moved on to another location, leaving the remnants and debris of their labors in their wake. At this time in our history, humans did not behave as conservationists of either the land or the animals who populated the continents. For instance, one theory called the *Pleistocene Overkill* suggests that the predation and indiscriminate hunting practices of early humans destroyed vast numbers of species of large mammals. During the climatic and geographic cataclysm of the last ice age, humans migrated from the African continent, where hominids originated, through Eurasia and over the Beringia land bridge between Siberia and Alaska into the Americas. As humans learned to build and sail boats, the Australian continent, the Pacific Islands, and South America also became populated through migrations. Archaeological evidence confirms that the first settlers on New Guinea and Australia reached these areas by boat approximately 40,000 B.C.E. The movement of the human population from one continent to another took place several times over centuries. For example, most experts agree that humans were in the Americas at least 11,000 years ago, but there is evidence to suggest that the Americas may have been populated by early humans as early as 21,000 B.C.E.

The retreat of the glaciers left the landscape of the Earth with vast, grassy plains, fertile river valleys, rivers and lakes with stocks of fish and aquatic life, and woodlands populated with trees and plants and abundant communities of wildlife. Historic evidence suggests that the earliest forms of rice and millet were cultivated in southeast Asia about 10,000 B.C.E.—roughly 12,000 years B.P. (Before the Present). Alfalfa, the oldest **forage** crop, reportedly was raised at about the same time in what is present-day Syria. Thus, many experts place this as the time when agriculture became viable. Others hypothesize that intensive agriculture began in the Middle East between 9000 and 7000 B.C.E.. All agree, however, that life was essentially short and brutal during this period. Characteristic of our species, early humans adapted in different ways to a particular environment, utilizing the plant and animal resources concentrated in that area. That adaptation did not mean that life was easier for them than for their nomadic counterparts. On the contrary, in many respects life was harder for early farmers. Dependent on the variables of climate and limited by what could actually be grown, farmers had poorer diets and suffered from more diseases while working harder. They also had to contend with pests who ate and damaged the growing crops. Yet, this somewhat more stable lifestyle was conducive to producing more offspring, thus their numbers grew rapidly. Larger populations of farmers could support more than merely those who worked the crops. However, their primitive methods of farming and animal management exhausted the land in just a few seasons, which forced these early settlers to move on to new territories, usually usurped from the nomadic hunters. This great antagonism between farmers and hunters was the impetus that eventually resulted in the replacement of the hunting/gathering lifestyle with that of the cultivators of the land. (The conflict of interest between farmers and cattle raisers continued into modern times in the western portion of the United States.) This change in lifestyle revolutionized the human condition and laid the foundation for modern civilization including commerce, industry, the arts, and science. Durable houses would be constructed and various tools and implements would be developed to ease the labor of the farmer and increase the production of crops. Crops would be grown not only to feed the farmers but also for the members of other communities, to maintain animal populations for human consumption and labor, to supply building materials and writing materials—in other words, everything needed by humans to supply their ever-expanding civilization and culture. Alfalfa is an example of a crop whose uses were manifold. It is an excellent feed for animals (as well as humans). It nourishes the soil and augments crop production, while possessing medicinal benefits as well.

Farming certainly spread with the migration of people across the continents, either on foot or by boat across the oceans, since our ancestors of some 30,000 years ago were first-rate sailors. Knowledge of agricultural practices, primitive and more advanced, often developed independently without the transfer of knowledge from one society of humans to another. This is believed to be the case in the Americas, for example. In ancient times crops were indigenous to a particular region or continent. However, when humans began to explore and conquer other lands in later centuries, nonindigenous plants were then introduced and grown with varying degrees of success. The following sections describe, by region, the expansion of farming, the conditions that existed, and the crops grown in that area.

The Levant, Anatolia, and Mesopotamia

Notwithstanding the belief that agriculture took hold in different regions with different varieties of cereals or grains, perhaps even simultaneously, fossil and artifact evidence strongly suggests that Earth's first farmers lived in the Near and Middle East. The region known as *the Levant* includes countries of the eastern shore of the Mediterranean Sea, namely Cypress, Egypt, Israel, Lebanon, Syria, and Turkey. *Anatolia,* or Asia Minor, is present-day Turkey. This region is different today than it was centuries ago when the climate was at its best. It was a much more fertile and lush region, and the growth and abundance of cereal grasses, the antecedents of later crops, can be traced back the farthest to these lands. As early as 9500 B.C.E. there is evidence that early humans gathered or harvested, but did not cultivate, wild grasses in Anatolia.

The Middle East region where farming was particularly successful is known as the "Fertile Crescent." This region comprises the lands that run north from Egypt through the former Palestine (now parts of Israel, Jordan, and Egypt) and the countries of the Levant through Anatolia to the hills between Iran and the south Caspian Sea to the fertile Tigris and Euphrates river valleys of *Mesopotamia* (present-day Syria and Iraq). Cereals and legumes, alfalfa, wheat, barley, peas, bitter vetch (a climbing plant), chickpeas, and fava beans were grown as early as 8000 B.C.E. Varieties of citrus plants were cultivated about 4800 B.C.E., and figs, grapes, pomegranates, and dates about 2500 B.C.E. However, according to an article that appeared in *American Scientist* (May–June 2001), alfalfa is believed to be the oldest forage crop. Evidence for the small seeded legumes (alfalfa) dating back to 10,000 B.C.E. has been found in present-day Syria.

Jericho, the ancient city of Palestine near the northwest shore of the Dead Sea, was, and still is, the site of a spring-fed oasis and the place where thousands of people have settled since 8000 B.C.E. (It is now near the present-day town of Jericho in the West Bank.) As early as 6000 B.C.E. the population of Jericho constructed water tanks and a massive stone tower, which is suggestive of both irrigation and a defense system. At Çatal Hüyük in Turkey, the largest Neolithic site in the Near East, inhabitants constructed a building of brick around 6000 B.C.E. as the rudiments of civilization developed. At the same time, agriculture was becoming a permanent and profitable economic institution.

Mesopotamia, from the Greek meaning "between the rivers," is often called the "Cradle of Civilization" and in many ways predates Greece as the fountain of culture. The soil in this region was particularly ideal for farming. The Tigris and Euphrates rivers flooded each spring, and the clay and silt that contained abundant nutrients were washed downstream and deposited in the southernmost part of the Mesopotamian region, which was known as Sumer. Sumerians developed the first recognizable civilization. They were an inventive people who in all likelihood created a trading system, primarily based on and sustained by their agricultural output, that established an interdependence among the countries of the Near East and, to some degree, India. The crops that were grown in great abundance in this region were sesame, millet, wheat, and barley, thought to have been their main crop. Mesopotamia was also home to a number of other tribes, including the Akkadians, Hittites, Assyrians, and Babylonians.

Egypt

Egypt is a country that is most defined by the Nile River. Before emptying into the Mediterranean Sea, the Nile River flows north through 6,400 kilometers (4,000 miles) of African territory. Sometime around 8000 B.C.E., Egyptian settlements and their civilization were established along the banks of this great river. Centuries ago, the Nile overflowed its banks each year when heavy rains fell at the river's watershed, causing the water level of the river to rise. (Obviously, this was before the Egyptians constructed dams.) This had a dual impact. First, the floodwaters irrigated the crops; second, they enriched the soil with eroded silt from the river. Early Egyptian farmers waited with great anticipation for the Nile to flood each year. In about 5000 B.C.E. irrigation systems were developed in Egypt that further increased the agricultural output.

The fertile soil along the banks of the Nile was ideal for growing grains, such as wheat and barley. In 1992 a team of archaeologists from several American universities gathered evidence that sorghum and millet, which are staple grains for humans worldwide, were introduced in the southern part of Egypt about 8,000 years ago. This was about 1,600 years before that particular region of Egypt learned of the domestication of wheat in the Near East. They concluded that this feat demonstrated that the **domestication** of sorghum and millet in Egypt took place independently of the domestication of wheat. Sorghum resembles a cross between a large wheat plant and an earless form of corn (maize). Millet is similar to a cattail. Both plants can be grown in hot and arid regions with a short growing season. About forty other crops were also grown in southern Egypt, including legumes, mustard, and an assortment of fruits, nuts, and tuber plants.

The Nile Delta, believed to contain the best farmland, is located in the northern part of Egypt. However, archaeologists believe that the Nile Delta did not exist before 6000 B.C.E. At about that time, the levels of the Mediterranean Sea changed rapidly, and along with eroded soil, the present delta was formed over the course of 1,000 years. Farming on this exceptionally fertile ground began about 5000 B.C.E. The Egyptians also cultivated cotton as early as 3500 B.C.E. and grew papyrus, used in the making of writing material, about 2000 B.C.E. Wild honey also was abundant and used in great quantities by ancient peoples as a sweetening agent. Though technically not an agricultural crop, raising and caring for bees was well established in Egypt by about 2500 B.C.E.

Indus Valley and the Indian Subcontinent

It is believed that agriculture was established in India at a later period, most likely 5,000 years ago (B.P.). However, there is some evidence that animal domestication in the Indus Valley and India itself was prevalent earlier than that, probably around 3700 B.C.E. The Indus is a river in south central Asia originating in the northern province of Tibet. It flows approximately 3,057 kilometers (1,900 miles) northwest through present-day Kashmir and southwest through present-day Pakistan into the Arabian Sea. By 3000 B.C.E. there were settlements and civilization on the alluvial plains of the Indus Valley that some historians believe were influenced and shaped by the migration of new peoples from the north. The city of Harappa, located on a tributary of the Indus, was one of the major settlement sites where large granaries were built. The Ganges River valley is another silt-rich area of India where large agricultural populations settled. Rice, which was and is a crucial staple of the Indian diet, was grown in the Ganges Valley. It is believed that rice was imported from China where it was grown with much success about 5,000 years ago.

The cultivation of cotton and the spinning of its fibers began about 2500 B.C.E. in the Indus Valley, while tea, bananas, and apples were cultivated about 500 years later in the same region and elsewhere in India. Sugarcane has been a consistently important part of the Indian diet. It was one of the earliest plants to be domesticated in India when it was first brought there from the island of New Guinea between 6,000 and 7,000 years ago during what is commonly referred to as the New Stone Age. One of the most ancient plants domesticated in India was hemp, a multipurpose plant. Oil is derived from its seeds, fiber from the stalks, and hashish, a narcotic, is obtained from the flowers and leaves of the hemp plant. Wheat, peas, sesame, barley, dates, and mangoes were also grown extensively in the Indus Valley.

Spices, most of which are indigenous to the Indian subcontinent and other regions of the Far East, were not cultivated per se, but were used in great abundance in the diets of ancient civilizations. Spices were also used as preservatives in food as well as in the process of embalming. (Peppercorns were found in the mummified body of Egyptian ruler Ramses III, dating back to the twelfth century B.C.E.) As explorers and navigators made their way eastward to the lands of the Far East and China, spices became coveted and popular commodities for the traders, who introduced them to the western populations of the European continent. Ginger, cinnamon, pepper, cloves, nutmeg, mace, coriander, and cumin were among the most prominent spices exported from this region.

China and the Far East

The huge expanse that is China is larger than the United States in land area and encompasses many varied climates and regions. The northern part of China is extremely hot and arid during the summer months, while the southern part is humid and prone to floods. In winter, the north is barren and dusty while the south is green and fertile. Three great rivers run across the country roughly from west to east. Today, they are called the Hwang-Ho or Yellow River, the Yangtze, and the Hsi. A great part of China is mountainous, except in its southernmost regions. Two prominent and desolate deserts, known today as the Ordos and the Gobi, cover huge expanses of the country. Despite the harsh topography, Chinese agriculture flourished in the river valleys centuries ago when the climate was more moderate and the land was fertile. Interestingly, this same harsh topography isolated China from the other developing cultures and civilizations of the Near East, Greece, and Rome; thus China's earliest perceptions of the world and how it worked were quite different.

Rice was cultivated in the Yangtze Valley over 9,000 years ago, and there is evidence that people cleared forestlands to make fields as early as 8,000 years ago. Millet was grown along the Yellow River 7,000 years ago. There is some dispute as to when soybeans were cultivated. It may have been as early as 4000 B.C.E. or as late as 1100 B.C.E. Barley and wheat were introduced somewhere around 1300 B.C.E., but wheat was more difficult to grow because of the arid nature of the land. Rhubarb, primarily used for medicinal purposes, was grown in China about 2700 B.C.E.

The Chinese also grew hemp for use in the making of paper. As the migration of ancient peoples continued, farming techniques and crop cultivation, especially of rice, spread to the Korean Peninsula, Japan, and other parts of the Far East with great success. (For example, using intensive agricultural techniques, Japan achieves higher yields [per acre] of rice than does China.)

Southeast Asia has supplemented the Indian subcontinent with a number of crops, probably introduced by people who emigrated from India sometime during the first millennium B.C.E. During later migrations, European colonizers introduced other agricultural products, such as sugar, coffee, tea, cotton, rubber, and maize.

Europe

At the end of the last ice age 10,000 years ago, the western and central section of the European plains (present-day western and northern France, Belgium, the Netherlands, the southern Scandinavian countries, northern Germany, and Poland) were marked by hilly moraines, which are boulderlike deposits from the last ice age, and long spillways that acted as drainage areas for the melting glaciers. The Northern European Plain contains the delta plain of the Netherlands that has always been rich and fertile. The East European Plain encompasses all of the present-day Balkans and Belarus, most of Ukraine, a large portion of Russia, and extends into Finland. Historians believe that as the human population increased after the last ice age, areas for hunting became increasingly overcrowded, thus hastening the great migration from the south into the plains of the European continent. These early nomads settled along the coastal areas, rivers, and lakes of the European plains, which were still heavily forested at that time. The earliest evidence of agricultural settlements appeared 8,000 years ago. However, about 5,000 years ago, despite the variances in climate and topography on the continent, farming was widespread in northern Europe, having reached Britain and the Scandinavian countries between 4000 and 2000 B.C.E. In Britain, prior to the Roman invasion, nuts, pulses, and chenopods, such as beets and spinach, were cultivated by its early settlers.

The primary and permanent agricultural settlements existed in areas that were sparsely wooded and had good drainage, until the invention of the plow, when settlements could be established in areas where the soil was hard and claylike. Wheat, barley, rye, legumes, and oats were among the crops that farmers grew on the European plains. The influence of Rome, including farming methods, was widespread in western

Europe during the height of the Roman epoch. However, the annual harvests, half of which were used as seed, were small by today's standards.

From prehistoric times until the Renaissance, there were essentially two "Europes," separated by geography. The "first Europe" consisted of countries on the Mediterranean coast, delineated by their cultivation of the olive tree. The population south of the "olive line" was literate, traveled, and was more urbanized. North of this delineation, the population tended to be backward and resistant to change unless and until it was conquered by outside forces. This happened with some regularity, as the Huns, Visigoths, and other barbarian tribes marauded the European populations. During the period from 146 B.C.E., when Greece fell to the Romans, until 476 C.E., when Rome fell to the Germanic tribes, farming in northern Europe continued and populations increased in Europe. From about 500 C.E. until 1200 C.E. Europe was divided into small fiefdoms or kingdoms, and most of the population was farmers. This era was known as the Dark Ages, since there is little historic evidence of great invention, discoveries, or scholarship. The plague or Black Death that reduced the population of Europe by 60 percent transformed the agrarian economy forever.

Greece and Rome (Etruscan)

The ancient Greeks were the architects of a great civilization on which many of our modern institutions have been built. The disciplines of philosophy, science, and medicine are based on the teachings of Greek noblemen. On the other hand, agriculture in Greece, at least on any large scale, was limited, primarily because of the small amounts of arable land available. The dearth of arable land actually was the reason that the ancient Greeks left their homeland in search of other territories, leading to the development of trade and commerce and the colonization of other lands. Greek farmers cultivated barley, lentils, and a grain called spelt, which is a variety of wheat, as early as 7000 B.C.E. Their farming techniques were dictated by the topography of the land as well as the climate. Other forms of wheat, beans, peas, vetches, chickpeas, alfalfa, millet, olives, turnips, and radishes, as well as a variety of fruit trees, were also grown. Greece is primarily known for its city-states and philosophers, such as Aristotle and Plato, and its poets and bearers of culture, rather than its agrarian achievements.

About 3,000 years ago the country now called Italy (which means "Calf Land") was divided into smaller countries, one of which was Etruria. The residents, called Etruscans, eventually conquered the other populations, who supported themselves by farming and animal husbandry. In time, they built the mighty city of Rome. (Rome was built around 753 B.C.E. It was named after Prince Romulus of Alba Longa, who chose the site after killing his twin brother, Remus.) The Roman civilization, unlike that of Greece, which was more scientific and scholarly, was oriented toward action and usurpation. The Romans used much of the knowledge developed by other civilizations and groups for their own benefit. They are credited, however, with the earliest development of what could be called the fertilizer industry. Roman farms produced enormous yields of wheat and rye. The farmers determined this was because of the manure deposited by the flocks of sheep and herds of cattle, and even horses, that were allowed to graze in these same fields. The nutrients from the manure, primarily nitrogen, enriched the rye and wheat plants, and the result was crops that were healthier and more abundant. Thus, it is reasonable to assume that this was the earliest use of fertilizer for agricultural purposes. Roman farming practices became widespread as Rome invaded and conquered other territories on the European continent and in the British Isles.

North America, Mesoamerica, and South America

Scientists and archaeologists posit that early humans arrived in the Americas (both the North and South American continents and Mesoamerica, which encompasses Mexico and Central America) sometime between 21,000 and 9000 B.C.E. There is no agreement on the exact time. They do agree, however, that they traversed the Beringia land bridge from Asia during the late Pleistocene Era and migrated throughout the North and South American continents. As with all travelers during this archaic period of our history, the Paleoindians, as they are commonly known, were fishermen and hunter/gatherers who followed herds of large animals, among them bison, mastodons, and mammoths, in their quest for food. It is widely believed that their hunting of these larger animals contributed to the extinction of some of these species on the North American continent. Thirteen thousand years ago during the last ice age, the climate could support a hunter/gatherer lifestyle with grasslands populated by these large, grazing animals. A dramatic warming period occurred about 10,000 years ago that caused the glaciers to melt and tropical forests to overtake the grasslands of Mesoamerica; thus there was a shift from a nomadic existence to a more settled one.

Agriculture in the Americas appears to have developed completely independently from that which evolved in the Near and Middle East and on the European continent. There is evidence that the earliest domestication and cultivation of potatoes and some other plants took place in South America as early as 8000 B.C.E. and, certainly by 7000 B.C.E., corn and other crops were widely grown in Mesoamerica. Agriculture appears to have come later to the North American continent, primarily from influences from Mesoamerica and South America. The Mississippi River drainage basin (present-day Illinois) contains archaeological evidence, dating back to 5000 B.C.E., which suggests that early humans inhabited this area during the summer months. The floodplains at this time in our history were covered with plants that had edible seeds, such as chenopods (goosefoot), sumpweed, and sunflowers. By 1000 B.C.E. these plants were deliberately cultivated, as evidenced by the physiological differences between the wild and cultivated varieties.

In the past, it was believed that maize (corn), a widely grown vegetable in Mesoamerica and South America, was not a major crop on the North American continent until sometime around 1200 C.E., and that beans and squash began to be grown in eastern North America at about the same time as corn, primarily because the climate at that time (**Medieval Warm Epoch**) was much warmer than today and the growing season was extended. Rather, a form of squash, called *cucurbito pepo,* was first cultivated by hunter/gatherers in what is now southern Mexico, approximately 5,000 years before the cultivation of corn. Recently, however, the journal *Science* (May 18, 2001) reported on research that suggests that ancient farms in what is the present-day state of Tabasco in Mexico actually cultivated cornlike plants along this coastal plain region in about 4000 B.C.E. Another issue of *Science* (April 27, 2001) details the work of other researchers who believe that maize originated in Mexico and spread southward throughout Mesoamerica and the South American continent. As evidence for this assertion, they point to the discovery of maize cobs, believed to be about 6,250 years old, in the Guila Naquitz Cave in Oaxaca, Mexico.

Mesoamerica is home to many ancient societies and cultures that used elements of one civilization to build on another. The earliest is the Olmec in what is present-day Central America. Four thousand years ago Olmec farmers grew potatoes and manioc (a starchy root vegetable), beans, squash, chili peppers, and cotton, in addition to maize. The more famous Mayan civilization of the Yucatan in Mexico, Guatemala, and Honduras, which began to appear in 100 C.E. and was at its height at about 900 C.E., used primitive agricultural methods to cultivate their crops since they did not possess plows or any other metal implements. They did, however, burn and clear the land to ready it for planting, and then moved on to another site after a few seasons, thus recognizing that the soil had been exhausted and was no longer fertile. The Mayans were successful in maintaining their civilization for centuries despite living in the most tropical parts of the Yucatan where insects, disease, the hot climate, and droughts were a constant challenge to their existence. According to an article appearing in *Science* (May 18, 2001), the collapse of the Mayan civilization around 900 C.E. most likely can be attributed to persistent droughts in this region.

Background and Breeding (Animal Domestication)

The exact time when early humans first domesticated animals is uncertain. However, it is generally believed that the domestication of animals preceded, or at the very least paralleled, the appearance of settled farming. Prehistoric humans primarily hunted the order of animals known as *Ungulata,* now subdivided into two orders called *perissodactyla* and *artiodactyla.* These are hoofed animals, among them deer, sheep, swine, horses, elephants, camels, and rhinoceroses. (Perissodactyls have an odd number of toes on each foot, while artiodactyls have an even number.) As prehistoric humans became increasingly more proficient as hunters of these wild beasts, they perhaps contributed to the extinction or demise of a number of species. Archaeologists, who have the knowledge and expertise to date their findings, can distinguish between the fossilized bones of wild and domesticated animals. Some 500,000 years ago humans cooked animals over stone hearths during their hunter/gatherer period. From the time of the early Neanderthals, the hides of animals were used for clothing and footwear. Bones, as well as horns and antlers, were fashioned into tools and weapons and other artifacts.

The evolution of early humans from a nomadic lifestyle into one of **sedentism** took centuries to accomplish and was driven by the need for an improved and abundant source of food. The dramatic changes in the Earth's climate that occurred at the end of the Pleistocene Epoch, along with the culture of deliberate hunting, and perhaps overhunting, decreased the availability of game animals. Early humans were not vegetarians by any stretch of the imagination. The animal protein diet that sustained them was only supplemented by the nuts and berries they foraged while they followed and hunted the herds of wild animals that would be killed to feed the group. For instance, archaeologists have found evidence in coprolites (fossilized excrement) that the diet of Neanderthals was 85 percent meat.

The word "domestication" is derived from the Latin word *domesticatim,* meaning "at home." In its usage to describe our relationships with animals and plants, domestication essentially means the initial stage of human mastery of wild animals and plants. Humans breed and cultivate animals and plants for specific purposes. In turn, the animals and plants adapt to the care and conditions provided for them. At some point in our history, most likely during the Mesolithic Period some 10,000 years ago, prehistoric farmers first attempted to domesticate animals and plants. There is some evidence to suggest that deer, antelope, and sheep may have been domesticated as early as 18,000 B.C.E. in the Middle East. Scientists, however, generally agree that the domestication of both animals and plants did not become an accepted activity until the Neolithic Period. Scientists also believe that animal domestication began independently in many regions, such as the Middle East, west and east Asia, and Central and South America, all of which are regions whose geography, climate, and biotic environments were conducive to propagating and sustaining large numbers of plant and animal life-forms.

The sedentary farming practices of the first Neolithic farmers gave rise to *pastoralism,* the practice of herding domesticated animals. The hunter/gatherer lifestyle of early humans can be considered as the training period for these early pastoralists. The need to constantly move from one place to another in search of a food source prepared them for the task of moving large groups of animals from one feeding ground to another. As was true for the first agriculturalists, pastoralists were at the mercy of climate and the availability of grasses as well as their own continuing health and that of the livestock. Over time, this demanding lifestyle became more stable, even in dry or semi-arid regions such as those in the Near and Middle East. Eventually, a symbiotic relationship arose between the earliest settled farmers and the pastoral nomads, albeit a rather tenuous one. Farmers had food and other items to exchange or barter with the pastoralists who had hides, wool, meat, and milk. In time, this became less important for the settled farmers who learned not only to work the land for crops, but also to domesticate a number of animals that would provide them with dietary necessities. The pastoralists, on the other hand, did not have the advantages of sedentism in that they could not independently produce crops or material goods. The relationship between these two groups deteriorated as the need for more agricultural land often encroached on and overtook the grazing land needs of the large herds of cattle and sheep. Though somewhat lessened, this tension between the two groups still remains in many parts of the world.

Animals provided a number of vital components necessary for the existence and survival of their human masters, namely a steady food supply, clothing, shelter, protection, labor, hunting, transportation, pest control, waste management, and fertilizer for agricultural plantings. It is hard to imagine how humankind would have progressed and developed to its present level without the domestication of animals who themselves have undergone significant changes that distinguish them from their feral ancestors. Nevertheless, the history of civilization is replete with paradoxes concerning the treatment of animals. For instance, the Egyptians worshiped the cat as a god, while the Romans imported and then slaughtered for sport tens of thousands of exotic animals in their arenas.

Only a small number of species of animals were actually domesticated by early humans. In all likelihood, the humans' choice of which species to "tame" was limited to those animals they could control and, of course, those species indigenous to a particular region. (Nearly 5,000 years ago, the early Egyptians attempted to domesticate a number of species, including the monkey and hyena, but their importance and popularity was seemingly limited to that particular region and civilization.) Also, the concept of animals as beasts of burden was secondary to the need for survival. For hundreds of thousands of years, women were the "burden-bearers" as they carried children, branches, wood, and other life-sustaining materials in one fashion or another while the males in the community hunted wild game to feed the group.

In time, human ingenuity found a way to exploit various species of the animal kingdom by identifying the traits necessary for domestication, which are:

- ability (usually a natural trait) to be organized into a herd or flock
- fully developed at the time of birth
- absence of a strong breeding pair bond
- ability to tolerate a varied diet
- ability to coexist with humans
- limited ability for flight

Purposeful breeding (animal husbandry) of animals to achieve improved qualities and characteristics was a natural consequence of animal domestication. However, in its initial stage it was most likely an accidental or serendipitous result. Early humans were unaware of the fundamental biological principles that are universal no matter what the species, and they had no concept of reproductive physiology, genetics, or statistics. The reasoning behind animal domestication was primarily to provide food for the first community of settlers who, because of climate and/or a dwindling supply of game animals, relinquished the nomadic hunting/gathering lifestyle of their ancestors. For hundreds of thousands of years, humans utilized all parts of wild game for food, clothing, tools, and weapons. Neanderthals lived for some 250,000 years in this manner. Thus, it is reasonable to assume that early farmers exploited the domesticated livestock in similar ways.

Domesticated animals are smaller than their feral ancestors, which in the earliest days of settled farming was suggestive of an animal's poor diet, a consequence of a limited grazing area and food supply. This may have been an unintended benefit as the smaller animals were in all likelihood more manageable. As with early farmers, who after a time began to keep seeds from their best plants for future plantings, the settlers who first domesticated certain wild species of animals eventually learned to corral, herd, and breed the best offspring of these wild beasts. Selecting the animals with the most desirable traits for breeding was in all probability based on empirical observation—or just plain instinct. There is tremendous genetic diversity in the offspring of wild or feral animals. Natural selection determines those best to survive and reproduce, thus continuing the genetic characteristics best suited to the environmental conditions. Those who are not strong will not live and therefore do not reproduce. However, in domesticated situations, animals that could never survive in the wild can and do thrive and reproduce offspring that bear characteristics that do not appear in wild populations. Domesticated species reflect the uses for which the animal was bred. On the negative side (for the animal), this may result in a loss of speed or agility, elimination of horns once used by the animal for self-defense, a change in disposition and/or intelligence, and the inability to survive without human intervention. On the positive side (for humans), there will be an increase in milk production, greater numbers of offspring will survive, and the quality of the meat and wool will improve.

In the earliest stages of domestication, breeding was often mixed with the wild or feral species. It was only when the animals were completely separated from their wild cousins that the actual alterations in appearance and temperament began to take place. Neolithic farmers and pastoralists manipulated various species of animals to achieve an increase in the population of these animals, as well as the traits or characteristics that farmers believed to be beneficial. This manipulation controlled the availability of mating partners and eventually altered some of the genetic qualities and physiology of the domesticated species. Controlled breeding can produce species that are less aggressive and more docile, and thus more manageable in a contained environment. It is believed, for instance, that in about 3000 B.C.E. domesticated species in the Near and Middle East had achieved complete differentiation from the wild species of cattle and sheep. These standard breeds produced superior meat, milk yields, and wool. Any interbreeding between the domesticated and wild species was considered a detriment to the stability of the herd. Eventually, a number of populations of wild species were eliminated because of environmental changes and availability of food sources—even overhunting. These early domesticated species were in most respects protected by their human masters from the catastrophes—except for disease—that could endanger their survival.

Principal Domesticated Species and Why They Were Chosen

Sheep and Goats

Both are hardy animals that are capable of existing on land that is less than ideal for grazing. They live outdoors at all times and give birth successfully without human intervention. They were especially plentiful in the Levant and Anatolia. The first evidence of domesticated sheep was found in Iraq in about 9000 B.C.E. It is believed that all sheep are descendants of the wild Asiatic mouflon that were prevalent in the mountains of Asia Minor. At some later point in history, humans discovered they could use the entire animal, as it provided wool for clothing, milk, cheese, and meat. After death, the sheep's hide was also used. Goats, domesticated sometime between 8000 and 7000 B.C.E., provided the same benefits as sheep, particularly their hides, which were exceptionally durable and long-wearing. Modern-day domesticated goats are descendants of the pasang *(Capra aegagrus)*, which was found in ancient Persia. In fact, recent evidence that was detailed in the journal *Science* (April 27, 2001) indicates that goats may well have been the first animal to be domesticated. Fossil evidence found in areas of the Fertile Crescent have led researchers to believe that ancient humans killed the largest male goats for food, while ancient farmers and herders killed the smaller and younger males and kept the female goats for breeding purposes. As these ancient tribes of humans migrated, they took their goat herds with them as a source of meat, milk, and wool with which to trade.

Goats were a primary source of milk and were usually raised in regions where it was difficult for cattle to thrive. One or two goats could provide a sufficient supply of milk for a family. Goathide was a favored source for water bags. However, unlike sheep, when given the opportunity, goats can revert quite easily to a feral or wild state.

A number of breeds of sheep were strictly food sources. Others were bred for wool or fur production. For example, the Central Asian Karakul sheep was bred specifically for its fur production, known as Persian lamb. When the Moors, a Saracen tribe from North Africa, invaded Spain in the eighth century C.E., they introduced a new breed of sheep, the Merino, renowned even today for its fine wool. Today, there are more than 200 distinct breeds worldwide, all products of the first domestication efforts of Neolithic humans.

The Angora goat is an ancient domestic breed native to the Angora region of Asia Minor (present-day Turkey). Its coat is the source of mohair, one of the oldest textile fibers, which for thousands of years was produced exclusively in this region before being imported to the European continent sometime in the mid-eighteenth century. Cashmere, another luxury textile, comes from the hair of the Kashmir goat, a descendant of the Siberian ibex, and native to the Kashmir region in India. Mistakenly, some soft, silklike wools are referred to as cashmere, but only the textiles made from the hair fibers of the Kashmir (or Cashmere) goat can truly be called cashmere. The popular pashmina shawl that is manufactured in Kashmir, India derives its name from the ancient word *pashm,* meaning underfleece.

Cattle and Oxen

The wild species of cattle or aurochs, known as the genus *Bos taurus,* were considerably larger than the domesticated species (six to seven feet high at the withers), primarily because of the need to better adapt and survive in Earth's harsh environment tens of thousands of years ago. There is some dispute as to the date of the first domesticated cattle. Some archaeologists believe the first cattle were domesticated in about 6000 B.C.E. in Greece and were raised only for their meat and hides. Others believe it took place much earlier, approximately 10,000 years ago. Recent research that was reported in *Science* (April 27, 2001) states that genetic similarities found in European and Near Eastern cattle *(Bos taurus)* populations strongly suggest that European cows descended from the Near Eastern herds. Researchers believe this supports their theory that ancient farmers who migrated westward from the Near East actually were responsible for the domestication of cattle on the European continent. Regardless of the time frame and location, at first cattle provided milk only for their own calves. The ability to produce large amounts of milk for human consumption was a consequence of purposeful breeding. In time, cattle provided humans with milk, cream, and cheese, as well as large quantities of meat and cowhide with which to make clothing and other accessories. An ox, which is a castrated bull, provided labor. Upon the discovery of how to harness these animals, oxen pulled plows and wagons for their human masters and were instrumental in the successful advance of agriculture worldwide. Oxen had another distinct advantage. Unlike horses, they could subsist on somewhat inferior fodder, particularly in ancient Greece and Italy, where the pastureland was limited. Also, after they had outlived their usefulness as beasts of burden, they could be eaten. (The eating of horsemeat, whether because of religious or cultural reasons, has never been widespread, except possibly in prehistoric societies.) Interestingly, cattle are not native to the North American continent.

After their initial purpose as a food source, cattle were then bred specifically for milk production. The increased supplies of milk spawned the production of other dairy products such as butter and cheese, further widening the types and availability of food for human consumption. Around 4000 B.C.E. the Sumerians in the Middle East became the first dairy farmers when they discovered how to make butter by churning milk. A wild auroch bull was a large big-horned animal that grew up to seven feet at the withers. Neolithic farmers, quite unintentionally, changed the genetic physiology of this animal by purposefully not domesticating it, rather allowing the animal to remain wild and removed from domesticated herds of cattle where the animal could not survive to reproduce the undesired traits. (According to a Breeds of Livestock project done at Oklahoma State University in 1995–96, the last surviving auroch lived until the seventeenth century on a hunting reserve in Poland, where it was reportedly killed by a poacher.) Bulls, more aggressive in temperament than cows, were always fewer in number in the herds, as they could mate with a large number of females who would then produce offspring. The farmer could choose which of the herd to use for milk, which for food, and which for labor. Oxen (castrated bulls that are at least two to three years old) were used almost exclusively as draft animals to pull plows and carts.

Swine (Pigs, Hogs)

These animals, whose ancestor was the wild Eurasian boar (*Sus scrofa*), were native to the Near East and the European continent. Domesticated pigs were somewhat smaller than their wild counterparts, although it is believed that breeding was not the reason for the domesticated animal's smaller size. (Wild and domesticated pigs are members of the pale-odont artiodactyl family called *Suidae,* which is presumably the source of the popular hog-calling term "sooey.") The domestication of swine dates back to between 8400 and 8000 B.C.E. in the Anatolia region. The Chinese, Romans, and Greeks, however, did not domesticate the swine until about 5000 B.C.E. Controversy remains even today regarding whether swine or sheep were in fact the first animals domesticated as a food source. The argument that the pig was the first is bolstered somewhat by the fact that pigs (or swine) were indicative of a settled farming lifestyle, primarily because pigs are difficult to herd and move over long distances. They do possess an adaptable nature and have a high rate of reproduction. The pig, mainly a scavenger, attached itself to our primitive ancestors. Pigs and hogs (pigs that weigh more than 55 kilos [120 pounds]) will eat practically anything. They found an unlimited food supply in the debris and offal from slaughtered animals that humans discarded and abandoned when they moved from settlement to settlement. The entire animal is edible while requiring a relatively small amount of food input.

Swine or pigs have always been bred to be a source of food. The wild boar, believed to be the ancestor of the majority of current breeds, was a larger animal in size than today's breeds. Its tusks, used for digging and defense alike, were enormous. Evidence exists in archaeological records that some 9,000 years ago early farmers successfully bred swine with smaller tusks or molars. Over the course of centuries, tusks or molars on domesticated pigs have been completely bred out.

Dogs

Many anthropologists believe that dogs were the first animals to be domesticated for something other than a food source. There is evidence dating back to 11,000 B.C.E. of the transformation of the wolf into the dog. Wolves were attracted by both the campfires and food scraps left by early nomadic humans; thus, they were relatively easy to tame or domesticate. Although they were eaten by certain clans or groups of early humans, they were bred from their ancestors, dire wolves *(Canis dirus),* to be guardians or protectors, herders, hunters, and vermin catchers—even as pets as early as the Paleolithic Period. The wolf is a descendant of a true carnivore that lived 60 to 65 million years ago, called *Canoidea* (doglike animals); thus, the derivation of the term *Canis familiaris* for dog. Dogs, considerably more massive in size than even some of today's large modern species, were domesticated some 2,000 years before agriculture was an accepted way of life. Dogs have powerful jaws and are swifter than humans. As such, they were tremendous assets to nomadic hunters who used them to track and entice prey. Since wolves were prevalent on all continents, the process of domestication occurred independently in several regions of the ancient world—including the Near and Middle East, Europe, the Americas, and Australia.

The dog possesses an innate ability to subordinate itself to a human master, a trait inherent in their wolf ancestry. Wolves live in hierarchical groups and learn from pups to adulthood to be subservient to the pack leader. Humans domesticated the wolf and successfully developed a number of different breeds, all of which had a specific purpose—from the most benign lap dog to the fiercest hunter of lions or guardians of encampments. In Egypt, for example, around 3500 B.C.E., dogs resembling the shape and size of greyhounds were kept on leashes by their human masters. At about the same time in the Near East, a mastiff-type breed was developed to hunt wild game. The Romans were the most successful dog breeders. Many of the most popular breeds of dogs today were bred thousands of years ago for reasons that bear no resemblance to their current status in today's society. For example, terriers were bred to be rat catchers, thus the term "rat terrier," while the standard or large poodle, thought to be rather patrician in bearing, was actually bred to be a herding dog.

Horses and Donkeys

Horses and donkeys, along with zebras, are the descendants of paleodont perissodactyl mammals in the *Equidae* family, commonly known as equids. These early animals, smaller in stature and sturdier in build than many of our modern breeds, were among the last breeds to be domesticated, probably between 5,000 and 6,000 years ago. As with most domesticated species, at first they were kept mainly for meat and milk. However, the donkey, known in its feral state as a wild ass, was used primarily as a transport animal and on occasion as a plow animal, since the meat of the donkey is unpalatable for human consumption. Also, the donkey is considerably smaller than the horse and thus is incapable of carrying larger loads or even an adult human over long distances. Their one great advantage over the horse is the donkey's tolerance of hot and arid climates where the vegetation is scarce and rough, making them more popular in those dry regions where farming was practiced out of necessity. The horse, on the other hand, is considered a prestigious animal that has played an indispensable role in the history of human civilization. Once humans realized that the horse could be harnessed and mounted, transportation and warfare were transformed. At first, humans used horses like donkeys, that is, as pack animals, eventually harnessing them to pull carts, wagons, and plows. However, it was the speed, as well as the size, of the horse that proved to be great assets to the tribes who made war and conquered each other. Cavalrymen could ride horses into battle or into a village under siege and easily achieve an advantage over foot soldiers. In about 2000 B.C.E. horses were effectively harnessed and pulled chariots into battles as well as races that took place in ancient Mesopotamia, Egypt, Greece, and territories of the Roman Empire. Horses were the prized possessions of barbarian tribes, kings, and the nobility, who often outfitted the horses, as well as themselves, with armor as protection in times of warfare. Horses provided transportation and were carriers of goods all over the European continent and the British Isles. The Celts were particularly skilled as horsemen and as such presented a challenge for their would-be Roman conquerors. In the country of Persia (present-day Iran), the horse was instrumental in the success of the first postal system. For some 6,000 years, the horse had been the mainstay of agriculture, transportation, and warfare. This is not an overstatement when considering that it was only with Nicholas Otto's invention of the internal combustion engine and Rudolf Diesel's invention of the compression ignition, both of which happened in the latter half of the nineteenth century, that "horseless" transportation became more than mere possibility. Steam engines were invented in the eighteenth century and a number of steam-driven automobiles were built and sold, but the automobile powered by a gasoline internal combustion engine has only been in existence for a little over a hundred years. Until these inventions were improved, marketed, and sold in large quantities, humans relied on the horse to aid them in their work, to get them to their destinations, and to help win their wars.

The wild species of horses and onagers (wild ass or donkey) interbred randomly and the resulting offspring did likewise for thousands of years. Mating between members of different species is not common, and the *mule* is probably the best known and most successful exception. This hybrid results when a mare and a male donkey mate. (When a stallion mates with a female donkey, the offspring is known as a hinny.) The offspring from either cross are almost always sterile. However, mules are the preferred pack animals as they possess a number of qualities that are superior to either the horse or the donkey. First, the mule's hide is tougher than a horse's and is less likely to be damaged by the friction of a harness. They have greater endurance, need less water to survive over longer periods, and are more sure-footed, particularly on rough terrain. Mules, generally less excitable, are believed to be more intelligent than horses, although somewhat more stubborn and temperamental. The ancient Greeks and Romans used mules extensively to transport all manner of goods.

Camels

It is believed that dromedary or one-humped camels were domesticated independently in both Arabia and Turkmenistan (a country in west central Asia east of the Caspian Sea) about 6000–5000 B.C.E. Bactrian or bi-humped camels are native to central and southwestern Asia, including the southern regions of China. Despite an often disagreeable temperament, body odor, and meat that is unpalatable to humans, camels have a number of characteristics that have made them valuable to their human masters. First, they can tolerate extremely hot weather. (Actually, they are ill-adapted to cold weather or high humidity and are susceptible to parasitic diseases. They are defenseless against mosquitoes and other insects when exposed to these types of conditions.) Second, they can go without water for seven to eight days, even under the hottest of conditions. (However, they may lose as much as 100 kilograms [220 pounds] of water or about one-fourth of their body weight.) Third, they can carry weights of up to 150 kilograms (330 pounds) and pull weights of up to 750 kilograms (1,650 pounds) for 25–40 kilometers (approximately 16–25 miles) over an eight- to 10-hour working day. This durability in the extreme desert conditions of Asia made possible the caravan trade in this region and opened up the almost impenetrable desert to the possibilities of civilization. Camels can be milked when they are nursing their own young, and their hair or wool is used for natural yarn and knitted wear.

Other Domesticated Species

Poultry (Chickens, Geese, Ducks)

Then as now, these poultry breeds were a consistent source of protein for humans. They can easily adapt to confined areas and are extremely efficient in converting high-energy grains into meat and eggs. The Egyptians, for example, domesticated geese around 2500 B.C.E., and force-fed the birds to increase their size and palatability. Prior to this, however, the first domesticated hens and cocks were used for sport—in particular, cockfighting.

Llamas

The four species of *Llamoids,* which are members of the camel family, are native to the Andean mountain region of the South American continent. The exact date of their domestication is unknown, but it is generally believed to be between 5,000 to 4,000 years ago. However, animals domesticated in South America, as well as those in Mesoamerica, were utilized primarily as a food source or for wool. Llamas were not as strong as oxen, or even horses, and while they could carry goods, their ability to do so over long distances was limited. Notwithstanding this limitation, the culture of the early settlers in these regions was such that animal power was secondary to that of humans. Thus, well into the first century C.E. farmers plowed their fields rather than utilizing the animals who lived in close proximity. For centuries llamoids have been bred specifically for their fur or hair. There are four members of this species: the llama, the alpaca, the guanaco, and the vicuna. The alpaca and the vicuna are the best-known producers of fine wool.

Cats

An approximate date for the domestication of the cat is unclear, because the cat—even today—defies the characterization of domestication as it is applied to most other animals. The domesticated cat (*Felis catus* or *Felis domesticus*) is a descendant of the wild cat, *Felis sylvestris.* It is believed that about 8000 B.C.E., with the beginning of settled agriculture, wild cats were attracted to and preyed on the mice and rodents that fed off the grains that were stored in some fashion by early farmers. Their usefulness as vermin catchers did not go unnoticed by these early settlers, and indeed the cat was exploited for just this purpose. The Egyptians' worship of the cat as a god has been somewhat overblown to the extent that the Egyptians considered other animals—among them crocodiles, cobras, and dung beetles—to be gods as well. The belief that the cat was held in singular high esteem by the Egyptians can be attributed to the Greek writer Diodorus, who in 59 B.C.E. recounted the death by lynching in Egypt of a visiting Roman diplomat who had accidentally killed a cat. Although it might be considered sacred by his human master, the cat has always retained its ability to catch and kill rodents; thus in times past keeping the animal in one's own home was a virtual necessity.

Guinea Pigs

On the North and South American continents, guinea pigs were domesticated as a food source.

Yaks

Native to the steep and cold mountainous regions of China, yaks are exceptional pack animals and are often called "ships of the plateau." They were domesticated for their hair, as well as their meat and milk.

Elephants

As far back as 7000 B.C.E., elephants were used as pack animals in India and Burma, as laborers to move heavy objects and timber, as well as to transport armies. Elephants are also the primary source for ivory, a commodity for which they have been exploited and slaughtered over the centuries.

Honeybees

The honeybee was domesticated at the end of the Neolithic Period to provide a sweetening agent for food, as well as for wax and bee venom, which was applied for medicinal purposes. Bees were also used quite effectively in warfare when their hives were thrown into the midst of enemy troops.

Indian Water Buffaloes

Existing today in both wild and domesticated states, Indian water buffaloes were first domesticated on the Asian continent thousands of years ago and then introduced on the European continent in Italy about 600 C.E. as a draft animal. They are large with a somewhat clumsy build and are used primarily as beasts of burden, with some milk and butter production, as well as for meat. (Their cousin, the Cape or African buffalo, has never been domesticated.)

Development of Animal Equipment

Early humans had effectively limited the ability of domesticated animals to wander, and over time had altered their temperament and physiology. The next and logical step was to gain more control and to harness the power and energy of those domesticated animals in order to ease the burden of their human masters. The same innate ingenuity that prehistoric humans demonstrated in fashioning tools for hunting and farming would prevail in human domination over animals. The ability to adapt to the harshest environments and to dominate other species are among the reasons why humans have survived and evolved as the dominant mammalian species. Although exact dates cannot be known, ancient humans invented what would be practical accessories or implements for their captive animals. Evidence gathered at archaeological sites worldwide indicates that many of these inventions were produced independently in various regions, thus we can only give approximate dates for their introduction and subsequent usage. Some sources ascribe the invention of almost all riding or equestrian equipment to the people who lived on the Eurasian steppes, the landmass between the European and Asian continents. The reasons for this attribution are twofold. First, the population and character of the wild horses that lived on the steppes were large and powerful. The horse, as a domesticated animal, enabled the also hearty humans of the steppes to traverse great expanses of this region. And human ingenuity utilized the available resources to its fullest—at least to the extent technologically possible at the time. As people migrated in all directions, these inventions obviously went with them, to be modified and augmented by the inhabitants of other regions. What is not in dispute is the fact that these inventions provided humans with tools that maximized the inherent strength of the ox and the prowess of the horse. (These inventions would also be used on certain other domesticated species, but none as successfully or widespread as the ox and the horse.)

Yoke

A yoke, which has been in existence nearly as long as animals have been domesticated, is merely a wooden framework consisting of bars or poles that joins draft animals at their head or necks for the express purpose of pulling some kind of implement or conveyance, such as a plow or cart. In its simplest form, the yoke was lashed together by leather straps or ropes attached to a longer set of straps or ropes that the driver used for control. In the earliest centuries of farming, yoked animals were difficult to control because the ancient yokes pressed against the windpipe of the animal, choking it. In an improved form, yokes were fashioned into U-shaped pieces, called *headstalls,* also made out of leather or rope, that encircled the neck of the animal or animals, most often two but as many as four. This type of yoke was somewhat less constrictive to the breathing of the animal, although still uncomfortable. Onagers or wild donkeys, and to a much lesser extent horses, were yoked in this manner. However, oxen were actually the preferred draft animal primarily because of their strength and physical conformation. Oxen have a natural hump at the withers, the high point of the back located at the base of the neck between the shoulders. Since the yoke fits across the horns or the back, this hump acted as the central point of thrust for the animal's front legs. The placement of the pole against the animal's throat continued to be problematical. However, the invention of the horse collar solved the problem of choking, and the faster and more intelligent horse displaced, in many but not all instances, the slower, clumsy ox in the field and on the road. Nonetheless, in some economically disadvantaged countries, particularly on the African and Asian continents, oxen continue to be yoked together in much the same fashion as they were in the early centuries of the first millennium C.E.

Horse Collar

There is some dispute regarding the origin of the horse collar, but most historians credit the Chinese with its invention, probably sometime between 200 and 100 B.C.E. Some type of horse collar, perhaps of Asian or Teutonic origin, was introduced on the European continent in about 500 C.E. The horse collar, made either entirely of leather, or leather and metal, surrounds the horse's head. Leather straps, called *traces,* are attached to both the collar and to the plow or wagon. While ancient yokes pressed uncomfortably against the animal's windpipe, the horse collar allowed the horse to pull across its chest and shoulders to move the load in much the same manner as an ox. The horse collar was modified and improved over the centuries to include more padding as well as eyepieces for the horse and was in general use on the European continent by 1100 C.E. Its invention made the horse a valuable animal. Agricultural practices advanced, and trade and transportation were accelerated when horses could be used almost entirely for plowing and for hauling heavy loads.

Harness

Crude harnessing systems of one form or another may actually have been in existence since 15,000 B.C.E. Stone Age cave paintings found in La Marche, France seem to depict images of bridled horses. And obviously, some type of harness was needed to keep the ancient yokes in place and to control the animal. However, the Chinese are believed to have developed the most effective and sophisticated harnessing system for horses, sometime before 500 C.E. Primarily because of their country's history of warfare and its vastness, the Chinese depended upon the horse for combat and transportation, and as a result were superior equestrians. The Chinese harnessing system was the first to successfully utilize the power of the horse without inhibiting the animal's ability to breathe. The harness consists of two rigid metal pieces *(hames)* that rest on the horse collar and are fastened at the top and the bottom by straps. For wagon harnesses, straps *(traces)* are positioned along the horse's sides and connected to the cart or wagon. Other straps may be secured around the animal's body to balance the load. *Reins* are the long straps used to control the animal. They are passed from the bridle through the loops in the hames and then back to the driver. A more elaborate system of harnessing is employed when a horse is harnessed between a wagon's shafts.

Another part of the harness system is the *bridle,* which consists of a headstall, bit, and reins, all used to control the horse. The *bit,* which fits in the horse's mouth and to which the reins are attached, exerts pressure and thus control by the driver. The first bits were probably made of rope, while the earliest evidence for the use of a metal bit dates back to 1500 B.C.E. Bits were also made of bone and antlers. These date back to sometime before 1000 B.C.E. The ancient Romans and Greeks experimented, quite unsuccessfully, with various harnessing systems, most of which were designed for oxen and not horses. For the most part, the Greek and Roman harnesses were awkward and inefficient. In addition to their choking the animal in much the same manner as the ancient yoke, these harnesses tended to press against the horse's carotid artery, causing further injury. Greek and Roman bas-reliefs and sculptures of horses often depict horses with heads held high. Rather than indicating the horse's sense of dignity and prestige, this in fact may have been an act of self-preservation in an effort to keep from choking.

Horseshoe

The invention of the horseshoe is somewhat contradictory. Some historians credit its development to the ancient Druids who inhabited the British Isles and France over 2,500 years ago. However, there is no hard scientific evidence to support this claim. Others maintain that the Romans were the first to fashion a horseshoe sometime after 100 B.C.E. (Actually, it was a mule shoe.) This may be a more accurate assertion since this procedure was referred to in the writings of the respected Roman poet Catullus (c. 84–54 B.C.E.). The plausible explanation for all the controversy is that like many ancient discoveries and inventions, the horseshoe was probably "invented" nearly simultaneously, and almost certainly independently, by a number of ancient civilizations. For instance, there is evidence that the horsemen of the Eurasian steppes experimented with various types of horseshoes, none of which were the traditional nailed-iron type. The nailed-iron horseshoe was most likely invented sometime in the fifth century C.E. somewhere in Asia and then introduced on the European continent by the conquering tribes from the East. The horseshoe, which is made out of an iron plate, is forged over a hot fire by the blacksmith or farrier who bends the iron plate to fit the outline of the horse's foot and then secures it by nailing the plate onto the animal's hoof. The dense and mostly insensitive hoof is the horny sheath that covers the bottom part of the foot of animals classified as perissodactyls and artiodactyls (horses, oxen, deer). When done correctly, this procedure does not cause the animal pain and protects the foot from injury on rough and rock-covered terrain. The invention of the horseshoe advanced trade as it allowed the animal to proceed at a greater and safer pace over longer distances. The art of warfare was also affected by this invention. An example of this is the Crusades that took place during the eleventh to thirteenth centuries. The large Flemish horses that the European Crusaders took with them into battle wore these iron horseshoes primarily to protect their weak feet. (Flemish horses, though massive in size, were raised in damp lowland areas in Europe. Consequently, they had flat and rather weak feet.) The sparks that flew off the horses' hooves as they plunged into battle was quite intimidating to those being vanquished. This invention also created the craft of the smith—commonly called blacksmith—which flourished for centuries.

Saddle

The origin of the leather saddle is more clearly delineated than other inventions dealing with the horse. Historians generally accept that the Sarmatians, a nomadic tribe of Iranian extraction who migrated from central Asia to the Balkans and southern European Russia, developed the leather saddle sometime in the third to fourth centuries C.E. (Sarmatian women, who rode into battle along with the male members of the tribe, are reported to be the inspiration for the Amazons of Greek mythology.) For centuries after the domestication of the horse, the animals were ridden bareback or with blankets or cloths, often making it difficult for all but the most accomplished of riders to remain astride. The saddle, which allowed the rider to keep a firmer seat on a moving horse, was secured around the animal's midsection by a leather strap called a *girth*. The Europeans, particularly the French, made a number of modifications or improvements to the saddle. In particular, a war saddle first designed in about the sixth century C.E. was further refined in the eleventh or twelfth century C.E. to incorporate a wraparound *cantle* or ridge at the rear for a more stable seat and a *pommel (saddlebow)* in the front to protect the rider's genitals. A double girth was also added, which further secured the rider with the horse. Human civilization has been shaped by warfare. The history of the saddle, from about the fourth century until its refinement as an indispensable apparatus in combat, coincides with the rampages of the barbarian tribes who battled each other and finally conquered the Roman Empire in 476 C.E., and culminates with the knights of the Crusades and the feudal conflicts in medieval Europe. The addition of the stirrups further enhanced the saddle as a powerful weapon of ancient warriors. Saddles have also been designed to accommodate the camel—both the dromedary (one-humped) and the Bactrian (two-humped)—and the elephant. Canopied elephant saddles, which are of necessity quite large, are called *howdahs.* Historians place the development of both these saddles in Asia and Africa in about the third century C.E.

Stirrup

It is believed that nomadic tribes who lived on the steppes in China are responsible for the invention of the stirrup in about the second century B.C.E. These tribes were superior horsemen and warriors who fashioned the first stirrups from looped pieces of rope, primarily to aid them in mounting their horses. The etymology of the word "stirrup" is from the Old English word *stizrap,* which combines the words "climb" and "rope." Stirrups are hung from either side of the saddle and are attached to the back of the animal. As mentioned, the original stirrups were made of rope, then leather. Bronze or iron platforms to support the sole of the rider's foot were added by about 300 C.E., also in China. In addition to facilitating the mounting of the animal, stirrups also further supported the rider, who then could exert greater control of the horse, as well as greater force with his weapons in time of combat, with less concern about falling off. The stirrup contributed to the rise of improved and successful combat tactics, as well as **feudalism** on the European continent, since knights and horse soldiers were able to carry armor more securely while wielding heavier and larger weaponry. (Attila the Hun is said to have introduced the stirrup on the European continent.) However, the first use of the stirrup in actual combat took place in France in 732 C.E. when invading marauders from North Africa (Saracens) with stirrups attached to their saddles attacked the French foot soldiers. Remarkably, the Saracens were defeated, but the stirrup became a necessary piece of equipment for the feudal knights who waged territorial wars in all regions of the European continent, as well as in their invasion of the Holy Land during the Crusades of the eleventh to thirteenth centuries. Another aftereffect of the invention of the stirrup was an increase in the breeding of horses specifically for combat purposes.

Summary

The contemporaneous development of agriculture and animal domestication transformed early humans and their world, and ultimately their future. While life was harsh for early farmers and often devoid of sufficient food and nutrients, the population continued to increase. They were settled and could produce more offspring, and feed greater numbers of the community. Over time, humans adapted to the climate and terrain and learned to manipulate plants and animals to their benefit. In essence, humans have been biotechnologists and practitioners of animal husbandry for thousands and thousands of years. Our earliest ancestors employed seed selection from the most desirable plants and then stored and planted them for later cultivation in order to produce the most productive variety. They instinctively and selectively bred the healthiest animals, and learned how to eliminate undesirable traits and characteristics while ensuring the most desirable would pass from offspring to offspring. They knew about fermentation, for example, in the making of bread, spirits, and cheese. Agriculture led to commerce, in which crops were bartered, then later sold. Domesticated animals were not only food sources but work and pack animals that helped to increase the output of crops and their distribution. Navigation and exploration of other lands for the purpose of colonization and expansion of wealth then followed. Agriculture afforded more leisure time in which cultural pursuits could be nurtured. Writing, reading, and the development of crafts and the arts—even politics—all evolved from the agrarian rather than nomadic way of life that early humans adopted 12,000-plus years ago. The bounties of agriculture and the quality of breeding have greatly improved over the centuries with the crossbreeding and hybridization of crops and animals. These efforts, at first, were trial-and-error experiments based on empirical evidence. Science continues to research methods of more effectively crossbreeding our domestic plants and animals to produce a larger and healthier supply of plants and animals to feed the world's ever-increasing population. The challenge as to how this knowledge, first applied by ancient humans, will be used by modern humans will not be dependent on science alone, but on society, and in particular on moral, economic, and political decisions. In the meantime, it is important to acknowledge that all of the inventions, discoveries, and advancements that presently define us as a species are the extensions and by-products of our early ancestors' agrarian achievements.

2

ASTRONOMY

Background and History

It is believed that humans practiced astronomy, which is the scientific study of celestial objects in the universe, about 5,000 years ago. Prior to that, humans were hunters and later, farmers, who settled into societies and settlements in an attempt to survive Earth's harsh environment. The heavens provided a semblance of order for them, inasmuch as the Sun rose in the same direction each morning and appeared to move westward to settle in the opposite direction each night. And the Moon always **waxed** and **waned** within a set time period. The same group of stars also appeared to move around a fixed point in the heavens. While primitive humans did not articulate these regularities in a formal manner (i.e., understanding vernal or autumnal equinoxes or recognizing constellations and the north celestial pole), in time they became dependent on this order for the planting and harvesting of their crops, the establishment of religious rituals, and later for navigational purposes. In other words, they recognized the cycles of nature and adapted their lives accordingly. Archaeological evidence dating from circa 4500 B.C.E. indicates that Neolithic humans built long barrows (burial mounds) that were aligned with the rising and setting of bright stars and from which they could ostensibly view the stars' movements. Later, as in the case of Stonehenge in southern England, dating back to between 3000 and 2000 B.C.E., monuments were constructed that were in line with the Sun and its **solstitial** positions, as well as with the Moon's cycles, presumably to predict eclipses.

The establishment of sites from which the celestial cycles could be observed and celebrated through rituals formed the basis of astrology, which is the foundation of astronomy. For example, the ancient

Mesopotamians built **ziggurats,** massive towers from which they viewed the heavens. One of these ziggurats found in Uruk, an ancient city in Mesopotamia, presumably dates back to about 5000 B.C.E. It is important to recognize that the historical scientific development of astrology and mathematics, particularly geometry, parallel each other. It is also important to recognize the discoveries and inventions of ancient mathematicians. They contributed to our understanding of the Sun, Moon, and planets and enabled succeeding generations to accurately measure time and distances within our universe. And as with most human inventions and discoveries, theories on astronomy were espoused by a number of ancient civilizations independently of the others, and later shared and expanded. Some of these theories were wrong, such as the theory of an Earth-centered universe that was accepted for about 2,200 years until the sixteenth century, when Copernican theory finally refuted the geocentric system. Other theories, amazingly, were correct: for example, Thales of Miletus (624–548 B.C.E.), using geometry, correctly estimated the Sun's diameter and correctly explained the nature of solar eclipses. The fact that many of these ancient astronomers/mathematicians were wrong in their theories should not negate their contributions to the overall body of knowledge in astronomy.

The ancient civilizations of Babylonia in Mesopotamia, Egypt, China, India, and Greece approached the "science" of astronomy from different perspectives, as did the Mesoamerican civilizations. Astronomy in the Islamic countries of the Near and Middle East and Europe did not flourish until centuries after the fall of Rome in 476 C.E. Islamic astronomy was influenced by the forces of Mohammadanism in the seventh century C.E. that later impelled the religious movement onto the European continent. Islamic scholars translated the works of Greek astronomers and mathematicians into Latin, the language of the educated of medieval Europe. After this, the Europeans dominated the field of astronomy.

Babylonian Astronomy

Babylon was the capital of the ancient country of Babylonia in Mesopotamia, often called the "Cradle of Civilization" (present-day Syria, Turkey, and Iraq). A number of tribes lived in this area, among them the Sumerians, the Akkadians, and the Hittites. However, Babylonians made some of the most important discoveries and achievements in numerous fields. In their astronomical discoveries, for example, Babylonian astronomers used only arithmetic—not geometry—to approximate celestial events. Babylonians identified zodiacal constellations, developed star maps and calendars, and recognized the cycles of the Moon. Beginning in about 1800 B.C.E. and continuing over the next 1,200 years, the Babylonians worked on and developed a calendar based upon the Sun's movements and the Moon's cycles. Sometime after 400 B.C.E. they arithmetically calculated, with amazing accuracy, the exact time and date of the first day of each month. Originally, they had designated the first day of each month as the day after the new moon, that is, when the Moon's crescent was first observed after sunset. Prior to this discovery they could only predict this date through observations. The Babylonians also calculated arithmetically the positions and movements, including the retrograde motions, of the five visible planets (Mercury, Venus, Mars, Jupiter, and Saturn), as well as the Earth and the Sun, which the ancients also considered to be a planet. (Retrograde means a direction of motion *opposite* to that of the Earth's axial motion, or the motion of the planets around the Sun.) Although the pseudoscience of astrology arose from Hellenistic (Greek) influences, the Babylonians studied and developed it extensively by practicing inductive divination, that is, predicting the future based on the motions of the stars, planets, Sun, and Moon. Thus, they are the civilization most associated with astrology. Babylonian astronomers also accurately predicted both solar and lunar eclipses.

Greek Astronomy

Hellenistic astronomy was more a myth-based astrology than what we now consider to be astronomy. However, Greek astronomers, who are among the most famous in the annals of the science, utilized Euclidian geometry rather than an arithmetic or numerical system to approximate the movements of the Sun, Moon, stars, and planets. Theories related to lunar and solar eclipses and the nature of the universe date back to the sixth century B.C.E. in ancient Greece. Greek astronomy was founded on the principle that all celestial observations were made from the fixed perspective of humans on the stationary and motionless planet (Earth) that was at the center of a rotating sphere to which the stars were "fixed." The movement of the planets, from the Greek word *planetes* meaning "wanderers," was explained against this backdrop with a high degree of accuracy. In the fifth century B.C.E. the Pythagoreans advanced one of the first important astronomical theories and subsequently erected the first physical model of the solar system. Using geometric progressions, Pythagoreans, who believed that the universe could be explained in mathematical terms, postulated the theory of a central fire around which all the stars, planets, Moon, and Sun revolve.

Sometime in the fourth century B.C.E., Greek astronomy began to change character as a result of the teachings of Plato (428–347 B.C.E.), who subscribed to the theory that celestial bodies moved in perfect or uniform circles. In other words, Greek astronomers were encouraged to develop "accurate" theories using only geometric combinations of symmetrical circles. As such, the geometric models that were developed were not considered to be accurate but simply tools to predict planetary positions and movements. Later Greek astronomers challenged Plato's theory and conceived of a homocentric model of the universe that incorporated a set of concentric spheres that produced all celestial motion. Aristotle (384–322 B.C.E.) developed a beautiful, symmetrical model of the universe consisting of a geocentric grouping of spheres (see Figure 2.1). His theory influenced astronomers for centuries—despite the fact that it was wrong. It is important to remember that, with

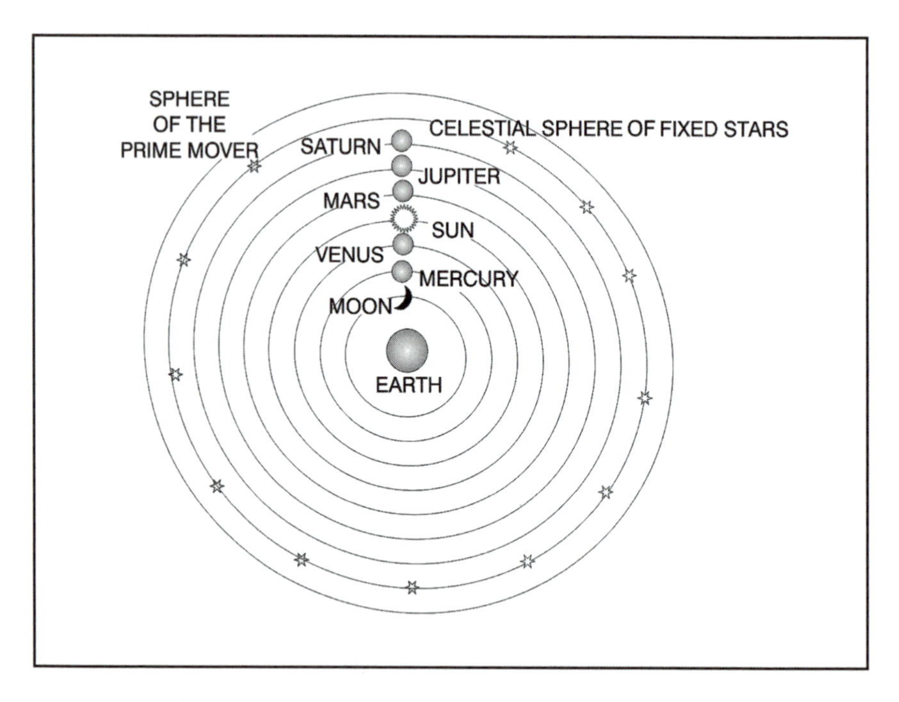

Figure 2.1 Aristotle's Geocentric Universe
Aristotle believed the universe was spherical and finite, with Earth at its center. Since he believed that an infinite thing cannot have a center, and the universe has a center—infinity does not exist. *Aristotle's universe* contained 10 spheres, with each sphere represented by the then five known planets, the Moon, Sun, the stars, and finally the prime mover.

a few exceptions, all of the ancient astronomers had problems with their calculations and tables because they adhered to the false belief that the heavens revolved around a stationary Earth (geocentric system). Greek astronomy expanded eastward to the Middle East, India, and the Islamic countries. It also expanded onto the European continent sometime after the conquest of Greece by the Romans in the second century B.C.E. Afterwards, while the Romans forced many of the Greeks into slavery, they did recognize the intellect of Greek scholars and utilized many of them as teachers.

Chinese Astronomy

The ancient Chinese are renowned for their many discoveries and inventions. China has always been a country of mystery, seemingly isolated from the rest of the civilized world, and with the reputation of being somewhat reluctant to share those ideas and discoveries. Nevertheless, Chinese astrologers and astronomers were keen observers of the heavens, and their astronomical calculations led to the invention of numerous devices to measure time. The Chinese or Eastern approach to astronomy was different than that of Western civilizations, including those of the Mesopotamians. Western civilizations focused on the ecliptic system, which holds that the plane of the Earth's orbit around the Sun is primary. Chinese astronomy was based on an equatorial system that focused on the constellations around the north celestial pole and the celestial equator. There is archaeological evidence from about 1281 B.C.E. that the Chinese adopted some of the Babylonian principles of divination using the Moon and stars and subsequently interpreted them using their own plotting of constellations in the equatorial system. The Chinese practice of astrology borrowed heavily from Western civilizations. However, the Chinese amended these ideas to fit their own concepts of celestial mechanics. Also, their interest in celestial matters was not especially scientific and did not incorporate the deductive character that is seen in Hellenistic astronomy. The first astronomical observatories were reportedly built in China in about 2000 B.C.E. Over 3,000 years ago, Chinese astronomers accurately predicted lunar and solar eclipses, which was about 500 years before Western astronomers were able to do so. In the last century B.C.E. the Chinese revised their belief in a flat Earth and began to think of the universe as a giant "egg" with the Earth being the "yolk" at its center. The sky curved around the Earth like a shell. They continued to revise their theories for the next 500 years, and in 800 C.E. they arrived at the belief that space is infinite and the universe had existed for 100 million years.

Indian Astronomy

Like all ancient civilizations, the Indians looked to the heavens to reconcile their religious beliefs and to foretell the future. The character of Indian astronomy began to take shape in the fifth century B.C.E. with the Mesopotamian influence that emphasized astrological omens. Sometime in the first century C.E., Greek astrological writings, part of which contained mathematical astronomy, were translated into Sanskrit. Then Babylonian astronomy took hold in India after the Persian conquest of India in the fifth century C.E. Aristotelianism became popular in India sometime before the sixth century C.E. This proved to be troublesome for Indian astronomers, who accepted unequivocally Aristotle's model of the solar system, which consisted of a concentric set of spheres, and then blended it with Hellenistic epicycle astronomy. The Indians were primarily interested in astronomy for religious and astrological purposes and did not contribute significantly to the body of knowledge in this field. They did, however, prepare calendars and other time scales for religious observances.

Egyptian Astronomy

Although the archaeological history of Egypt dates back 5,000 years, there are no technical records or writings dealing with Egyptian astronomy until the first millennium B.C.E., after Egypt's conquest by Persia. It was at this time that the influence of the Near Eastern astronomers became important. Prior to this, however, the Egyptians used simple astronomical methods to measure time and to develop accurate calendars, as well as to align their buildings with the stars. They constructed elaborate temples and sphinxes to worship the Sun and the stars, and to regulate both their civil and religious lives. Egyptian astronomers charted stellar movements and grouped them into constellations that were different from those of Western astronomy. They were also able to observe the planets of Venus, Mars, Saturn, Jupiter, and perhaps even Mercury. As with all the other ancient civilizations, astrology and divination were the dominant forces in the study of celestial bodies. Egyptian astronomy began to flourish after Egypt's conquest by Alexander the Great in 332 B.C.E. The dynasty of the Ptolemies, who were of Macedonian Greek heritage, ruled Egypt after Alexander's death in 323 B.C.E. Many of the Egyptian astronomers were, in fact, of Greek heritage, and Egyptian astronomy during this period was, in actuality, Hellenistic in character. Ptolemy (Claudius Ptolomaeus) of Alexandria (ca. 90–170 C.E.), often considered to be the greatest of the ancient astronomers, was certainly the most influential. In fact, a vast majority of astronomers accepted his theory of a geocentric system of the universe until the middle of the sixteenth century, when Nicolaus Copernicus (1473–1543 C.E.) would discover that it was not the Earth at the center of the universe, but rather the Sun.

Mesoamerican Astronomy

Four great civilizations flourished in Mesoamerica (present-day Mexico and Central America) for about 2,000 years. Their demise coincides with the discovery of the Americas by Christopher Columbus in the fifteenth century C.E., as well as with dramatic changes in climate and rainfall. Thus, the term "pre-Columbian" is generally assigned when discussing the Olmec, Zatopec, Aztec, and Mayan civilizations. As with all ancient civilizations, the Sun, Moon, stars, and planets were essential in understanding and divining the events that took place in their lives. Astrology is a more apt description of the practices of these ancient peoples, who nevertheless developed accurate calendars and built temples and pyramids that were carefully aligned with the celestial bodies. These were creative civilizations but extremely ritualistic. Both animal and human sacrifices were part of their ceremonies to worship the deities that they believed lived in the Sun, Moon, stars, and planets. Archaeological evidence confirms that, at the least, the Mayans used mathematical concepts to formulate astronomical events, such as the equinoxes, solstices, and eclipses. However, all these civilizations believed in a layered universe theory—13 layers comprised the universe with each layer containing one type of celestial body (Moon, clouds, Sun, stars, planets, comets, etc.), with God residing in the thirteenth layer. The Aztecs erected the famous temple complex at the ancient city of Teotihuacan in Mexico, believed to coincide with the movements of the Pleiades, also known as the "Seven Sisters," a group of seven (now six) stars in the constellation of Taurus. These Mesoamerican civilizations considered the Pleiades to be an important grouping of stars in terms of the prediction of meteorological and agricultural events.

Astrology

Astrology is defined as the use of astronomical phenomena (i.e., the position and aspects of stars and planets) to predict future events. Ancient humans believed that celestial events, in particular the planetary motions, directly impacted their own fortunes. They studied the heavenly bodies and tried to understand how the stars and planets, Moon, and Sun influenced their religious, personal, and civil lives. It is important to recognize that throughout most of human history, no distinction was made between astrology and astronomy. Thus, the foundation of modern-day astronomy was ancient astrology. It was only when Copernican theory of a Sun-centered universe was accepted as fact, and after Johannes Kepler (1571–1630 C.E.) and Isaac Newton (1642–1727 C.E.) accurately explained celestial motion, that astronomy became a systematic science. Although the Mesopotamians were the first to categorize and catalog celestial omens based on the alignment and configurations of planets and constellations, astrology is historically Hellenistic in its philosophy and character. Constellations are random or arbitrary yet conspicuous patterns or groupings of stars that have been recorded and mapped and subsequently given various names, primarily by the early Arabs and Greeks. Ancient astrologers in Mesopotamia, China, India, Egypt, Mesoamerica, and Greece projected patterns within these constellations that they believed represented animals, gods, or objects that could affect events on Earth. Thus the idea that these heavenly bodies could influence earthly life was conceived. These star patterns or signs could be found in the zodiac, which is a zone or band around the celestial sphere that extends approximately nine degrees on either side of the **ecliptic.** The paths of the Sun, Moon, and the principal planets of Mars, Mercury, Jupiter, Saturn, and Venus are encompassed within this zone. (The ancients considered the planets to be "wandering stars," whereas the stars were considered fixed.) Ancient astrologers believed that the planets, which are located in specific "houses" within the zodiac, are associated with human traits and events. The term "zodiac" originated with the Greeks, who called it *zodiakos kyklos,* meaning "circle of animals." The names of the signs of the zodiac were popularized in a poem written by the Greek poet Aratus sometime after 600 B.C.E., essentially so that they could be remembered and utilized by farmers. This is believed to be one of the reasons why the Greek names for the zodiacal signs were adopted rather than Arabic, Chinese, or Egyptian designations. (They are Greek names in the sense that the Greeks translated the names from the original Babylonian texts.) Astrology contributed to people's understanding of time, agriculture, and celestial events and influenced modern astronomy, cosmology, and mathematics. Given the fact that technology and instrumentation were limited, it is remarkable that our ancient ancestors were able to understand and catalog accurately a tremendous amount of astronomical information and events. Nonetheless, astrological predictions are often based on metaphysical phenomena. As such, there is no way to check the accuracy or validity of their outcomes. Thus, astrology, as currently practiced, cannot be classed as science, but only as pseudoscience.

The Geocentric View of the Universe

The practice of ancient astrology and astronomy emerged from the perspective of humans as the observers of celestial movements. From the earliest days of human existence until the middle of the sixteenth century, it appeared that humans lived on a motionless planet around which the Moon, Sun, stars, and planets revolved. Ancient astrologers believed the planetary movements within the zodiac accounted for events on Earth. Later astronomers developed epicyclic theories of celestial motion, some of which were amazingly accurate. The application of Euclidean geometry enabled astronomers to formulate more correct astronomical models. But most ancient astronomers believed, obviously incorrectly, that the Earth was a stationary planet. Notwithstanding this false belief, they continued to amass vast amounts of data concerning the lunar cycles, equinoxes, and planetary and stellar motion. They did all this using deductive strategies and empirical or observable evidence since science, as either a word or system, did not exist. Neither did scientific method. Following is a compendium of the most famous ancient astronomers and their various theories of a geocentric system of the universe, as well as their discoveries related to celestial mechanics.

Thales of Miletus (ca. 625–ca. 547 B.C.E.)

Believed to be the first philosopher of nature, Thales was also the first to depart from the metaphysical explanations of astrology/astronomy by attempting to understand the physical makeup of the Earth and everything on it. As a young man, he traveled to Egypt and the Near East to study geometry, a branch of mathematics concerned with points, lines, and surfaces in two dimensions.

He founded a school of natural philosophy in Ionia, a city on the Aegean Sea in Asia Minor, where he espoused his theory that the Earth was flat, and it and everything on it floated on a huge body of water, from which they originated and to which they would all return. The reason for this misconception is probably founded in Thales' observance of water in its three phases: gas, liquid, and solid. Most likely, Thales developed his theory through empirical observations of water as it changes from one state to another. Ice, when heated, turns to liquid water, and steam (gas) results when water is heated to a high-enough temperature. Another source of Thales' belief that Earth floated on water was the creation stories of Babylonians and Egyptians in which all things came from water. Nonetheless, he accurately measured the apparent diameter of the Sun, as well as accurately explained the nature of a solar eclipse. He also described the constellation Ursa Minor and its usefulness for navigational purposes.

Anaximander (ca. 611–ca. 547 B.C.E.)

Most likely a pupil of Thales, Anaximander is often referred to as the founder of astronomy as he was the first to formulate a cosmological view of the universe. He disagreed with Thales' assertion that the Earth floated on water, and proposed that the Earth was unsupported at the center of the universe. In other words, it just "hung" there. He correctly stated that the Earth's surface was curved (even though he mistakenly believed it to be cylindrical in shape) and that the heavens were a complete sphere encircling the Earth—not just a semisphere—in which the Sun, Moon, and stars moved in perfect circles.

Pythagoras (ca. 580–ca. 500 B.C.E.)

Most famous for establishing the Pythagorean Order of philosophy and for his Pythagorean theorem for right angles, he utilized mathematics and its relationship to the patterns found in nature to divide the universe into three components: (1) *Uranos,* or the Earth as a sphere at the center of the universe; (2) *Cosmos,* the heavens surrounded by fixed stars, also in a sphere or shell; and (3) *Olympos,* home of the gods. In addition to philosophy and mathematics, he taught astronomy at his famous school located in Crotona (Italy). He correctly determined that the orbit of the Moon was inclined to the Earth's equator, and was the first to discover that the evening and morning stars were the same celestial body: the planet Venus.

Anaxagoras (ca. 488–428 B.C.E.)

A follower of Thales of Miletus, he also incorrectly believed the Earth was a cylinder, the basis of all matter was the atmosphere, all things were composed of air, and that the Sun was a large, very hot stone. Nonetheless, he was the first to describe how shadows on the Earth caused by the Sun and Moon are responsible for the solar and lunar eclipses. He also correctly determined that the Moon, the surface of which is covered with plains, mountains, and ravines, is the closest body to the Earth.

Eudoxus of Cnidus (ca. 400–350 B.C.E.)

He was one of the first ancient astronomers to attempt to account mathematically for the irregular motions of the planets and still maintain the Earth as the center of the universe. He introduced geometry into the "science" of astronomy. His system required not only a motionless Earth, but also 27 crystal-like celestial spheres that were responsible for their motions. (The Sun and Moon each had three spheres; each of the five known planets had four spheres; and the remaining twenty-seventh sphere contained all the fixed stars; beyond that lay the heavens.) Eudoxus described mathematically the rising of the fixed stars and constellations over the period of one year. It is interesting to note that some of the instruments used by modern-day astronomers are based on many of Eudoxus's theories, as are some of the techniques used by today's meteorologists when predicting weather patterns.

Aristotle of Macedonia (384–322 B.C.E.)

In the estimation of many historians, Aristotle was one of the most influential humans to have ever lived. His philosophy, methods of reasoning, logic, and scientific contributions are still evident and valid. However, his astronomical theories were inaccurate. He believed that the Earth was the center of a spherical universe, and all celestial bodies moved in perfect circles in a perfect medium, which he called *aether.* Aristotle's teachings were widespread and respected. However, the Church of Rome challenged his theories on astronomy, primarily because of their astrological or metaphysical components.

Hipparchus of Nicaea (ca. 190–ca. 120 B.C.E.)

An astronomer and mathematician, Hipparchus is often referred to as the "greatest astronomer of antiquity." This designation may be questionable given the fact that he was a firm geocentrist and roundly dismissed any proposals that set forth a heliocentric (Sun-centered) universe. Nevertheless, he did reject the accepted astrological teachings of his time, as well as the theory of an immovable Earth, and formulated positions and motions for the planets and Moon. He is credited with changing the direction of Hellenistic astronomy from one that was qualitatively geometrical in nature to a science that was more empirical or experimental in character. From his observatory in Rhodes, he utilized geometrical calculations, as well as his own observations, and determined that it was the Earth that was actually moving and not the stars as earlier astrologers/astronomers had proposed. He is credited with several "firsts," among them, using latitude and longitude, as well as the compilation of the first star catalog of about 850 stars. Hipparchus was the first to discover the precession of the equinoxes, which is defined as the slow motion of the Earth's axis westward along the plane of the ecliptic at an annual rate of 50.27 seconds of arc. He also recalculated the inclination of the ecliptic to within five degrees of the now-accepted figure, as well as estimated the length of the year to a measurement within 6.5 minutes. Using the astronomical instruments of his time, he made the first realistic measurements of the Earth's distance from both the Sun and the Moon. Although Hipparchus lived several centuries before the birth of Ptolemy of Alexandria, Ptolemy considered his theories to be those of a contemporary and subsequently based many of his own theories on those of Hipparchus.

Ptolemy of Alexandria (ca. 90–ca. 170 C.E.)

Also called Claudius Ptolomaeus, he is considered to be the most influential of the ancient astronomers, since his theories of an Earth-centered universe and other astronomical concepts related to celestial bodies were accepted, essentially without challenge, for over 1,300 years. His works can be found in the 13-volume work called *The Almagest.* Ptolemy compiled, utilized, and expanded the data and research of other ancient astronomers, among them Aristotle and Hipparchus. As an example, Hipparchus's star catalog of 850 stars was expanded by Ptolemy to include 1,022 stars. From Aristotle, he accepted that there were two parts to the universe—the Earth and the heavens—and that the Earth's natural place was at the center of this universe. Ptolemy believed the Earth to be a sublunary region where all things are born, grow, and die. The heavens are composed of compact, concentric crystal spheres that surround the Earth. Each shell was the home of a heavenly body arranged in the order of the Moon, Mercury, Venus, Sun, Mars, Jupiter, and Saturn, followed by the fixed stars and the "prime mover," who was responsible for the movement of the whole system. From this background, Ptolemy claimed that all celestial bodies that revolve in orbits do so in perfect circles and at constant velocities. The stars, however, move in elliptical orbits at inconsistent speeds. This theory required the application of a complicated geometry that Ptolemy used to describe these motions (eccentric, epicycle, and equant), thus forming the basis of the *Ptolemaic system for planetary motion.* (The *eccentric, epicyle,* and *equant* are complicated geometrical constructions. The *eccentric* places the Earth outside the center of its circumference; the *epicycle* is the placement of the center of a smaller circle on the circumference of a larger circle; Ptolemy added the *equant* as more or less a combination of both of the other geometric constructions. Using these geometrical constructs, Ptolemy was able to account for the motions of celestial bodies—insofar as he was able to observe them.) As evidenced by *The Almagest,* Ptolemy was a prolific writer who formulated detailed charts of planetary movements (although not always accurately), observed and detailed the curvature of the Earth, and experimented with optics and refracted light. Ptolemy was so well respected that despite the inaccuracy of his astronomical charts and his incorrect belief in a geocentric system of the universe, astronomers for the next 1,300-plus years would base their astronomical findings on the Ptolemaic system.

The Heliocentric View of the Universe

Although the geocentric system of the universe was the dominant theory until the middle of the sixteenth century, a number of ancient astrologers/astronomers challenged this view—with obvious failure. The ancients could not satisfactorily explain why the relative positions of the stars appeared to be constant despite the Earth's changing perspective as it moved around the Sun. Early astrologers/astronomers, and later Ptolemy, ostensibly answered this by asserting that the Earth was a stationary object around which all celestial objects revolved. Nevertheless, several ancient philosophers, mathematicians, and astronomers, as far back as the fifth century B.C.E., proposed various heliocentric models of the universe. None of their theories received wide support.

Philolaus of Tarentum (fl. 475 B.C.E.)

A follower and author of Pythagorean cosmology, the Greek philosopher Philolaus asserted that the Earth was a sphere revolving around a mystical central fire, called "Hestia," that controlled the universe. His reasons for proposing this are unknown. As a follower of the Pythagorean model of cosmology, which was outside the realm of Hellenistic cosmology, Philolaus believed that the Earth moved in some type of orbit.

Hicetas of Syracuse (fl. fifth century B.C.E.) and Ecphantus of Syracuse (fl. fourth century B.C.E.)

Little is known of these two ancient Greek philosophers, both of whom are believed to be Pythagoreans. However, the writings of the Roman statesman and philosopher Marcus Tullius Cicero (106–43

B.C.E.) specifically mention Hicetas as a proponent of the theory of a central fire and of the movement of the Earth. When Nicolaus Copernicus proposed his model of a heliocentric universe, he credited several of the ancient philosophers and astronomers with giving him the courage to forge his theory, which was counter to the then-accepted Ptolemaic system. Two of those ancient philosophers were Hicetas, who "made the Earth move," and Ecphantus, who "made it rotate about its axis."

Heracleides of Pontus (ca. 390–ca. 322 B.C.E.)

Believed to be a student of the Greek philosopher Plato (ca. 427–347 B.C.E.), Heracleides is also believed to be the first Greek philosopher and astronomer to propose that the apparent westward movement of celestial bodies across the sky was caused by the eastward rotation of the Earth on its axis. This attribution stems not from his own writings, which have not survived, but from other seemingly reliable sources. The other astronomical attribution of Heracleides is his assertion that Mercury and Venus revolve around the Sun, which revolves around the Earth, although it is unknown whether he made the same assertion regarding the movement of the other known planets. He was obviously correct about the movement of Mercury and Venus, but incorrect about the Sun revolving around the Earth. (The famous sixteenth-century astronomer Tyco Brahe adopted Heracleides' mistaken theory about the Sun revolving around the Earth.)

Aristarchus of Samos (ca. 320–250 B.C.E.)

About two centuries after the deaths of Philolaus, Hicetas, and Ecphantus, Aristarchus expanded on their ideas and proposed that the Earth as well as the other planets revolved around a central object that he believed to be the Sun. Not only did he place the Sun at the center of the universe, but Aristarchus also explained the daily rotation of the Earth on its axis and its yearly revolution around the Sun. He believed that all the planets revolved around the Sun and that the Earth's movements caused the apparent movement of the "fixed" stars. He also considered the universe to be much larger than anyone—at the time—believed to be possible. One of his greatest achievements was his *method* for determining the sizes of the Sun and Moon, as well as their distances from the Earth. (However, his calculations were incorrect primarily because his observational techniques were insufficient to the task.)

Seleucus of Babylon (fl. second century B.C.E.)

A Babylonian king, Seleucus is believed to be the only named ancient astronomer to have followed Aristarchus's belief in the movement of the Earth. He lived during the time following the Roman conquest of Persia and the death of Alexander the Great in 323 B.C.E. Seleucus ruled Babylon as the founder of a new dynasty. He was a proponent of the Hellenistic culture and is often referred to as Seleucus the Chaldaean, suggesting that he was a practitioner of Babylonian astrology/astronomy. (Chaldaeans were a Semitic people, well versed in matters of the occult and divination, who ruled in Babylonia.) In this regard, Seleucus discovered the periodic variations in the tides of the Red Sea that he ascribed to the position of the Earth's Moon in the zodiac.

Compilation of Astronomical Maps and Star Catalogs

From the time shortly after the big bang, the stars, the planets, the Sun, and the Moon have shone in the heavens, and it was our most ancient ancestors who first visually charted these celestial bodies and then later invented instruments and devices with which to make more accurate measurements. Ancient *astronomical maps,* invaluable for navigation on both land and sea, were most often illustrated with mythical figures and objects where the outlines of the constellations were immediately recognizable. The stars are usually separated into two groups: (1) the stars that rise and set; and (2) the stars that remain above the horizon, which appear to circle a fixed point in the sky (the celestial pole). The ancients recognized that, when traveling northward, the fixed point was higher. When traveling southward, the fixed point was lower. Only later did ancient astronomers/mathematicians attribute this to the curvature of the Earth's surface.

Babylonians are believed to be the first civilization to assemble astronomical data and events into a formalized system. They tracked the position of the Moon among the stars, charting the appearance and disappearance of the brightest stars on the horizon, and projected imaginary objects and animals among the stars that appeared to be in close proximity to each other, which they called *constellations.* Among these constellations are the *Big Dipper, Orion the Hunter, Cassiopeia,* and *Draco the Dragon.* (Their apparent proximity is actually a line-of-sight effect since the stars that comprise any of the constellations are not necessarily related.) The Sun and Moon and the bright points of light that are the five visible planets (Mercury, Venus, Mars, Jupiter, and Saturn) appeared to move among the dimmer stars in the same path (called the *zodiac*) over the same annual time period. There is evidence on ancient cuneiform texts dating from about 1500 B.C.E. that the Babylonians organized the heavens into 12 sectors, naming each one after a proximate constellation. These are the signs of the zodiac. (Other archaeological findings suggest that the naming of the zodiacal constellations may have originated as early as 4000 B.C.E.) The names given to these signs are based on the Greek and Latin translations of the ancient Babylonian designations: *Aquarius,* the giant or water bearer; *Pisces,* the fish or tails; *Aries,* the hired man or ram; *Taurus,* the bull of heaven; *Gemini,* the twins; *Cancer,* the crab; *Leo,* the lion; *Virgo,* the virgin or the barley stalk; *Libra,* the balance; *Scorpio,* the scorpion; *Sagittarius,* the archer or the god Pablisag; and *Capricorn,* the goat or goatfish.

The Chinese are also believed to have charted and cataloged approximately 1,464 stars at approximately the same time as the Babylonians, grouping them into 284 constellations. The earliest Chinese *star maps* date back to at least 500 B.C.E., and perhaps as far back as 850 B.C.E. Only a handful of these configurations are the same as those of Western astronomy, namely the scorpion, the lion, Orion, and the northern or Big Dipper. Otherwise, the Chinese constellations are different. Ancient Chinese astrologers/astronomers also grouped the stars into 28 *lunar mansions,* each of which was believed to be a section of the celestial equator.

In Greek astronomy, recognition of star groups or constellations can be found in *The Odyssey,* written by Homer (fl. ninth century B.C.E.). Thus it can be assumed that ancient, unknown Greek astrologers/ astronomers identified the constellations of Ursa Major, Orion, the Pleiades (or Seven Sisters), and the Big Dipper. The earliest systematic recording of constellations can be found in *Phaenomena,* written by the Greek poet Aratus in the third century B.C.E. In this text, Aratus described 43 constellations and named 5 stars. The ancient Greeks used maps and globes to chart these celestial bodies. Unfortunately, they have not survived. Astronomical maps can be found in several of the royal tombs from ancient Egypt that date to about 2000 B.C.E. These tomb paintings also depict constellations. However, they are not considered to be accurate representations.

A *star catalog* in ancient times was a systematic record of stars, listed according to their position and brightness (magnitude). Star or stellar brightness is measured in magnitudes. A first-magnitude star is 100 times brighter than a star of the sixth magnitude. It is impossible for the naked human eye to view a star that is fainter than the sixth magnitude. Using this system of magnitudes, Hipparchus, the Greek scholar, compiled the first known star catalog of approximately 850 stars, completing it sometime between 129–150 B.C.E. A century later, Ptolemy of Alexandria expanded Hipparchus's catalog and compiled a listing of about 1,022 stars and 48 constellations. However, because a huge portion of the heavens could not be viewed from the location at which he made his observations (Alexandria, Egypt), Ptolemy's catalog was quite limited—even for its time. In addition, although he listed 15 first-magnitude stars, very few of the more numerous, but faint and barely visible, less-bright stars were included in his catalog.

Discovery of Eclipses, Comets, Novas, and Sunspots

Eclipses

The word "eclipse" comes from the Greek word *ekleipsis,* which means "abandonment." In Chinese, the word for eclipse is *shih,* meaning "to eat." Both of these terms were apt, since it must have seemed to our ancient ancestors as if, on occasion, the ever-present Sun had abandoned them, or in the case of the Chinese, that either a dragon was dining or a heavenly dog had eaten the Sun. In either case, the ancients viewed eclipses as a foreboding of trouble. Neolithic humans and those who came after had no concept of the movement of the Earth, and therefore would likely have ascribed mystical reasons for the solar and lunar eclipses that have always occurred with regularity. (See Figure 2.2.) A *lunar eclipse* takes place when the Earth, aligned with the Sun, casts a shadow through which the Moon passes. A *solar eclipse,* which is less frequent, occurs when the Moon passes across the face of the Sun when the Earth, Moon, and Sun are all aligned. A *total solar eclipse* occurs only at the time of a new moon, when the Moon is directly in the path of the Sun and completely blocks its image. Even though the Sun is approximately 400 times the diameter of the Moon, a total solar eclipse is possible because the Moon is much closer to the Earth than the Sun. A total solar eclipse lasts about 7.5 minutes. A *partial eclipse* is as it sounds and is of a much longer duration, and occurs before and after and on each side of the path of the total eclipse. An *annular eclipse* occurs when the apparent size of the Moon is too small for a total solar eclipse, resulting in a bright ring of sunlight surrounding the Moon.

The ability to predict the precise occurrence of lunar and solar eclipses was first discovered by the Chinese in about 2000 B.C.E., about 500 years or so before Western astronomers were able to do so. The Babylonians and the Assyrians, another tribe living in ancient Mesopotamia, also recorded lunar and solar eclipses beginning in at

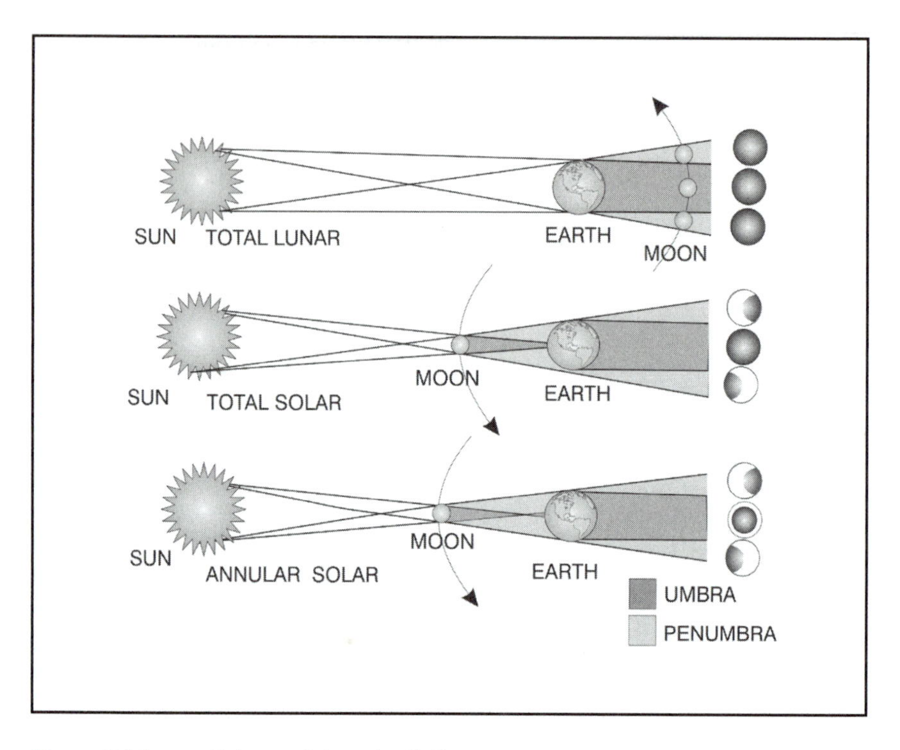

Figure 2.2 Lunar, Solar, and Annular Eclipses
As viewed by an observer on Earth, an eclipse is the complete or partial obscuring of one celestial body by another. A *lunar eclipse* occurs when the Earth, aligned with the Sun, casts a shadow through which the Moon passes. A *solar eclipse* occurs when the Moon passes across the face of the Sun when the Earth, Moon, and Sun are all aligned. An *annular eclipse* occurs when the apparent size of the Moon is too small for a total solar eclipse, the result of which is a bright ring of sunlight surrounding the Moon.

least 747 B.C.E. These careful records written on cuneiform tablets enabled subsequent Mesopotamian astronomers to predict accurate intervals, particularly between lunar eclipses. This led to the discovery by Suidas, the Greek lexicographer, of the *saros,* the 18-year natural cycle over which the sequences of both lunar and solar eclipses repeat themselves. (A more accurate figure is 6,585.33 days.) Hipparchus was the first astronomer/mathematician to use geometry and the occurrence of a solar eclipse to calculate the **parallax** of the Moon, and it was his work that enabled future astronomers to predict lunar eclipses with extreme accuracy. The research done by the ancient astronomers on eclipses is a prime example of how the work of one person built upon the work of others.

Comets

Comets are nebulous celestial formations that consist of rocks, ice, and gases that move around the Sun in large **elliptical** orbits. There are three parts to a comet: (1) the *nucleus* or the center, made of rock and ice; (2) the *coma,* which is composed of the gases and dust that form around the nucleus as it evaporates; and (3) the *tail,* which is made up of gases and spreads out from the coma. The word "comet" comes from the Greek word *kometes,* meaning "hairy," similar to "beard," and later called a "tail." There is some dispute as to who actually recorded the first sighting of what has become known as Halley's Comet. Some historians believe that in about 200 B.C.E. Chinese astronomers recorded the first sighting of this famous comet. However, writings found on ancient Babylonian cuneiform tablets that are now at the British Museum include sightings of comets approximately 76 and 152 years prior to those recorded by other civilizations, including the Chinese. Thus, it is possible that the ancient Mesopotamians may actually have been the first to sight Halley's Comet, which is visible every 77 years or so. The Greeks and Romans also recorded sightings of this famous celestial event.

Novas

Novas, also called "new stars" or "guest stars," appear suddenly, burn brightly for a short time, and then disappear. It was Chinese astronomers in about 1300 B.C.E. who made the first recording of a "guest star" that was visible for about two days, believed to be near the star now called Antares. When a star explodes or collapses, it is called a *supernova,* which is a tremendous explosive event that sends great bursts of electromagnetic radiation (light) into space. This usually occurs once every century or so. The Crab Nebula, which is visible today using power telescopes, is a remnant of the explosion that took place nearly a millennium ago. (A *nebula* is an immense and diffuse cloudlike mass of gas and interstellar dust particles, visible due to the illumination of nearby stars.)

Sunspots

About 2,000 years ago the Chinese also detected dark spots on the Sun's surface. They attributed these blemishes to shadows cast by flying birds. Today we know these are actually sunspots that range from 3,000 to 62,000 miles in diameter. Although the ancients did not comprehend exactly what they were viewing, the Chinese, in particular, were perceptive enough to refrain from looking directly at the Sun when observing both eclipses and sunspots. They used translucent crystals and thin slices of transparent jade to block the harmful rays.

Calculating the Earth's Circumference

Two Greek scholars are famous for calculating the circumference of the Earth: Eratosthenes of Syene (ca. 276–194 B.C.E.) and Poseidonius (ca. 135–ca. 51 B.C.E.). Conventional reasoning would assume that with the scholars having lived centuries apart, Poseidonius's calculation would be the more accurate, since he would have the benefit of more accumulated data and information than that available to Eratosthenes. However, this is not the case.

Eratosthenes, a philosopher and mathematician, had previously determined that on June 21, the date of the summer solstice, the Sun cast no shadow at the bottom of a water well located in Syene (also spelled "Cyrene"). Therefore, he knew that on this date and at this point, the Sun was at its zenith. At the same time, he measured the angle of the shadow from a stick (gnomon) he had placed upright in the ground at a location in the city of Alexandria, which he knew was 5,000 *stadia* from Syene. Both cities were located in Egypt, although some distance apart. (*Stadia* is the plural of stadium, which is from the Greek word *stadion,* the unit of measurement based on the length of the course in a stadium. Today, we know it is equal to approximately 185 meters [607 feet]. However, the figure that was used in ancient times is uncertain.) Knowing how many stadia a camel could walk in a day, Eratosthenes estimated the distance between the two cities by multiplying the distance a camel walked in one day by the number of days it took a camel caravan to make the journey. On June 21 in Alexandria, the angle of the stick's shadow was $7^\circ 12'$, which corresponds to about $^1/_{50}$ of a 360-degree circle. Multiplying 5,000 stadia by 50 equals 250,000 stadia as the Earth's circumference. His calculation (24,700 miles) was very close to today's accepted equatorial circumference of 24,902 miles.

Poseidonius, a Greek philosopher and astronomer and a student of the Stoic school of philosophy at Rhodes, also attempted to measure the circumference of the Earth by assuming that the cities of Rhodes (Greece) and Alexandria (Egypt) were on the same meridian and that the star Canopus touched the horizon at Rhodes at a meridian altitude of $7^\circ 30'$ or $^1/_{48}$ of the circumference of a circle at Alexandria. The distance between these two cities was 5,000 stadia. By multiplying 48 by

5,000, he figured the Earth's circumference to be 240,000 stadia, which was close to Eratosthenes' figure. However, at some point Poseidonius decreased the figure to 180,000 stadia after presumably recalculating the distance between Rhodes and Alexandria. It was Poseidonius's incorrect figure that was contained in Ptolemy's work that Christopher Columbus used more than 1,000 years later on his westward voyages to Asia. This error made his trip much longer than expected, but led him to his discovery of the Americas.

Note: Modern calculations continue to use Eratosthenes' geometric and algebraic methodologies to arrive at the current figures.

Calculating the Length of the Solar Year

The exact calculation of the length of the solar year was a rather late development in the annals of astronomy, for it did not occur until about the beginning of the tenth century C.E. Written calendars have been in existence since about 3000 B.C.E., and primitive calendars of varying forms have been around much longer—some archaeologists say as early as 13,000 B.C.E. Using the regularity of the phases of the Moon, as well as the regularity of the Nile River floods, ancient Egyptians believed the length of the year to be 354 days—close to the correct figure, but nevertheless $11^1/_4$ days short. It is also important to remember that the geocentric view of the universe was the accepted theory until the middle of the sixteenth century C.E., and astronomers and other scientists of the time did not accept or even know that the length of the year is determined by the amount of time it takes the Earth to make one revolution around the Sun, $365^1/_4$ days. One of the greatest of the Islamic astronomers, Abu Abdullah al-Battani, was the first to discover the accurate calculation of the solar year: 365 days, 5 hours, 46 minutes, and 24 seconds, which is remarkably close to today's figures. His discovery led to more exact recalculations for the dates of the spring and fall equinoxes. Accurate dates for the length of the year and for the spring and fall equinoxes were important then, and remain so today, primarily for religious purposes.

Calculating the Earth's Ecliptic

Both Eratosthenes in the second century B.C.E. and Al-Battani in the tenth century C.E. worked on calculating the Earth's **ecliptic.** The ecliptic can be thought of as the apparent yearly path of the Sun as the Earth revolves around it. In other words, it is the angle of tilt of Earth to the solar orbital plane in the sky. This is the major reason why the Earth has four seasons, not because of how close or how far the Earth may be to the Sun. Eratosthenes calculated the ecliptic by applying similar measurements to those he used when estimating the Earth's circumference. His figure was 23°51′20″; Al-Battani's was 23°35′. Today, the plane of revolution of the Earth around the Sun's ecliptic is measured at 23.5°.

Tides

Tides are produced by the gravitational forces of the Moon and the Sun, which pull the solid Earth out of shape by approximately 12 inches. The Sun's gravitational attraction of the Earth is approximately 40 percent of that of the Moon because the Sun is much farther from the Earth than it is from the Moon. At the same time the Earth's gravity is attracting the Moon, the Moon's gravity is attracting the solid Earth. It also attracts water toward the Moon, causing high tide. Conversely, as the Moon is attracting the Earth and water on the side nearest the Moon, the water on the opposite side of the Earth is also at high tide attracted, but to a lesser extent. Thus, low tide results on the side of the Earth opposite the Moon. During a full moon, when the Moon and Sun are in a straight line on opposite sides of the Earth, their combined gravitational pull creates an extra-high tide known as a *spring tide.* An extra-low tide or *neap tide* is produced when the Moon and the Sun are at a right angle to each other, counteracting their respective gravities. (Isaac Newton developed the concept of gravity in the seventeenth and eighteenth centuries C.E. Newton's law states that two bodies of $mass_1$ and $mass_2$, separated by a distance *d,* will exert an attractive force on each other proportional to the square of the distance separating them. The force of gravity is expressed as $F = Gm_1m_2/d^2$, where F is the force, G is the proportional constant for gravity, m_1 and m_2 are the masses of the two bodies, and d^2 is the square of the distance between them.)

It is believed by historians that the explorer Pytheas of Marseilles (ca. 360–ca. 290 B.C.E.) was the first of the Greek astronomers to correctly relate and record the movement of the tides with the phases of the Moon. He did this while sailing out of the Mediterranean Sea into the Atlantic Ocean to the British Isles in about 300 B.C.E. It was at this time that he experienced coastal tides for the first time. Two centuries later, Poseidonius also correlated variations in the tides with the phases of the Moon. (An interesting point: Galileo rejected this theory in the seventeenth century C.E.) However, the Chinese were in all likelihood the first to recognize the connection between the tides and the cycles of the Moon. The tides along the eastern shores of the Asian continent have a tremendous range, and thus were the subject of close observation and hypothesizing. By contrast, the tides along the Mediterranean coast were less noticeable.

Instrumentation

The most important and obvious of instruments used by astronomers today is the telescope, a device that was not invented until the early seventeenth century C.E. All the astronomical theories that had been correctly proffered by ancient astronomers through Copernicus were developed without benefit of this most essential tool. First and foremost, the ancient astrologers/astronomers were observers who used philosophical and mathematical principles to explain nature and the universe. The Greeks, in particular, were outstanding mathematicians who applied geometry and trigonometry and, to a lesser extent, algebra, to seek solutions for and develop theories dealing with celestial matters. Thus, as far as Greek instrumentation is concerned, many of the instruments are, in fact, geometers' tools that were adapted to conform to the ever-expanding field of astrology/astronomy. Other civilizations constructed instruments that would essentially aid them in their observations of the stars, planets, Moon, and Sun. For example, ancient Egyptian astrologers/astronomers invented a forked stick with an attached cord or string to follow the movements of a set of 36 specifically chosen stars, which they called the "Indestructibles." The cord or string hung to the ground at a right angle. Ostensibly, this enabled the astrologer to view the stars through the tines of the forked stick and record their changing positions. (This device does not have a specific name attributed to it.) Another astronomical device used to determine the width of the Sun was merely a block of wood with an eyehole. The block of wood slid back and forth within a wooden frame. Thus, the viewer could make his estimations as to the size of the disk of the Sun.

The most complete record of ancient astrological and astronomical instruments is contained in Ptolemy's *The Almagest.* As with many of the ancient discoveries, the exact year of the invention, as well as the country and name of the inventor, are often unknown. A detailed listing of all the ancient astronomical devices is beyond the scope of this book. However, a brief description of the most unusual and most important follows.

Gnomon

This device, also called a shadow-pole, is named after the Greek word *gnomon,* meaning "one who knows." It is simply a vertical stick that casts

Figure 2.3 Chinese Sighting Tube
This lens-less tube blocked out all other sources of light, thus enabling ancient Chinese astronomers to observe a single object.

a shadow and which is placed in the ground in a flat and open area. (Eratosthenes used this device when he measured the Earth's circumference.) At sunrise and early morning, the shadow is quite long; as the morning progresses and the Sun continues to rise, the length of shadow decreases. At high noon, the shadow is the shortest because the Sun is at its highest point in the sky. The gnomon's shadow increases as the Sun moves westward, until it finally sets. Ancient astronomers utilized this tool to observe the summer and winter solstices and the vernal and autumnal equinoxes.

Sighting Tube

The ancient Chinese developed a telescope-like device that was fashioned in the shape of a tube, but without lenses. (See Figure 2.3.) It blocked out all sources of light, except for a tiny opening, which allowed them to observe a single object, for example, a faint star, in the night sky. It enabled the viewer to track and to accurately chart the movements of the planets and stars above the horizon.

Astrolabe

It is generally accepted that Greek astrologers, in either the first or second centuries B.C.E., invented the *astrolabe,* an instrument that measures the altitude of stars and planets above the horizon. Some histori-

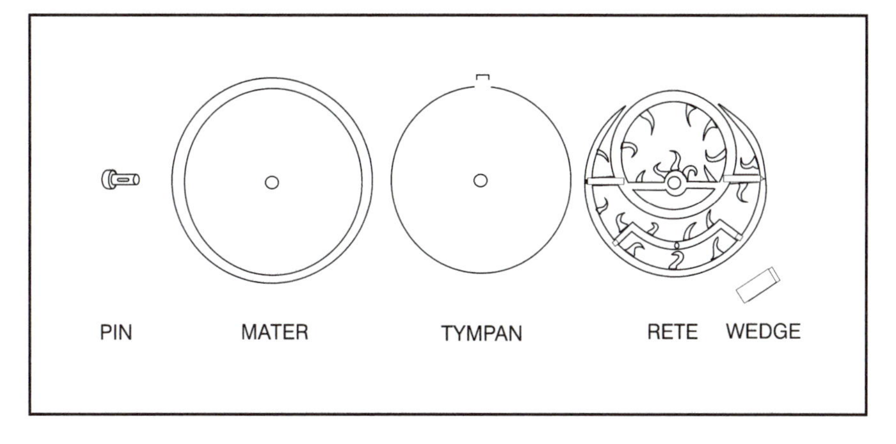

Figure 2.4 Ancient Plane Astrolabe
The *astrolabe,* invented by the Greeks either in the first or second centuries B.C.E., was the first instrument to measure the altitude of a star or planet above the horizon.

ans attribute its invention to Hipparchus, although there is no written verification for this assertion. (The word is a combination of two Greek words, *astro* meaning "star" and *labio* meaning "finder.") There is also some question as to the extent to which ancient Greek astrologers and astronomers actually used the astrolabe, often referred to as "the mathematical jewel." In addition, the term "astrolabe" has been applied to a number of different instruments. But it is the plane astrolabe that has been the most popular and useful. (See Figure 2.4.) It consists of one or more flat, circular brass disks pierced at their centers. The disks pivot around a pin at the center point. One of the disks (rete) is a star map that shows the positions of the brightest stars, as well as the paths of the planets and the Sun; the other (plate) displays the fixed local coordinates—meridian, the horizon, and the lines of altitude at 5°, 10°, and so on, above the horizon. The disks are encased in a brass circular form with the scale of hours written on its rim. The Greek mathematician Hypatia of Alexandria (ca. 370–415 C.E.), believed to be the only female scientist of the ancient world, also designed an astrolabe, as well as other instruments. The astrolabe was especially helpful in determining latitude or the horizontal position of the Earth. Over the centuries, there were variations in the style and characteristics of the astrolabe, but its basic design remained essentially unchanged. Astrolabes were widely used by astronomers in the East and the West, where it was a symbol of their knowledge and skill, and by ancient seafaring navigators. The astrolabe is the forerunner of the *sextant,* an instrument designed to measure the elevation of the Sun above the horizon at noon, as well as latitude. It was also a testament to the ingenuity of the Greek scholars who designed it by combining geometry with astronomy.

Armillary Sphere

Early astronomical instruments, armillary spheres were three-dimensional models of a skeletonlike celestial globe, usually made of brass, with circles divided into degrees for angular measurement. They were used to measure the positions of the stars and planets and the relationships among the principal celestial circles. *Armilla* is the Latin word for "ring." Primitive models were most probably invented in either ancient China or Greece in the last centuries B.C.E. or at the beginning of the first millennium C.E. The Chinese astronomer Zhang Hen (ca. 78–ca. 139 C.E.) designed an armillary sphere that contained three rings. One represented the equator, one showed the paths of the Sun, Moon, and planets, and the third delineated the poles of the Earth. The pole of the sphere pointed northward, and a tubelike device was attached at the center through which the astronomer viewed and mapped the stars. For example, using the armillary sphere, it was possible to locate a particular star's direction and distance from the North Pole. The Chinese continued to improve and retool their design of the armillary sphere well into the thirteenth century C.E. The Alexandrine Greeks designed an armillary sphere that contained nine rings, also known as the *meteoroskopion,* which represented the great circles of the heavens, among them the horizon, meridian, equator, tropics, polar rings, and an ecliptic hoop.

The great astronomer Ptolemy designed several armillary spheres, including the *astrolabon,* which is considered to be his most important. (See Figure 2.5.) This instrument consisted of a group of metal rings that were pivoted one with the other, nesting within a larger circle that was affixed to a heavy base. The innermost rings contained an **alidade** with pinholes through which the astronomer could view celestial bodies. The astrolabon allowed the astronomer to measure angles, such as the ecliptic axis and the polar axis, directly, rather than engage in complicated and lengthy calculations necessary for astronomical theories. Islamic astronomers utilized armillary spheres and the astrolabon extensively. It was their design of these instruments, introduced onto the European continent after the Moorish invasion, that were the models for those used by medieval European scholars and astronomers.

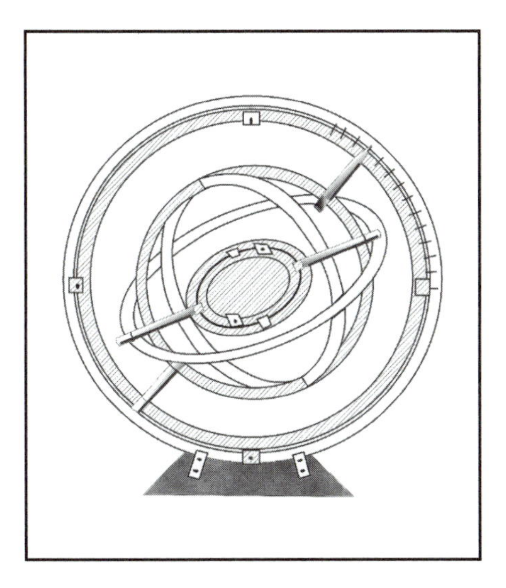

Figure 2.5 Armillary Astrolabon Invented in the first century C.E. by the Greek astronomer Ptolemy, the *armillary astrolabon* was a variation of the armillary sphere, a three-dimensional model of a celestial globe used to measure positions of the stars and planets and the relationships among the principal celestial spheres.

Summary

Astronomy and cosmology as we think of them today are sophisticated and far-reaching sciences that almost routinely astound and spellbind us with discoveries of star systems and planets, and yet-unexplained phenomena existing at unfathomable distances. All we know of our universe today stems from the innate curiosity of our most ancient ancestors, who first gazed into the skies and eventually questioned the relationship between themselves and the distant objects that could be seen far above them—albeit in the most primitive terms. If we ponder how far we have come on this earth as a species and how much knowledge we have accumulated in such a short period of time, geologically speaking, we must stand in awe of the discoveries and inventions made by ancient mathematicians, philosophers, and astronomers who used deductive reasoning as well as intuition to determine humankind's place in the universe.

3

BIOLOGY, BOTANY, AND ZOOLOGY

Background and History

As defined in modern-day terminology, biology is that branch of the natural sciences that studies living organisms and systems (cellular and acellular). It encompasses several specialized areas of living organisms, such as cytology (the study of cells), zoology (the study of animals), botany (the study of plants), and ecology (the study of environmental systems). Although ecology will not be discussed in this chapter, it is important to understand what ecology is. *Ecology* is the scientific study of the interrelationships of organisms and their physical, chemical, and biological environments. It is a study of systems. As a science, it draws on many other disciplines, such as physics, chemistry, geology, climatology, oceanography, economics, and mathematics. Ecology is often confused with environmentalism; the latter is often considered a political movement rather than a science.

However, the term "biology" actually originated about 1802 with Jean Baptiste Monet, Chevalier de Lamarck (1744–1829), when he was developing his theories of evolution. The fact that biology, as a body of science, was not identified until the beginning of the nineteenth century does not mean that biology, as a body of knowledge, was unknown. Quite the contrary, since the beginning of time, humans have been "practicing biologists." The roots of agriculture and animal domestication were planted millions of years ago with the recognition that both plants and animals were part of the natural cycle of life that could be bred, cultivated, and propagated. It is reasonable to assume that, by trial and error, ancient humans selected a variety of plants that sustained nourishment without adverse affects. In other words, they knew which plants they could and could not eat safely. They knew where certain plants could be grown and which plants would aid in the healing of a wound or some other ailment. The same was true for the animals they hunted and killed. Trial and error played some part in the hunting of certain animals over others. In other words, ancient humans knew which animals provided the greatest amount of meat to feed the community, as well as hides for clothing and shelter, and bones for weapons and other implements. They had no concept of kingdom, phylum, class, order, family, genus, or species. But they did know that animals mated with animals of the same group, just as humans mated with other humans. The awareness of life processes and the surrounding environment were not apparent with the appearance of the first hominids. Rather, the passage of millions of years was needed in order for *Homo sapiens sapiens* to evolve into the dominant species capable of marshalling the resources of nature to survive and control Earth's often harsh environment.

As was discussed in chapter 1 ("Agriculture and Animal Domestication"), the advent of settled farming transformed the history of civilization. The increase in the human population and the eventual improvement of its health and lifespan can be attributed to better nutrition and a more stable food supply. Approximately 10,000 years ago—and perhaps as early as 12,000 years ago—Neolithic farmers began to cultivate the first seeds for crops. With each century and millennium, the crops and harvests improved and the herds of domesticated animals increased, all without the benefit of an organized taxonomic system to guide ancient humans. The absence of an accepted formal classification system for living organisms did not mean an absence of curiosity. Theories and myths concerning the origin of life abounded among ancient peoples. Historically, humans have always attempted to explain what they do not understand. Our innate and intense desire to understand ourselves and our surroundings was and is primary to our survival and dominance as a species. And if the theories and myths that ancient humans developed about the world around them seem naive and ludicrous by today's standards, they seemed completely reasonable and acceptable to civilizations that were responsible for some of the earliest inventions and discoveries that mankind has ever known.

The Greeks of Ionia

The history of the biological sciences has its roots in (1) the Iron Age (ca. 1400–1200 B.C.E.), (2) the development of the Phoenician alphabet, and (3) the civilization of ancient Greece. And in order to understand why the biological sciences began when and where they did, it is necessary to provide some historical background on the events that led to the major biological discoveries of the ancient world. The Sumerians of Mesopotamia were the first to develop a form of writing known as *cuneiform.* This was a system of wedge-shaped characters, usually inscribed on clay tablets, which dates back to about 2300 B.C.E. The Egyptians also had a form of writing that dated back to 3000 B.C.E., commonly called *hieroglyphic script,* which was more adapted for transcribing on papyrus. Using these two systems of writing, a millennium later the Phoenicians developed an alphabet upon which the Indo-European and Semitic scripts were based. The discovery of iron smelting led to the development of iron weaponry and subsequently to the continued but more efficient vanquishing of one civilization by another. Warfare and invasion overtook Asia Minor, Mesopotamia, Egypt, and the mainland of Greece, which subsequently fell to its own barbarian tribes. The civilizations that flourished were those that exploited iron in all its forms and utilized alphabetic writing, primarily in seagoing exploration and commerce. The "Hellenes" (the term "Greek" was not used until centuries later by the Romans) were illiterate, as were all civilizations, until the introduction of the Phoenician alphabet by their own traders sometime in the eighth century B.C.E. Also, regardless of the region, ancient Greeks shared a common language and conversely a common culture, both important components in the development of classical Greek civilization. The Hellenistic tribes that were most prominent at this time included the Ionians, the Dorians, and the Aeolians. The Greeks took to the sea primarily because of the limitations of the mainland of Greece. It simply could not sustain the growing population. The Greeks also held themselves to be different from other civilizations and were openly contemptuous of what they considered to be barbaric behavior, even though they themselves descended from barbarian tribes. Religion was also an important part of Hellenistic identity. The Greeks were proficient geometers and mariners who appropriated ideas and traditions from other civilizations, mainly Egyptian and Mesopotamian, and inculcated them into their own unique and often imperious philosophy of life and science.

In particular, the seafaring Ionians, who traded regularly with other ancient and settled countries, such as Egypt, Phoenicia, and India, were exposed and receptive to foreign influences. A wealthy Ionian merchant and the son of a Phoenician woman, Thales of Miletus was the first of the natural philosophers who hypothesized about the Earth and its origin. During his lifetime in the late seventh and early sixth centuries B.C.E. he professed that the Earth, a cylinder or disk, had its origins in and floated upon an immense body of water. He believed that water provided a continuous "cycle of existence" that passed from the sky and air to earth and to the bodies of plants and animals and then back to the air and sky, and so on. Two other Miletians and followers of Thales, Anaximander and Anaximenes, continued the study of natural law. Anaximander added a fourth element, fire, to the three elements of air, earth, and water that the ancient Babylonians, Egyptians, and Greeks believed comprised the known world. He also taught that these four elements separated out of a primal substance as ordered strata: earth, water, air, and fire. His poem *On Nature* is often credited with being the first written work on natural science and encompasses what is believed to be the first written theory on evolution. Anaximander believed that life had come from water and that in the beginning humans were like another animal, namely fish. Conversely, Anaximenes (ca. 570–500 B.C.E.) taught that air rather than water was the essence of life, and that the Earth was created from a formless substance called "aperion." Obviously, these were all misconceptions, but they were the genesis of the biological or life sciences. Another Greek philosopher, Xenophanes of Colophon (ca. 560–478 B.C.E.), mounted the first systematic challenge to the **anthropomorphic** view of nature that was widespread at that time in our history. Humans began to question and examine the diversity and variety of the world around them and to formulate answers that were based, not on myths, but on the very elements of the Earth itself.

The Greeks continued to trade with Persia, which under its Emperor Darius I (522–486 B.C.E.) made ever-increasing advances to the West. As a result of these Western influences, the nature of Hellenistic, mainly Ionian, natural philosophy changed. It was no longer left to the merchants or sailors and members of the leisure class to ponder the origins of life. Rather, "thinking" became an acceptable profession. One of the most influential of the "thinkers" was Herodotus (ca. 484–425 B.C.E.), who came from another Ionian city, Halicarnassus. Herodotus is considered to be the father of anthropology. Using primarily empirical observations, he composed a work entitled *History* that is considered to be the first thesis on the science of humans.

Athens—the Center of Ancient Learning

After the defeat and subjugation of the Ionian towns by the Persians in 530 B.C.E., the city-state of Athens assumed the leadership of the other Hellenistic cities against the Persians. The Ionian town of Miletus was completely destroyed in about 527 B.C.E. The Athenians ultimately defeated the Persians on land in 490 B.C.E. and at sea in 480 B.C.E. Afterward, the city-state of Athens embarked upon a period of power and affluence. This continued until the advent of the Peloponnesian Wars in 431 B.C.E., which resulted in the defeat of Athens by the Spartans in 404 B.C.E. The ancient Greek philosopher Socrates (ca. 470–399 B.C.E.) lived during this time of upheaval and chaos. His influence was substantial, quite positively in the study and practice of ethics, morals, and politics, but quite negatively in the study of natural science. As a result, his rejection of natural philosophy (natural science) was so pervasive that it discouraged the study of science and nature for decades. On the other hand, his belief that the body was a temporary abode for the soul, which lived on after death, served as the justification for dissection of the human body by Greek anatomists and physicians. As is always the case, other scholars and philosophers challenged Socrates' beliefs, namely Plato (ca. 428–348 B.C.E.) and Aristotle (384–322 B.C.E.).

Plato characterized his relationship with Socrates as that of a friend, while historians believe he was actually a disciple or pupil. Regardless, Socrates had a formative influence on the young Plato, who believed it was necessary to formulate a theory on the nature of the universe that could be subordinate to ethics, politics, and theology. He was concerned with "forms" and their hierarchical arrangements. While he is often considered to be the most important figure in Western philosophy, Plato also influenced the development of science in that he believed in the mathematical character of nature. As a matter of fact, an inscription over the vestibule of Plato's Academy in Athens read: "Let no one enter here who is ignorant of Geometry." Plato believed in a perfect universe. However, in order to rationalize the imperfections of nature, he developed his belief system of dual realities, that is, an *Essential universe,* which was the perfect universe, and the *Perceived universe,* which was an imperfect facade that was evident to us but that would not affect the perfect Essential reality that lay beneath. Plato believed that every living organism (humans, animals, and plants) possessed this dual reality. Thus, this belief system affected any assumptions that the ancients formulated about the species of living things.

A student of Plato, Aristotle distanced himself from his teacher. Aristotle openly questioned the philosophy of dual realities espoused by Plato and formulated his own philosophy of nature, which stated form and matter are always joined. Aristotle, in the estimation of many historians, was one of the most influential humans who ever lived. He was a philosopher concerned with classes and hierarchies rather than a scientist concerned with observations and evidence. For Aristotle, and for most Greek philosophers, observation was the accepted means of investigation. Whatever happened spontaneously was considered to be natural. Experimentation that altered the natural state in order to illuminate that which was hidden was considered "unnatural" and therefore could not be valid. However, Aristotle's philosophy, methods of reasoning, logic, and scientific contributions are still with us and continue to be respected.

Taxonomy

One of the basic characteristics of humankind is the ability to classify things in the environment. This aptitude was necessary for survival. For example, when gathering berries, it soon became apparent and necessary to distinguish the "good" berries from the "bad" varieties. Experience taught our early ancestors to generalize the types of foods, both plant and animal, suitable to sustain life—or those which were dangerous and should be avoided. Once they arrived at generalized "classes" of foods, early humans were able to "stereotype" from a general group to a more specific group. In other words, they knew which foods were desirable versus undesirable. This ability to generalize, classify, and stereotype has been a part of human nature since the beginning of time.

Historians almost universally attribute the beginnings of biology, in particular, and science, in general, to the work done by Aristotle. His approach to nature was primarily *teleological,* from the Greek word *telos* meaning "object" or "end." In other words, the components of all living things (animals and plants) can only be explained by what functions they perform in and for that particular organism. Aristotle was a prolific writer whose topics, in addition to biology and botany, covered physics, astronomy, meteorology, economics, politics, and ethics to name a few. However, his entire philosophy was most influenced by his examination of the natural world and his biological observations. His studies on animals transformed the belief systems of his time. It was not until the early medieval and Renaissance periods in Europe that Aristotle's works were translated into Latin, thus shaping the development of what would be termed "modern biology." Further investigations and discoveries in the nineteenth century, especially Darwinism, corrected and expanded what were the generally accepted biological principles of the time.

Taxonomy is derived from the Greek word *taxis,* which means "arrange." However, like "biology," "taxonomy" was not a term that was used by Aristotle or his contemporaries. In fact, ancient scientists had no scientific nomenclature to access, relying only on the classical languages of the day (Greek and Latin) to explain and describe their discoveries. Taxonomy and classification of living organisms are used synonymously and refer to the arrangement of living organisms into a hierarchical group based on their similar biochemical, anatomical, or physiological characteristics. More recently, **DNA** is used to "sharpen" the classifications. Today, we know that the taxonomic levels, from the most general to the most specific, are as follows:

kingdom → ***phylum*** (for animals); (***division*** for plants) → ***subphylum*** (intermediate taxonomic level) → ***class*** → ***order*** → ***family*** → ***genus*** → ***species.***

As an example, humans are members of the ***animal kingdom*** characterized by the ***chordata phylum*** (at some point in human development we had a notochord or primitive backbone from which the spine developed) and the ***vertebrata subphylum*** (an animal with a backbone) and the ***mammalia class*** (a warm-blooded animal, usually but not always with hair, that feeds its offspring with milk from the mammary glands of the female) and the ***order of primates*** (mammals with specific physiological traits, e.g., five digits on all limbs, generally with an opposable thumb and big toe) and the ***hominoid family*** (animals that resemble humans) and the ***homo genus*** (e.g., bipedalism, increased cranial [brain] capacity) → and the ***sapiens species*** (from the Latin word *sapient,* meaning wisdom).

In the fourth century B.C.E. Aristotle wrote *Historia Animalium* (History of Animals), a series of 10 books, in which he formulated the first taxonomic or classification system to be developed in the biological sciences. (Some historians dispute this characterization, rather believing that *Historia Animalium* is merely a volume of uncategorized data.) Aristotle proposed that species could be arranged in a hierarchical order that included the lowest or least complex (inanimate matter) to the highest or most complex (humans). Aristotle wrote the following passage in *Historia Animalium* that sets up his theory of animal and plant life:

Nature proceeds by little and little from things lifeless to animal life, so that it is impossible to determine the exact line of demarcation, nor on which side thereof an intermediate form should lie. Thus, next after lifeless things in the upward scale, comes the plant. Of plants one will differ from another as to its amount of apparent vitality. In a word, the whole plant kind, whilst devoid of life as compared with the animal, is yet endowed with life as com-

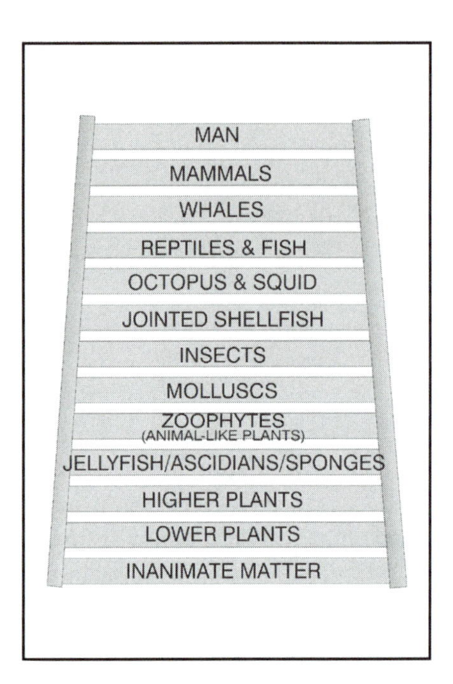

Figure 3.1 Aristotle's Ladder of Nature
Alternately called *Scala Naturae* or *Chain of Being,* Aristotle's *Ladder of Nature* was the first attempt to arrange each species on an ascending "ladder." Aristotle was the first to use the terms *species* and *genus,* as well as classify living organisms from the simplest (inanimate at the bottom) to the most complex (humans at the top).

pared with other corporeal entities. Indeed, there is observed in plants a continuous scale of ascent toward the animal.

Naturalists and other historians have characterized Aristotle's arrangement or classification of living organisms as the *Scala Naturae* (also called the "Ladder of Nature," the "Great Chain of Being," and the "Ladder of Life"). (See Figure 3.1.) Aristotle did not believe in any sort of evolutionary theory. Rather, he believed in a perfect world and universe in which each species is permanent and unchanging and has an allotted place on an ascending "ladder." He originated the terms "species" and "genus," the Latin translations of the Greek words *eidos* (species) and *genos* (genus) that were used by Aristotle in his writings. (There is some controversy in the actual translation of Aristotle's work, since Aristotle occasionally used the word *eidos* to refer to form or formal material, and *genos* as matter.) Species refers to a group, the members of which have a set of common attributes; genus refers to a larger group, the members of which possess a similarity as to size, shape, and outward features. To Aristotle, nature proceeds from tiny lifeless forms to larger animal life, and thus living organisms were arranged in three classes: (1) vegetable, which possessed a nutritive soul; (2) animals, who were able to move and thus had a sensitive soul; and (3) humans, who had intelligence and thus a rational soul, and who also possessed souls of all the types of creatures. Reproduction identifies those giving live birth (**viviparous**) as being mammals and humans, while those laying eggs (**oviparous**) are subdivided into birds, reptiles, fish, and insects. His *Scala Naturae* listed inanimate matter at the bottom, progressing upward from lower plants and higher plants, both of which are capable of self-nutrition; to minor water organisms that have an ability to feed themselves; to shellfish, insects, fish, reptiles, whales, and mammals, all of which have varying degrees of movement, consciousness, and desire; and finally on the top rung of the ladder, humans, who encompass all the attributes of other animals plus the ability to think and reason. As with most of his contemporaries, Aristotle, the philosopher, was concerned with the four elements of air, earth, water, and fire, and their individual characteristics were significant. He valued hot above cold, moist or wet above dry. Blood was both warm and moist, and thus was a principal characteristic with which to differentiate animals.

In addition to the four classes of matter (air, earth, water, and fire) Aristotle believed that a fifth medium, called *aether,* existed. He felt this was necessary to explain the movement of celestial bodies. In other words, he believed that since all celestial bodies move in perfect circles, there must be a perfect medium for this to occur. As an example, Aristotle and others of his time felt that the Sun's heat could not reach the Earth without some form of "matter" transporting it.

Aristotle's taxonomic system, though imperfect, was nevertheless important for future biologists. This may seem somewhat contradictory given the fact that both Aristotle's work on a classification system and other zoological studies were widely ignored after his death in 322 B.C.E., and he held an unyielding position that a higher form of life could never arise from a lower form of life. Nevertheless, Aristotle's categorization of living organisms into a hierarchical chain expedited the work of future scientists who, through further observation and investigation, formulated contemporary evolutionary theories. In the thirteenth century C.E. Albertus Magnus (ca. 1220–ca. 1292 C.E.) attempted to facilitate many of Aristotle's concepts into Christianity. However, it was not until the eighteenth century that an accurate taxonomic system for both plants and animals was actually devised by Carolus Linnaeus (1707–1778 C.E.). It is still in use today. And of course, Charles Darwin's book *On the Origin of Species by Means of Natural Selection,* written in 1859, proffered his theory on organic evolution. Darwin himself referred to Linnaeus and Georges Cuvier (1769–1832 C.E.), a French biologist who formulated theories on animal classification and evolution, as "mere schoolboys" in comparison to Aristotle. In other words, Aristotle laid the methodical foundation upon which it was possible for subsequent classification and evolutionary theories to be built.

Zoology and the Classification of Animals

Among all Aristotle's writings, the earliest are those concerned with biology. These were written in large part during his residency in Asia Minor from 347–342 B.C.E., when he was employed for a time as a tutor to the young prince Alexander of Macedon, more commonly known as Alexander the Great. Aristotle's study of animals began with his admiration and respect for the natural world, as well as the writings of other ancient philosophers, alchemists, and astronomers who recorded their own interpretations of the mechanics of the universe. He also accumulated information from the workmen of the time (fishermen, hunters, etc.) along with his own observation of animal behavior and structure. He recognized similarities and parallels of anatomy and other characteristics between humans and animals. He understood that animals provided humans with essential components for survival, and he appreciated animals for their aesthetic beauty. Above all, Aristotle's motivation was simply to understand why things were as they were. To that end, Aristotle studied animals, and in particular sea creatures, in his pursuit "to know." *Historia Animalium* is the longest of his writings and each of the 10 books covers specific topics as follows:

Book I—an overview and comparison of animals, suggestions for classification into groups, and description of characteristics and habits

Book II—description of the parts of animals; common attributes; animals arranged into groups, such as viviparous, oviparous, quadrupeds or four-footed animals, birds, fish (only red-blooded animals are described in this book).

Book III—internal organs are described, as well as the essential component body parts, such as bone, marrow, cartilage, nails, hoofs, horns, beaks, hair, scales, sperm, etc.

Book IV—description of bloodless animals: cephalopods or mollusks, crustaceans, insects, etc.

Book V—reproduction systems

Book VI—continuation of reproduction systems of several classes of animals, such as birds, fish, quadrupeds; external influences on reproduction

Book VII—human reproduction; description of humans from birth to death

Book VIII—nature and habits of all animals as observed by Aristotle, including food sources, migrations, disease, climatic influences

Book IX—continuation of Book VIII; relationships of animals to each other: companionship, hostility between species, etc.

Book X—most probably a continuation of Book VII; also infertility in humans

Aristotle, who is considered the father of the science of zoology, classified approximately 540 animal species according to the variations in their forms and performed extensive animal dissections. He did not perform human dissections, since human dissection was not accepted practice until circa 280 B.C.E. when the Greek surgeon Herophilus (ca. 335–280 B.C.E.) performed the first public dissection at the University of Alexandria. (However, it is widely thought that as early as 500 B.C.E. the Greek philosopher and physician Alcmaeon of Croton performed dissections on human and animal cadavers.) Rather, Aristotle relied on other animals for his information and subsequent conclusions. As he stated, "the inner parts of man are for the most part unknown, and so we must refer to the parts of other animals which those of man resembles and examine them."

Aristotle's topological-species theory (or in modern terms, his taxonomic or classification system of animals) states that an ideal form is a living group in which each member resembles each other, but the group is distinct in structure from members of other groups. In other words, each living thing has a natural built-in pattern that, through reproduction, growth, and development, leads to an individual type (species) similar to its parents. Aristotle believed that all living species reproduce as to type (e.g., humans beget humans, cattle beget cattle), but he considered a possible exception to this theory when applied to the lowest of the species. He recognized that this categorization could not be done simply or easily, owing to variations in individual species. For example, the dolphin resembles a fish and lives in the oceans, but it develops its offspring internally, nurses its young with milk, and is warmblooded. Thus, he decided to classify cetaceans with other mammals rather than with fish, which, visually at least, seemed to make more sense. Others, in fact, did group dolphins with fish, at least throughout the medieval period. Aristotle was, thus, ahead of his time.

Aristotle maintained that there were two main groups of animals: first, vertebrates with red blood, either viviparous or oviparous; and second, invertebrates without red blood, oviparous, **vermiparous,** producing buds, or reproducing spontaneously. Within those two main groups were 12 levels of "perfection" based on embryological exemplars. The highest levels consisted of animals that were *warm and moist* and gave live birth (e.g., all mammals). The next levels consisted of animals that were *moist* but not *warm,* and that laid eggs that formed inside the female (e.g., sharks). Next came the *warm, dry animals,* such as birds, that laid eggs that were complete (i.e., the eggs maintained their size and shape before the animal hatched). After the *dry animals* came the *cold animals* that laid incomplete eggs (e.g., frogs that laid eggs that changed shape and size before tadpoles hatched). Below the *cold animals* were the cephalopods (octopi), crustaceans (crabs, lobsters), insects, worms, mollusks, sponges, lower plants, and finally inanimate matter. It is important to note that his "Ladder of Nature" contained no sharp delineation between species, since Aristotle created this chart based only on "assumption" and empirical observation. As Isaac Asimov stated in his book *A Short History of Biology,* "Aristotle was not an evolutionist.... the preparation... set up a train of thought that was bound, eventually, to lead to the evolutionary concept."

Aristotle also wrote four other volumes on animals, namely, *Parts of Animals, Movement of Animals, Progression of Animals,* and *Generation of Animals,* in which more detailed and descriptive studies of the characteristics of a number of species are recorded.

The history of biology (zoology and botany) after Aristotle is guided in great measure by the Roman Empire and by the emergence of Christianity. Although it did not employ trial and error or experimentation in its conduct, Aristotelian philosophy affected the character of Hellenistic science inasmuch as it provided the foundation for a systematic method or approach to biological examination. However, the adoption of Aristotle's biological teaching and beliefs was not widespread. In fact, only his student Theophrastus (ca. 372–287 B.C.E.) continued Aristotle's work and expanded on it, primarily with the study of plants and their classification. After Aristotle's death, the practice of the "biological" and "zoological" sciences shifted from Athens in Greece to Alexandria in Egypt and to Rome. There the focus was on the practice of the healing arts, anatomy, and medicine.

The Roman philosopher and naturalist Gaius Plinius Secundus (Pliny the Elder) (23–79 C.E.) also played a role in the classification of animals and the zoological sciences, albeit a somewhat controversial one. He is the author of *Natural History* or *Historia Naturalis,* a 37-volume series that is essentially a plainly written "encyclopedia" of accumulated knowledge covering a wide range of subjects, from astronomy, biology, and metallurgy to black magic. He relied heavily on the writings of others, including Aristotle and Theophrastus, always careful to credit other ancient writers and to cite appropriate sources in an effort to bolster his reputation as a scholar of scientific research. *Natural History* contains some 500 individual sources of reference. However, he was a proponent of folklore and magic and oftentimes distorted or mistakenly interpreted translations of other scholarly works. Thirteen of the 37 volumes are devoted to biology (zoology and botany). However, some historians maintain that Pliny was completely opposed to any organized system of classification. Pliny's biological belief system maintained that nothing existed in and of itself but only in its relationship to humans as a source of food or medicine or as a "danger" that potentially could be interpreted as a moral lesson. These principles were in keeping with the beliefs of early Christianity and, therefore, helped to give Pliny credibility. Nevertheless, he traveled extensively, made his own observations of the natural world as he saw it, and attempted to make genuine contributions to the body of scientific knowledge of his time. While exploring the volcanic eruption of Vesuvius in 79 C.E. he died after inhaling toxic gases from the massive explosion. Pliny was held in the same regard as Aristotle, at least through the fifteenth century C.E., when his writings came under open attack by Niccolo Leoniceno (1428–1524), a teacher of mathematics and philosophy at the University of Ferrara. Leoniceno also translated ancient Greek and Latin texts, especially those dealing with medicine and life sciences. Afterward, Pliny's influence lessened, and by the end of the seventeenth century C.E., it had all but faded from scientific consideration.

In the second century C.E., the Greek physician Claudius Galen (ca. 129–ca. 199 C.E.), who is often considered to be the last of the ancient Greek biologists, became an influential force in the fields of biology and medicine. He was not a philosopher or theorist in the sense of Aristotle, or even Pliny, but a practitioner and anatomist who derived his information from the gladiatorial arenas of ancient Rome. As a physician, he was able to observe firsthand the innards and anatomy of wounded and slain gladiators. Although not a Christian, he believed in a single deity and that humans and animals live for a purpose. At the time, this belief reinforced the increasingly popular Christian view of life and, as such, accounted for his popularity, which lasted for centuries. It was not until the sixteenth century that most of Galen's concepts were challenged and his mistakes corrected. The acceptance for almost 1,500 years of his concepts in biology and medicine has become known as the "Tyranny of Galen."

Botany and the Classification of Plants

Although Aristotle wrote prolifically about animals, he composed only one comparatively short treatise on plants, entitled appropriately *De Plantarum* (On Plants), in which he applied the same principles to plants that he had to animals. To Aristotle, a living organism's "perfection" was determined by its physical structure, which was subsequently ruled by its nature and character. Although plants grow and reproduce, they are simpler organisms and, as such, possess a "vegetative" soul or psyche that is vastly less complex than that of animals. However, long before Aristotle wrote about their structure and psyche, ancient humans "classified" plants. They were the sources of food, medicine, shelter, clothing, tools, and weapons for our most ancient ancestors, who categorized plants based on their function, usefulness, and danger. The development of settled farming was a natural outgrowth of biological classification inasmuch as the decision to plant and cultivate certain types of plants and grains was based on a kind of intuitive classification system that was practiced by even the most primitive of humans.

With respect to the formal classification of plants, however, it was one of Aristotle's pupils, the Greek philosopher Theophrastus of Eresus, who was regarded for many centuries as the authority on the **morphology** and **physiology** of plants. He is still referred to as the father of botany. After Aristotle's death in 322 B.C.E., Theophrastus succeeded his master as head of the Lyceum in Athens, a school that was founded in opposition to Plato's Academy. Theophrastus, who studied with Plato before his association with Aristotle, was known for his autonomy of thought. Notwithstanding this independence, he continued Aristotle's research into living organisms, utilizing many of the same principles and techniques as his teacher, including a rejection of any and all evolutionary theories. Theophrastus's botanical works were exceptionally well arranged. They were accepted by and influenced botanists for centuries. His classifications, as well as many of the plant names assigned by him, have survived into the modern-day botanical lexicon primarily because they contain numerous critical and precise observations and conclusions.

Theophrastus is credited with writing over 200 treatises on botany. However, only two of these books have survived, *De Causis Plantarum* (On the Causes of Plants) and *De Historia Plantarum* (On the History of Plants). In these two main books, one dealing with plant structures or morphology and the second with their functions or physiology, Theophrastus described and classified over 500 plant species, arranging them in three classes: trees, shrubs, and herbs. In addition, he was the first to establish a relationship between the structure of flowers and the resulting fruits of plants. His terms, *carpos* for fruit and *pericarpion* for seed vessel, survive into modern botanical terms. However, his main theory was the one that distinguished between *monocotyledon* and *dicotyledon* seeds. He discovered this after examining grass and wheat seeds and then noticing that they had only one seed "coat," thus classifying them as monocotyledons. On the other hand, bean seeds, having two seed "coats," were classified as dicotyledons. Cotyledons are the first shoots of the new plant arising from the germinating seeds. He considered these to be "coats" or "covers" of the seeds. Theophrastus also described the differences between flowering plants (angiosperms) and cone-bearing plants (gymnosperms). He was also aware of sexual reproduction in higher plants (e.g., palm trees), a phenomenon of which the ancient Babylonians were also cognizant. However, this belief was not generally accepted at this time in history.

Although Theophrastus's works were essentially unchallenged for nearly 1,800 years, there were other ancient botanical texts written by "botanists" who were interested in the properties and benefits of plants. In the first century C.E., the Greek botanist and surgeon in the Roman army, Pedanius Dioscorides of Anazarba (ca. 40–90 C.E.), reportedly authored numerous texts on plants, with particular emphasis on their pharmacological or medicinal characteristics. The only one of these writings to which his authorship can be attributed without question, however, is *De Materia Medica*. This book describes 600 plants, placing them into three main categories: (1) aromatic; (2) culinary; and (3) medicinal. The descriptions, which deal mainly with the habitats of the plants and their particular purposes, are short, with little emphasis on the characteristics of the plants themselves. Although much of the information was inaccurate, it was an important body of work for several reasons. First, it was an alphabetical classification, thus it was easier to access the information contained therein. Secondly, it was illustrated in the style of the Greek physician Crateuas (120–63 B.C.E.). Crateuas collected plants for medicinal purposes and wrote about their specific properties. He made colored, artistic representations of those plants within his texts. Crateuas's texts and illustrations have not survived in their original form. However, it is widely believed that these illustrations were included in Dioscorides' first-century-C.E. text *De Materia Medica,* and thus fairly accurate copies of the original illustrations have been preserved. Lastly, *De Materia Medica* was considered the definitive botanical and herbal text for over fifteen centuries.

Reproduction

All living things have a *life cycle* that extends from birth to death. Ancient humans were aware of this and accepted the transitory nature of life, probably with less resistance than humans do today. Our ancient ancestors had little control over the hazards and diseases that resulted in death on a daily basis, and only with time and the accumulation of knowledge were they able to improve the quality of their lives, as well as increase their spans.

Primitive humans were obviously aware of themselves and their immediate environment. Human intelligence began with the mental activities of conception, creative imagination, and symbolic logic. And the "seed" of primitive reason and logic was language—the symbolization of thought. As language developed, things (nouns) and actions (verbs) were then understood as concepts by early humans. We can assume with a fair amount of certainty that the questions that early humans had concerning "life" and its origin were more metaphysical in nature than actual. In other words, they questioned how their world and all its resources came to be. In answer to this, all civilizations developed their own concepts and myths concerning creation. On the other hand, the basics of the reproductive cycle of animals and humans (and later plants) were fairly obvious to ancient humans, that is, the mating instinct of males and females of all species resulted in offspring. (However, the connection between the act of copulation and the birth of an offspring was not readily apparent to our most primitive of ancestors. This realization came later in our evolutionary period as the human brain enlarged and intellect grew.) Ancient humans were unaware of zygotes, spermatozoa, DNA, RNA, and all the other biological characteristics and functions necessary for the production and reproduction of living organisms. Although folklore and myths about human and animal reproduction certainly existed within specific civilizations, it was the ancient Greek philosophers who, because of their inquisitive, restive, and intelligent natures, first explored the physiologic reasoning behind the processes of conception and birth.

Spontaneous Generation

Given the biological knowledge that we have today, the concept of and belief in spontaneous generation seems absurd. However, to ancient humans, even someone like Aristotle, who is generally believed to be a man ahead of his time, the idea that some living organisms could develop from nonliving material was perfectly plausible. There were no microscopes to detect bacteria, cells, and viruses. The maggots that appeared on rotting meat, garbage, and manure seemingly came out of nowhere, as did the rats that "emerged" from the piles of rags that covered cheese and moldy bread, and the frogs, salamanders, clams, and crabs that arose from mud. Without the resources and technology that were developed after the Renaissance, spontaneous generation (abiogenesis) was accepted as fact. Also, the character of Greek science, which influenced the ancients' approach to the inexplicable, was based on observation rather than experimentation. Aristotle mistakenly believed that flies and low worms were generated from rotting fruit and manure. The first challenge to this concept did not come until the early part of the seventeenth century C.E. when the British physician William Harvey (1578–1657) hypothesized that vermin do not appear spontaneously but come from eggs and breeding. The seventeenth-century Italian researcher Francisco Redi (1626–1697) also challenged the theory of spontaneous generation. He conducted controlled experiments that led him to believe that maggots and flies, for example, do not generate spontaneously but rather develop from eggs that are deposited. Despite this, he continued to believe that certain types of worms and flies could appear spontaneously. Thus, the complete refutation of spontaneous generation did not take place until the nineteenth century C.E. when Louis Pasteur (1822–1895) discovered the distinction between organic and inorganic matter and the existence of microorganisms.

Embryology

Even as ancient biologists and philosophers struggled with the theories concerning the origin of life and believed that life could, in fact, emerge from nonliving matter, life itself flourished in all forms—from the single-celled organisms that live in slime to the baby that grows into a functioning, intelligent, and literate human being. By nature, humans are curious and questioning and it is probably safe to assume that humans were cognizant of many facets of animal and human biology—at least in the empirical sense. Thus, their observations and experiments, such as they were, were limited to the usually fully formed offspring of various species. The one exception was the egg, since it was evident that the progeny that eventually broke through the hardened shells of eggs were not fully formed at the time they were first laid. Rather, an incubation period was needed for their full development. Thus, they were the perfect specimens with which to focus this natural curiosity about the formation and development of life. Embryology is defined as the science or study of the formation, early growth, and development of living organisms. The word "embryo" is derived from the Greek word *embruon,* which means "to grow," and the science of embryology can be traced back to the Ionian philosophers of Greece. They were the first to formulate a philosophy called "rationalism," which essentially stated that the world could be understood through the process of reason rather than merely relying on supernatural divination. This mode of thinking (rationalism) influenced the biological sciences in that animals in particular and life processes in general were studied for their own sake.

Historians usually attribute Alcmaeon of Croton, the Greek physician, with performing the first dissection on a human cadaver, as well as with dissecting and describing the structure of chicks that were in various stages of growth within eggs. This took place about 500 B.C.E. Whatever conclusions he drew from these egg dissections were necessarily based on description and comparison and did not dispel the controversy concerning *preformation* and *epigenesis* that existed until the eighteenth century C.E. *Preformation* was a biological theory, now invalid, that stated that all the parts of the embryo or future living organism are fully formed and enlarge during its stages of development. In other words, the egg contained a miniature individual called a *homunculus,* or animal that fully developed into the adult stage within the appropriate environment. On the other hand, the theory of *epigenesis* stated that the embryo, at the beginning, is indistinct and that formation of the specialized individual occurs within a series of developmental stages. Aristotle was a proponent of epigenesis and based his taxonomic system on embryological differentiation, using the terms "complete eggs" and "incomplete eggs." For example, higher animals were warm and/or moist creatures that could either give live birth or lay eggs that developed within the female. These animals laid "complete eggs." On the other hand, cold and earthy animals, such as amphibians, laid "incomplete eggs" that changed shape and size before hatching. Aristotle made his conclusions after he had dissected and studied fertile chicken eggs at different stages of embryonic development. The result was his concept of reproduction that stated that an invisible "seed" of the most rudimentary structure was imparted by the male to join a female egg to produce an offspring of the same species.

Summary

The history of ancient Greece, Rome, and the Islamic countries abounds with the exploits of ancient "scientists" (usually physicians and philosophers) who questioned and challenged the accepted beliefs of their day. Their methods were primitive, their investigations were flawed, and their conclusions were often incorrect. Yet, a great deal of what we know and utilize today in biology, botany, and zoology was, in fact, unearthed by these ancient scientists. For that very reason, we cannot discount their efforts and contributions. A more in-depth history of the discoveries and experiments in ancient biology that pertain to the fields of anatomy, physiology, and pharmacology can be found in chapter 7 ("Medicine and Health").

4

COMMUNICATION

Background and History

Although humans may be atop the evolutionary food chain, the ability to communicate is not unique to our species. Within their particular groups, all life forms transmit some commonly understood signal or method of communication, if only for the purpose of reproduction, which is, after all, the primary function of all living things. (It might be said that even uni- and multicellular organisms that reproduce asexually—characterized by the lack of transfer of genetic material (gametes) between individuals—are, at the least, communicating with themselves.) As we understand it today, the degree and level of communication among members of a species, aside from reproduction, are dependent on taxonomic classification, or the species' placement on the "ladder of nature." However, the evolution of our species from the small African hominid *Australopithecus* five million years ago to the subspecies that most closely resembles modern humans, *Homo sapiens sapiens,* 40,000 years ago, establishes that our earliest ancestors possessed physical and cognitive abilities that were, and still are, lacking in other mammals.

Speech, awareness of self (consciousness), and the ability to reason are the three characteristics that differentiate us from all other mammals. Many anthropologists believe that humans have, to some degree, possessed the power of speech from the time of the emergence of *Homo sapiens* as a distinct species 250,000 years ago. Other researchers and geneticists, however, place the timeline of speech at 200,000 years ago. While many animals have "speech" (vocal sounds) in the most primitive sense, in that the tonal quality or timbre of a chirp, a roar, a bark, or a growl (to name a few) can convey a sense of danger, foreboding, or contentment, and some social animals (e.g., dolphins, elephants, and whales) have their own "encoded languages," enabling them to communicate in a highly developed manner with other members of their groups, their cognitive abilities are limited. Only humans possess speech along with memory and language, which allows us to impart and record a shared history. All animals exhibit instinctive behavior, and what humans may attribute to "memory" in other animals is, in fact, imitation that is either spontaneous or learned.

While speech is inherent in humans, language is learned. The ability to speak is not an invention or discovery, but results from the genetic attributes of our species. And, at the very least, most discoveries, inventions, and advancements, whether cultural or technological, were made possible because of human language. Humans were able to communicate and disseminate information and experiences not only from one person to another but from one generation to another. The hard lessons of survival that were learned hundreds of thousands of years ago were shared first orally by our early ancestors, primarily so that succeeding generations need not repeat the mistakes of the former. As the physical and cognitive abilities of primates (hominids) evolved, so too did the characteristics that would dominate all other species. Hominids, of which *Homo sapiens sapiens* (modern humans) is the only **extant** species of the *Homo* genus, manufactured weapons and tools to hunt and domesticate prey and cultivate plants, and developed a societal culture that utilized cognitive strategies not only to ensure survival but also to improve upon and surpass existing conditions.

Speech is a precondition for spoken language, but writing is not necessarily a component of language. After all, anthropologists assume with some validity that, for many thousands of years, our earliest ancestors communicated orally with a language system that continually expanded and became more complex and inclusive. However, at some point in our history, early humans took the first step toward literacy when pictorial symbols—also referred to as pictograms, pictographs, and petroglyphs—were etched, carved, or painted onto the surfaces of cave walls or rocks. This was the advent of writing, for these illustrations became the historical record of life that dates back to the Middle Paleolithic Era (100,000 to 30,000 B.C.E.). As reported in the October 2001 issue of *Nature,* advanced radiocarbon dating techniques indicate that the world's oldest known animal paintings are located in the Chauvet cave in France. French scientists believe that the 400 black-and-ochre paintings that depict 14 different animals are approximately 30,000 years old and are equal in complexity and skill to the cave paintings found near the French Pyrenees that are believed to have been done between 12,000 to 17,000 years ago. There is, however, new evidence that suggests that early humans may have carved fairly sophisticated symbols some 70,000 years ago. As reported in the January 2002 issue of *Science,* an anthropologist working at a South African site discovered cave paintings and decorated tools that are more than twice as old as those found in Chauvet in France. These pictograms and carvings illustrate the inherent abilities of our species that rapidly advanced, in terms of geological time, from the most primitive examples of writing into the more sophisticated forms—hieroglyphics, cuneiform, and the alphabet.

The very human desire and ability to chronicle and share information with other members of our species, characterized by cognition, speech, language, and writing, also led to the development of literacy. There is little benefit in recording pictures or symbols if other members of the group are unable to access or interpret the messages contained therein. Interpretive writing dates back to about 3500 B.C.E. These ancient carvings and paintings are the forerunners of stone and clay tablets, papyrus, scrolls, paper volumes, and the printing press. It is interesting to note that many journalists, political leaders, and historians consider Johann Gutenberg (1400–1468 C.E.) the most influential person of the last millennium. Gutenberg is often credited with "inventing" the printing press. This is a misconception, because printing presses and movable type were invented by other, earlier Eastern civilizations, namely, the Chinese and the Koreans. Gutenberg's "invention" was really an improvement on the concept of movable type so that one piece of type for each letter could be reused after a page was printed. In essence, Gutenberg's printing press was the precursor of the principle of mass production that was used well into the twentieth century, and for this reason, ascribing the status of "most influential of the last millennium" may be well placed. Knowledge may be power, but the ability to communicate that knowledge is even more powerful. This, after all, may be one of our species' greatest attribute.

Language

Speech is not a required component of communication, nor is it necessary for some forms of language. For example, sign language, which is presumably older than speech itself, can be as primitive and crude as grimaces and bodily gestures or as refined as it is today in the use of advanced sign language for the hearing-impaired. Also, the Braille system, which is technically not a language but a writing system, is utilized universally by the blind and visually impaired. Both of these systems, which have been adapted to 40 or more languages, have enabled those with hearing or visual disabilities to live fuller and more productive lives. While a detailed history of all 6,000 languages is beyond the scope of this book, this section will trace, in general terms, the development of spoken language, particularly those languages whose ancient roots either formed the basis for some of the dominant languages of today or those languages that, while extinct, are important in the development of ancient writing or alphabet systems.

The larynx in hominids (humans) is lower in the throat and farther from the soft palate than in other primates. This indicates that the position of the larynx altered as a result of the evolution of the hominid form. Cranial formation and position and the manner in which the skull was balanced on the neck changed. The neck became elongated, and the lower positioning of the larynx allowed for the distinctive speech of humans. Over time, early hominids developed greater control of their air passages (throat, nose, lungs) as well as of their tongues, lips, and teeth. These anatomical changes imply that hominids, specifically *Homo sapiens,* were able to form and sustain more complex sounds and speech patterns. Recent research at Oxford University in England ascribes speech to a genetic mutation, specifically the *FOXP2* gene, that influenced the mouth and facial movements needed for speech. Anthropologists and geneticists, however, generally agree that the origin of man's first language, as well as culture, is contemporary with the evolution of larger brains in *Homo sapiens.*

Language is taught, primarily by one generation to another, most often by parents who deliberately teach their children to speak in the language that they themselves speak. Cultural tendencies are transmitted and taught in much the same way as language—instructively. This is how groups, societies, and civilizations were formed. The transmission of a universally shared and understood vocabulary induced similar behavioral patterns within a group. The result was the creation of a cultural identity that was unique and comfortable. A spoken language that was understood by all members of a group imparted information that was primarily instructive. Secondarily, it aided in the development of new skills, techniques, and opportunities that over a period of time, albeit thousands and thousands of years, improved the standard of living for humans.

Definition and Types

Webster's defines language as "human use of voice sounds... in organized combinations and patterns to express and communicate thoughts and feelings." All languages have shared commonalities. Using a set of grammatical rules, they first combine words or wordlike components into sentence structures; second, they differentiate between nouns and verbs in sentence structures; and third, they have the ability to use subordinate clauses in a sentence structure. In other words, a sentence can incorporate one or more meanings or thoughts. (An example: He fell off his horse, and he broke his arm.) However, these commonalities are structured differently in specific languages.

Historical linguists categorize languages as belonging to three broad groups: *inflectional (fusional), analytic (isolating),* and *agglutinative (attaching* or *chaining)*. Latin is a prime example of an *inflectional* language, in which a change in the form of a specific word can indicate distinctions in tense, person, gender, number, mood, voice, and case, and changes within the stem of that specific word can illustrate another type of inflection (e.g., go, going, gone). *Isolating* or *analytic* languages use word-form variations sparingly, if at all, and sentences are structured by the order or grouping of specific words. Classical Chinese is an isolating language. In *agglutinative* languages, prefixes and suffixes that connote grammatical functions and relationships are attached to nouns and/or verbs in chainlike formations. This results in polysyllabic and polymorphemic words that are considerable in length and difficult to pronounce. Examples of *agglutinative* languages are the ancient, extinct Sumerian language, as well as the languages of Anatolia (present-day Turkey and Iran). (Another group, called *synthetic,* encompasses both inflectional and agglutinative languages. Some languages, such as Native American languages and Eskimo, are referred to as *polysynthetic* in that a whole sentence may consist of one word that is usually in the form of a verb. For example, the vernacular of the Pennsylvania Dutch in the eastern United States uses the word "ain't" as a one-word sentence that can be a declaration, an exclamation, or a query.) Modern English is considered to be an amalgamation of all three types: inflectional, analytic, and agglutinative.

Dialects

Languages develop over far-reaching geographic areas. On the other hand, dialects develop within small, regional localities. This has been true since ancient times; specific dialects have been "identifiers" regarding a person's place of birth as well as his or her social status within the community. The dynamics of dialectal differences within a geographical region where a common language was spoken depended primarily upon the frequency and size of the migration of the population. If migration was sporadic and random, the dialectal differences were greater, obviously because there was little contact between the residents of localized areas. If the migration was large and frequent, then the differences tended to diminish, and the dialect of the migrating population became dominant over the broader geographic area, or became a mixture. Also, the dialect that became prevalent was, in general, one that signified or identified a higher social status. In ancient times a prime example of a single and dominating dialectal force is *koine,* which in the Hellenistic era replaced all Greek dialects. *Koine* was the dialect of Athens, the city in Greece where philosophy and learning were centered. On the other hand, tremendous dialectal diversity existed in the British Isles, primarily because the population was far less transitory within this particular geographical region. (The tendency among the British to "stay put" was less evident during the worldwide expansion of the British Empire of the eighteenth and nineteenth centuries. Nevertheless, even today, there are greater numbers of dialects within the British Isles as compared to the dialectal differences that exist in the United States, a country far greater in geographical area than Great Britain.)

History

Thus far, science has been unable to determine what existing language represents the oldest human language. Most linguistic scholars concede that this determination, in all likelihood, will never be known. However, we do know that the most rapid changes in the development of human history occurred in the Middle East about 45,000 years ago with advances in tools and weaponry and an increase in population growth. Scientists believe that agriculture and animal domestication also originated in this region, while some type of spoken language was most likely prevalent among all populations of hominids. The Neolithic Period, or New Stone Age as it is commonly called, was a time during which "specialization" of trades and occupations became significant. For example, people became farmers, potters, and weavers, and vocabularies associated with these endeavors developed. And although some populations of indigenous peoples believe and respect their elders' oral recounting of past events, history primarily relies on written accounts, which have proven to be more credible and consistent. Thus, the history of language and the invention of writing are necessarily intertwined.

The earliest written language was Sumerian, the now-extinct language of Sumer, the ancient country of Mesopotamia (present-day

Iraq). It flourished about 5,000 years ago. However, around 2000 B.C.E. Akkadian supplanted Sumerian as the spoken language of the region, but not the written. Sumerian was the written language of Mesopotamia until the beginning of the Christian era, which coincided with the demise of the spoken Akkadian language. Akkadian was the language of the Semitic Arabian nomads who migrated throughout the Near and Middle East, where they took up residence as tradespeople and farmers. The spoken form of Akkadian was essentially a trade language, while written Sumerian tended to reflect the business, cultural, legal, royal, and private records of the now-extinct civilizations of Sumeria, Babylonia, Assyria, and Mesopotamia. The spoken Sumerian language was relegated to a small portion of the population in Sumer, while other Semitic peoples migrated throughout Mesopotamia and influenced the culture as well as the language. Sumerian continued to be taught in the scribal schools of the Middle East during the first and second millennia B.C.E., primarily because it was the language in which the religious and cultural history of the region had been transcribed. (A parallel can be drawn in the teaching of Latin in the European universities in the Middle Ages, even though Latin had ceased to be a living language centuries before.) Various dynasties formed as a result of this migratory influence, including the most famous, the Babylonian.

Writing and the Alphabet

Cave Paintings and Hand Stencils

The oldest evidence of "writing" can be found in the cave paintings of the Paleolithic Period. Paintings discovered in the Chauvet cave in France are believed to be 30,000 to 32,000 years old; new findings in South Africa date back some 70,000 years. These discoveries call into question the belief held by some archaeologists that hand stencils are actually the archetypes of the first writing or art to be done by early humans. Hand stencils, as the name implies, are created by blowing paint over a hand that is placed palm-side down on an unpainted surface. In 1985 French diver Henri Cosquer discovered the underwater entrance to a limestone cavern on the French Mediterranean coast. Cosquer Cave, as it is now known, contains 55 hand stencils, as well as drawings and engravings of animals and geometric symbols. French archaeologists, applying carbon-14 dating techniques on the paint, have determined that early humans who used organic sources did them some 25,000 to 26,000 years ago. The paintings and drawings of animals in Cosquer Cave were created about 10,000 years after the hand stencils.

Pinpointing the age of these and other archaeological finds becomes more precise each decade with the technological advances in dating methods, such as thermoluminescence and radiometrics. There are other sites where ancient cave paintings, pictographs, and petroglyphs have been found, including Lascaux, France; Mazouco in Portugal; and numerous sites in the southwestern area of the United States. These drawings represent prehistory, when early humans were hunters who shared the land with fearsome creatures. The drawings, particularly those depicting the killing or subjugation of animals, are the earliest written chronicles of life and survival during the Stone Age. As important as the drawings may be in illustrating the environment that existed thousands of years ago, they also are a testament to our species' creative abilities to not only draw complex and faithful representations of animals and events, but also to discover the materials, such as charcoal and black oxide of manganese, with which to do so.

Cuneiform

Notwithstanding the discoveries of paintings and pictographs in various sites around the world, scientists and archaeologists have bestowed the title of "Cradle of Civilization" on Mesopotamia, because it is generally believed that settled agriculture and animal domestication, as well as the foundations of writing, began in this region, which was home to numerous tribes of ancient humans. Controversy now surrounds the origins of writing, however. As reported in the June 29, 2001 issue of *Science,* researchers are revising their formerly held beliefs that ancient clay tablets found in the long-buried temple of Uruk, the ancient city of Mesopotamia, are the oldest examples of cuneiform writing. New credence is being given to the belief held by other scientists and archaeologists that writing was "invented" by more than one group or civilization independently of others, rather than by a single individual or group. For instance, generations of historians believed that Egyptian hieroglyphics resulted from the Mesopotamian influence of cuneiform. The 1989 discovery of another Egyptian writing system challenges this belief while raising new questions about whether the opposite was true—or neither. Limitations of radiocarbon dating and the reluctance of a number of museum curators to release artifacts for new testing, which has the potential for further degradation, are hampering research efforts into this controversy. Another controversy involves objects, called *tokens,* that were discovered decades ago in Iraq, Syria, and Iran. Tokens are ceramic pieces in varying geometrical shapes, or hollow clay spheres or cylinders that contained markings on their exteriors, and are from

4,000 to 9,000 years old. Some archaeologists believe that each of these tokens was used for commerce and represented a distinct object or measurement, such as an ox or a measure of grain. They believe that the utilization of tokens was the first form of communicating information and was the basis for the cuneiform system. Thus, tokens were the first form of organized writing. Other archaeologists remain skeptical. What these controversies do confirm, however, is that information is constantly evolving as generations of new researchers armed with more precise tools and methods with which to gather data continue to search for the true answers to ancient mysteries.

While it may never be known with exact certainty whether the Sumerians were the earliest settlers or the actual inventors of cuneiform writing, cuneiform texts that are more than 5,000 years old can be traced back to the Sumerians, a people of unknown ethnic origins. The word "cuneiform," which describes the writing system used in the ancient Middle Eastern regions of Sumeria, Akkadia, Babylon, and Assyria, was first used in the early part of the eighteenth century by Engelbert Kampfer. The word itself is a combination of Latin and Middle French and refers to writing that utilizes wedge-shaped characters. Sumerians used a square-ended stylus, made out of either wood or reed, to inscribe various symbols onto the surface of soft clay tablets. The tablet was then sun-dried or baked and the text saved for perpetuity. The clay tablets found at Uruk contain lists of merchandise along with a set of numerals and personal names. At first the symbols were pictographic; that is, an ox's or donkey's head represented the entire animal. Since pictographic representations could not convey complex thoughts, the system evolved into a combination of pictographs that conveyed syllables. For example, combining the signs for "mouth" and "water" meant "to drink." Further refinements of cuneiform made it possible to express the sound but not the actual meaning of the symbols. For example, combining the symbols for "bee" and "leaf" meant "belief." Linguistic scholars refer to these examples as the *rebus principle*—words or syllables represented as pictures.

Further improvements in Sumerian cuneiform led to the concept of *phonetics.* As an example, the Sumerian word for "hand" was *su,* and the phonetic syllable *su* could then be used in any necessary context. The next invention in cuneiform was the use of *logograms (ideograms).* A logo-gram is a single character or symbol that represents an idea or object without having to use the particular word to phrase it. The $ and & signs are examples of logograms. With the dissolution of the spoken Sumerian language, the Akkadians then developed their own Semitic form of cuneiform that borrowed heavily from the Sumerian. It was a complicated language inasmuch as it retained the Sumerian logograms while using the Akkadian pronunciation for corresponding words. Nonetheless, cuneiform writing did evolve into a more phonetic language that could express complex and sophisticated grammatical concepts. There were numerous adaptations of cuneiform writing during the first and second millennia B.C.E. Old Akkadian cuneiform was used to compose literature, such as that written about 2300 B.C.E. by the world's earliest known poet, Enheduanna, daughter of the Akkadian king Sargon. The famous *Law Code of Hammurabi,* which is the most complete existing collection of Babylonian laws, was written in Old Babylonian cuneiform during the reign of King Hammurabi (1792–1750 B.C.E.). Although the code is a chronicle of primarily Sumerian laws, Hammurabi's intent was to integrate all Sumerian and Semitic traditions and populations. The source of the Hammurabic Code is now housed at the Louvre in Paris on a **stele** carved from crystalline diorite. It was discovered in Iran at the ruined city of Susa, the capital of the ancient country of Elam (present-day southwestern Iran). During the middle of the second millennium B.C.E., traders in Asia Minor recorded their transactions in Old Assyrian cuneiform, while writers in the Middle and Neo-Assyrian periods of cuneiform transcribed legal and other documents that were kept at the library at Nineveh nearly 3,000 years ago. The first postal system, using clay tablets and clay envelopes, can be traced back to the Assyrian and Babylonian civilizations. Clay was heavy and bulky but able to survive indefinitely, unlike the papyrus and paper records that came afterward.

The influence of cuneiform writing beyond the borders of Mesopotamia began in approximately the third millennium B.C.E. when the Elamites adopted this writing system. From there the Indo-European Persians used cuneiform to devise a pseudoalphabetic form in which to transcribe the Old Persian language. Invasion and conquest of one tribe over another were also the mechanisms for the spread of cuneiform from Asia Minor to Syria and the mountainous regions of Armenia, as well as other areas of the Near and Middle East, where its usage became universal. Late Babylonian and Assyrian cuneiform survived as written languages until the beginning of the first millennium C.E.

Hieroglyphics

The term "hieroglyphics" first appeared in the writings of the Greek scholar Diodorus Siculus in about the first century B.C.E. when he used the Greek word *hierogluphikos,* meaning "sacred carving," to refer to Egyptian writing. Hieroglyphics refers to a system that utilizes pictograms, called *hieroglyphs,* to represent complete words, syllables of words, and/or sounds. While the term is generally associated with Egyptian writing, for the past hundred years or so it has also been used to describe the similar style of pictographic writings of other civilizations, including the Mesoamerican and South American civilizations of the Mayans and Incas. The hieroglyphs found on other continents give further weight to the belief that writing was an independent invention of more than one civilization. Scholars have long believed that Egyptian hieroglyphics grew out of the cuneiform system of writing of the Sumerian civilization. Recent finds have led to some reinterpretation of this belief, but no definitive determination has been made as to the origin of hieroglyphics in ancient Egypt. Sumerian influences on writing are a logical assumption, given that Egypt and Mesopotamia, only 1,000 kilometers (622 miles) apart, were ancient trading partners.

Hieroglyphics developed in stages, as was the case with all writing systems. At first, only the simplest symbols were developed. Then the number of symbols increased, making it easier to read. It was essentially unchanged for several thousand years until about 500 B.C.E., when the number of symbols again increased dramatically. The demise of hieroglyphic writing coincided with the rise of Christianity. The oldest Egyptian hieroglyphic writings date back to approximately 3100 B.C.E. and, technically, are classified as such only when they are written on the monuments, temples, and tombs of the Egyptian monarchs. Hieroglyphics were ostensibly difficult to transcribe because the hieroglyphs (pictures) were always cut in **relief** into stone, metal, or other hard surfaces. The Egyptians, however, developed other forms of writing that were copied onto papyrus, in which an individual sign denotes a specific syllable or even a consonant, and reliance on pictures decreased. These were part of the *hieratic* system that the ancient Greeks also used for the transcription of religious texts, and the cursive *demotic* writing system that was employed in the transcribing of ordinary documents. However, all three forms of writing, hieroglyphic, hieratic, and demotic, were identical typologically. In other words, all three were phonetic in character, with common characteristics and predominant features of grammatical structure.

Accurate translations of Egyptian hieroglyphics were made possible upon the discovery of the Rosetta Stone in 1799. The Rosetta Stone, a slab of black basalt that was found in the Egyptian delta town of Rosetta (Rashid), contains inscriptions written by the priests of Memphis in 196

B.C.E. that commemorate the accession to the throne of Ptolemy V (205–180 B.C.E.). The inscriptions are in two languages, Egyptian and Greek, as well as three writing systems, hieroglyphics, demotics, and the Greek alphabet.

Linear B

Less well-known than cuneiform, hieroglyphics, and the alphabet is *Linear B,* a phonetic writing system based on sound structure that developed in Mycenae, the ancient Greek city, sometime around 1400 B.C.E. Linear B represents the old form of the Greek language, and like cuneiform, it was transcribed onto clay tablets. The system, written from left to right, contained approximately 90 symbols that represented syllables, ideograms, and a decimal-based numbering system. As with all forms of early writing, it was used primarily for commerce and business records. The linear system was most probably the antecedent of the alphabet. Also like cuneiform, it did not survive, because claylike materials for writing were not universally available. Thus, another system, more easily transcribable, was needed.

Alphabet

Linguistic scholars and historians are divided in their opinions as to exactly what ancient civilization invented the alphabet. However, they are in general agreement concerning two aspects of its invention. It first appeared in the Middle East sometime during the second millennium B.C.E., in what is now present-day Syria and Israel, by an ancient Semitic group of people who had contact with the Egyptians, most probably the Phoenicians, but possibly the Hebrews or Canaanites. Secondly, both cuneiform and hieroglyphic writing influenced the inventor or inventors of the alphabet. The history of this region and its neighbors provides the reasons for this belief. Settled agriculture began in the Middle East, and the Syro-Palestinian region was the geographical center of what became known as the "Fertile Crescent," an agriculturally rich area with expansive trade routes. During this time, the ancient Bronze Age civilizations of Egypt, Mesopotamia, and Crete in the Aegean were rife with political corruption and in the midst of their demise as powerful forces. Contemporaneously, the ancient Middle Eastern countries of Israel, Phoenicia, and Aram began to exert political and commercial influence. Arabia, a country to the south of the Fertile Crescent, became an important trade route between the East and the countries of the Mediterranean. In the West, the nation of Hellas, later known as Greece, developed. Hence, the term "Hellenistic" came to be used when referring to Greeks and their influence. Cuneiform, hieroglyphics, and the hieratic and demotic forms of Egyptian writing, which were all used to record legal and historical documents, of necessity needed to be simplified in order to accommodate the needs of the ever-expanding trade and commerce of the region. Also, the taxation of goods, commerce, and property by the governments of these burgeoning nations became increasingly important, and a system of measurement (accounting) was a concurrent invention, along with a more universally understood alphabet.

Regardless of the alphabet's origin in the Middle East, the Phoenicians played a major role in its diffusion. A fierce Semitic group with close blood ties to both the Arabs and the Hebrews, the Phoenicians were sea traders who settled on the coast of present-day Syria and Lebanon. The Phoenician alphabet was a linear system incorporating 22 letters that expressed only consonants. In about 1100 B.C.E. its use spread to the populations of the countries of Syria, Palestine, and Arabia. At about the same time and using the same Semitic sources as the Phoenician alphabet did, the Arabic alphabet came into existence. A common belief is that the alphabet "follows the flag," and thus the Mediterranean trade routes of the Phoenicians were the logical transmitters of this new writing system. The successful trading practices of the Phoenicians among the countries of the Mediterranean and the resultant settlements of new populations in this area were the primary motivators for the dominance of the Phoenician alphabet, rather than the Arabic. In about the ninth century B.C.E. the Greeks, who centuries earlier had utilized the extinct Linear B writing, adopted the Phoenician alphabet. This gave rise to the cultural and literary ascendancy of the ancient Greeks, whose influence is still evident in modern times. From the Greek alphabet came the Etruscan alphabet, and from the Etruscan came the Roman (Latin) alphabet. In about 100 B.C.E., the Roman alphabet consisted of 23 letters. The Modern English alphabet consists of 26, after the addition of the letters *j, u,* and *w* during the medieval period.

Just as with language, a full history of all the alphabets that existed prior to 500 C.E. is beyond the scope of this book. However, the following is a brief synopsis of the development of the alphabetic systems that have been most influential. (See Figure 4.1.)

PHOENICIAN C. 2000 BCE	GREEK C. 800 BCE	ETRUSCAN C. 700 BCE	LATIN C. 100 BCE	MODERN WESTERN C. 1000 CE-PRESENT
𐤀	Α (ALPHA)	𐌀	A	A
𐤁	Β (BETA)		B	B
𐤂	Γ (GAMMA)	𐌂	C G	C
𐤃	Δ (DELTA)		D	D
𐤄	Ε (EPSILON)	𐌄	E	E
𐤅	Ϝ (DIGAMMA)	𐌅	F	F
𐤆	Ζ (ZETA)	𐌆	Z	G
𐤇	Η (ETA)	𐌇	H	H
𐤈	Θ (THETA)	𐌈	Th	I
𐤉	Ι (IOTA)	𐌉	I J	J*
𐤊	Κ (KAPPA)	𐌊	K	K
𐤋	Λ (LAMBDA)	𐌋	L	L
𐤌	Μ (MU)	𐌌	M	M
𐤍	Ν (NU)	𐌍	N	N
𐤎	Ξ (XI)	⋈	X	O
𐤏	Ο (OMICRON)		O	P
𐤐	Π (PI)	𐌐	P	Q
𐤒	Ϙ (QOPPA)	𐌒	Q	R
𐤓	Ρ (RHO)	𐌓	R	S
𐤔	Σ (SIGMA)	𐌔	S	T
𐤕	Τ (TAU)	𐌕	T	U*
	Υ (UPSILON)	𐌖	U V	V
	Φ (PHI)		Ph	W*
	Χ (CHI)		Ch	X
	Ψ (PSI)		Ps	Y
	Ω (OMEGA)		Ō	Z

* ADDED DURING MEDIEVAL PERIOD

Figure 4.1 Evolution of the Modern Western Alphabet (ca. 2000 B.C.E. to present)
The *western alphabet* has its origins in the Middle East. The seafaring people of Phoenicia spread their unique linear system to other countries on the Mediterranean, where it became incorporated into the Greek alphabet, then to the Etruscan and Latin alphabets, and finally into the modern Western alphabet.

Other Major Alphabets

Arabic

After the Roman alphabet, the Arabic alphabetic writing system is the second-most widely used. The origins of the Arabic alphabet are vague. Some scholars believe the language descended from the inhabitants of the ancient Arabian kingdom of Nabataea (present-day Jordan). Others believe that its derivations are Aramaic. It has been in existence since about 300 C.E., and it flourished as Islam spread throughout the Eastern Hemisphere. The Arabic alphabet contains 28 letters, all consonants, 22 of which are the same as those in the Semitic alphabet. The remaining six letters connote sounds that were not used in earlier languages or alphabets. In addition, the form of each letter is determined by the position of that letter in a word—initial, medial, or final. *Kufic* is one of the major Arabic scripts. It was developed about 680 C.E. primarily for inscriptions on stone and for writing texts of the Koran. *Naskhi* is a cursive script that is more prevalent and is used in numerous decorative displays. Both are written from right to left.

Hebrew

The *Square Hebrew* alphabet and script is believed to have derived from the Aramaic alphabet during the third century B.C.E. It became standardized sometime before the beginning of the first millennium C.E., and from this standardization arose the Modern Hebrew alphabet. The decline of spoken Hebrew, which was supplanted by the Aramaic language, impacted the ability to read ancient biblical manuscripts and documents. To counteract this deficiency, scholars imposed distinctive vowel systems on the Hebrew alphabet (i.e., the *Babylonian, Palestinian,* and *Tiberiadic*) so that the Bible could be read with understanding. Only the Tiberiadic has survived. There are four distinctive types of Hebrew writing, all of which can be traced back to the Square Hebrew writing: first, the square script, from which Modern Hebrew writing evolved; second, the medieval style; third, the rabbinic (Rashi); and fourth, the cursive script. Also known as daily handwriting, the cursive script was the basis for a number of regional varieties. The most popular is the Polish-German variety, which became the current or modern-day Hebrew handwriting.

Chinese and Japanese

Scholars believe that Chinese writing can be traced back to the Shang dynasty of circa 1520–1030 B.C.E. (This may be a further verification of the belief that writing was an independent invention of several civilizations.) Artifacts from this and later periods reflect that Chinese writing evolved from complex scripts to those with a more modern appearance. Chinese writing is a logographic system that utilizes pictorial symbols, called characters, to represent entire words. There are some 40,000 individual characters in the Chinese lexicon. In order to read basic Chinese, it is only necessary to recognize approximately 2,000 of these characters. To function at a higher level of literacy, however, it is necessary to be familiar with at least 10,000 Chinese characters. Chinese is the oldest continuously used written language.

Japanese writing is based on the Chinese scripts. Although ancient Japanese and Chinese civilizations had interacted with each other since the end of the last millennium B.C.E., the Japanese did not actually begin to write texts in their own language until sometime after 400 C.E. Despite the Chinese influences, the typology of the Japanese language is quite different than Chinese. For instance, Chinese is monosyllabic and *isolating;* Japanese is polysyllabic and *inflectional.* (See the Language section in this chapter for further explanation.) Written Japanese involves both a logographic and syllabic system. Ancient Japanese was, and still is, a difficult and complex language to decipher, primarily because of the use of homophones that are present in the system. (Homophones are characters representing words that often have several completely distinctive meanings. An example: the Japanese word "kan" has more than a dozen different meanings, including "sweet," "print," and "slow." Likewise, the English word, "fire," has several dozen meanings.)

Paper

There were several stages in the invention, development, and manufacture of paper. The soft clay tablets that were used to transcribe the ancient cuneiform texts in the Middle East were heavy and cumbersome and limited in their ability to reach large groups of people over a wide geographic area. Their great attribute was their longevity, evidenced by the remarkably well-preserved tablets that have been unearthed as late as the twentieth century C.E. (However, for a time, smaller soft clay tablets with clay envelopes were utilized in a Mesopotamian courier system.) The ingenuity of ancient humans and their desire to chronicle and dispense their experiences and information led to the development of more useful materials on which to record and maintain historical documentation. For instance, palm fronds were used as writing and painting surfaces in ancient India. The Chinese first used silk, wood, bamboo, and leather before their invention of paper sometime in the second century B.C.E. And the Egyptians are most closely associated with the use of papyrus for writing before they developed parchment and vellum.

Papyrus

Cyperus papyrus is known as the paper plant. It grew in great abundance along the Nile Delta region in Egypt, where Egyptian farmers cultivated and used it for numerous items, the most famous being the writing material known by the same name as the plant. The terms "papyrus" and "paper" should not be used synonymously. They are completely different materials, and the processes by which the final products are developed are also completely different. Papyrus, an aquatic plant with long, woody stalks, triangular in shape and up to 4.6 meters (15 feet) in height, grows in areas of slowly flowing shallow water. Ancient Egyptians cut the central fibrous core (pith) of the stalks (reeds) into strips. The strips were then placed side by side, while other strips were placed atop these at right angles. These strips were then dampened and pressed and allowed to dry, the result being a light cream-colored sheet of material. The dried sap of the plant was the adhesive that held the fibers together. In time, the Egyptians refined the process, resulting in a smoother writing surface. The original discovery of papyrus as a writing material dates back to at least 3500 B.C.E. At first, single sheets of papyri, approximately six inches wide, were used. Then papyrus rolls, called *biblions* by the Greeks, and the precursor to the bound book, were invented. On the papyri rolls, ancient Egyptian scholars known as scribes, using reed pens dipped in different-colored inks, wrote only on the sides of the papyri where the fibrous strips were laid horizontally, presumably because it was easier to transcribe on this side. The text itself was written starting from the left, in columns at right angles to the edge of the rolls. After the ink dried, the papyri (no more than 20) were pasted edge-to-edge and then rolled with the text on the inside. The reader unrolled the *biblion* as he went along. Afterwards, the document was rerolled. The drawback to this was that repeated readings and rerolling caused the papyri to crack, often rendering the document unreadable.

Because papyrus is an aquatic plant, it is vulnerable to humidity and therefore particularly fragile in damp climates. Ancient papyri found in Egypt and the Middle East that are nearly 5,000 years old have survived, primarily because of the dry climate. The desert sands, in a very real sense, protected these ancient documents that, when unearthed, were remarkably well preserved. Notwithstanding the fragility of papyrus, it was widely popular not only in Egypt and the Middle East but throughout Greece and the Roman Empire during the first centuries of the first millennium C.E. Papyrus began to be replaced by parchment and vellum as the writing materials of choice sometime during the third century C.E., particularly in the colder and wetter climates of the European countries. Its demise was almost complete by 800 C.E., after the manufacture and availability of paper became more widespread. Nevertheless, papyrus was still used, although sparingly, until about 1100 C.E. In addition to its use as a writing material, the ancient Egyptians used papyrus in the manufacture of ship sails, mats, ropes and cords, and fabric. Today, papyrus is grown primarily as an ornamental plant.

Parchment and Vellum

Parchment refers to a writing material made from the skins of certain animals, primarily sheep or goats, first used sometime in the second century B.C.E. The derivation of the term itself is thought to be from the ancient city of Pergamum in Asia Minor (present-day Bergama in Turkey). Some scholars believe that a rivalry between Pharaoh Ptolemy V of Egypt and Eumenes II of Pergamum was the impetus for its invention. Ptolemy placed an embargo on the exportation of papyrus to Pergamum because he feared that Eumenes' library in that city would surpass in size his own collection at Alexandria. Supposedly, Eumenes' response to the embargo was the invention of parchment. The Greek and Latin words for parchment mean "stuff from Pergamum." Thus, there may be some credence to this story. The use of animal skins for writing materials was not new. Both the Hebrews and the Egyptians had learned to tan hides to make leather over 5,000 years ago. A fragment of an ancient Egyptian document written on leather and dating back to the twenty-fourth century B.C.E. is still in existence. In addition, a number of the famous Dead Sea Scrolls, some of which are over 2,000 years old and were discovered in the mid-twentieth century C.E., are written on leather as well as papyrus. The reason the ancient Egyptians and other ancient civilizations opted to use papyrus rather than leather for writing materials was twofold: papyrus grew in abundance, and it was cheaper to process in quantity. Nonetheless, the improved, albeit more expensive, refining of leather to make parchment resulted in its increased popularity. With improvements in the cleaning, stretching, and scraping of the animal skins, along with whitening with chalk and smoothing with pumice, it was now possible to transcribe on both sides of the parchment sheet, also known as a manuscript leaf. Parchment sheets were also durable and could be cleaned and then reused. The reused documents, upon which traces of the erased writings were visible, are called *palimpsests*. Parchment also lent itself to binding in a form that became known as a *codex* (*codices* in the plural). These were really the first forms of the modern bound book, and they supplanted the papyrus rolls that heretofore had been in existence. (For a time, however, some of the very early codices continued to use papyrus.) At first, codices were stitched together on one side. Later codices were folded to form a double-leaf folio, or into *quartos,* where the pages of the codex were folded into quarters. These resembled modern-day bound volumes. Codices contained large numbers of pages that were transcribed on both sides. The reader could select any page to read without having the burdensome task of rolling and rerolling. And the parchment itself did not crack or deteriorate, so that the manuscript was well preserved despite frequent use. While we think of the codex as a development in Egypt, the Middle East, and Europe, the pre-Columbian civilizations in Mesoamerica, at about the end of the first millennium C.E., independently developed their own sets of codices, using pictographs and ideographs to depict astrological, calendrical, and religious celebrations and speculations.

Vellum is a finer grade of parchment that is made exclusively from the skin of young calves and/or lambs. The derivation of this word comes from a variation of the Latin and French words for calf, *vitulus* or *vitellus* and *veau.* Many of the finest manuscripts of the early centuries of the first millennium C.E., including biblical and religious texts, were transcribed onto vellum. By 1100 C.E. the quality of vellum further improved, making it especially soft and supple.

Since parchment and vellum, as well as papyrus, were expensive to manufacture, *wax tablets* were used for daily text entries and correspondence. (For example, the skins of about a dozen sheep were required to create a parchment codex of 200 quarto pages.) These wax tablets were hollowed-out pieces of wood into which melted wax, usually black in color, was poured. Entries were then made onto the hardened surface. After the invention of paper and the proliferation of paper mills, and most particularly after the Gutenberg printing press in the mid-fifteenth century C.E., the use of parchment and vellum ceased. Today, the words "parchment" and "vellum" refer to paper of exceedingly high quality, as neither is manufactured from animal hides.

Paper

Aside from the facts that the word "paper" is derived from the word "papyrus," and that both were used for writing materials, the two are quite different. Paper is a Chinese invention from at least the second century B.C.E. The oldest extant piece of paper that contains writing dates back to 110 C.E. It was discovered in 1942 under the ruins of an ancient watchtower at Tsakhortei, China. This is significant because when the Chinese invented paper, it was not for use as a writing material. Rather, paper was used for clothing, blankets, wrapping material, and lacquerware, as well as for personal hygienic use (i.e., nose "tissues" and toilet paper). Exactly when the Chinese first utilized this invention for writing purposes is unclear, as bamboo, silk, and wood were the materials on which Chinese characters were inscribed. Ts'ai Lun, a councilor to the Chinese Imperial Court of the Han Dynasty who lived during the second century C.E., is believed to be the inventor of paper. He is reported to have made a pulp or mash from the bark of a paper-mulberry tree *(Brousometia papyrifera)*, rags, fishnets, and hemp, with which he mixed water. This solution was poured over closely woven fabric mats through which the excess water drained. The dried sediment suspended on top of the fabric mat became the paper. The quality was crude and coarse, so it is reasonable to assume that the newly invented material would have been unsuitable for writing material, in any case. This is an interesting story, but in fact the development of paper may have actually been a more gradual process. The Chinese, as was their position with all their inventions, were reluctant to share their knowledge, not only with "outsiders" but with their own countrymen as well. Thus, a true history of the development of paper may never be known.

What is known, however, is that the spread of paper followed the Silk Road, the ancient trade route that linked China with the countries of the West. It was so named because silk, a Chinese invention, was exported westward, while precious metals, wool, and other manufactured goods traveled eastward. Another commodity that traveled in both directions was knowledge. However, in the case of paper, the product rather than the process was the first export, but not for centuries after its initial discovery.

Printing

Literacy and all that accompanies the accessibility to and acquisition of knowledge was the foundation for the renaissance of science and discovery. However, the invention of modern printing and all that it implies was not a discrete process. Rather, it was a progression of innovation and redesign, with varying influences and importance, not the least of which was the invention of paper. It is, however, an indisputable fact that Gutenberg's "reinvention" of a process that was first invented in the Far East at least four centuries before did completely transform civilization. As with most inventions, the transformation came in stages.

Seals

Printing or imprinting has been in existence for thousands of years. The stone, wood, brick, and metal seals used with soft clay or melted wax on ancient official documents, some of which date back to the fourth millennium B.C.E., were forms of imprinting, as were the brick stamps that the kings of Babylonia used to engrave their names on buildings. In the fifth century C.E. the Chinese devised a new method of imprinting seals: using inked characters on a wooden seal and then pressing onto paper, similar in process to the modern-day rubber stamp.

Engraving

Although prehistoric engravings have been found on cave walls, rocks, and fossilized bones, the Sumerians were assumed to be first to perfect the technique of duplicating images. They engraved cuneiform writings on stone cylinder seals that they then rolled over soft clay, leaving relief impressions. The stone cylinder seals or rollers could be reused to make the same impression on another clay tablet, thus multiplying the images. This invention was in essence a primitive printing press or copying machine. Engraving became more sophisticated with the invention of the graver or burin, used to cut the design into a metal plate, preferably copper, to produce a high-quality, intense image.

Rubbings

Some historians hypothesize that the Chinese may have invented a crude form of printing as early as the second century C.E., the rubbing, to reproduce Chinese calligraphy. This ancient process, as the name implies, entailed placing paper over a carved surface (wood or metal) and then rubbing the paper with a mixture of wax and carbon black, or blotting with ink, thus raising the image on the carved surface onto the paper surface. However, the earliest authenticated wooden block rubbings can only be dated back to the eighth century C.E.

Postal Systems

Courier or postal systems are, at the least, 4,000 years old. The ancient Egyptians and Mesopotamians maintained a postal service, albeit for their wealthier citizens, in about 2000 B.C.E. The Assyrians of Mesopotamia had established successful commercial trade routes as far as Anatolia (present-day Turkey) that enabled the merchants, via caravans, to send all manner of correspondence (legal, accounting, personal) to their trading partners. The correspondence was written in cuneiform on clay tablets, approximately three inches square, and enclosed in clay envelopes on which the name and address of the recipient was likewise written in Assyrian cuneiform script. In about 1900 B.C.E. the pharaohs established a courier system of relay stations in Egypt that was expanded throughout the next several centuries to include outposts in Mesopotamia, Palestine, and Anatolia. Egyptian excavations have unearthed letters that date from about 1450 B.C.E. that were addressed to the famed King Tutankhamen. However, these letters were not composed in Egyptian hieratic or demotic writing, but rather in Babylonian cuneiform. China's postal service dates back to the Chou Dynasty, circa 1027–1021 B.C.E. Like the Egyptians, the Chinese also utilized posthouses or relay stations, a system that the Mongols further developed after their conquest of the Chinese in the thirteenth century C.E. The Chinese postal service was renowned for its speed and innovation. For instance, feathers attached to a letter were indicative of importance and/or urgency. The Chinese government–sponsored postal system was abolished in 1402 C.E. when the then-Emperor Yung Lo turned the system over to private individuals.

In the mid-sixth century B.C.E. the Persians established the most sophisticated postal service system during the reign of Emperor Cyrus the Great. This system used horse riders who operated along the 1,600-mile "Royal Road" that ran from the ancient Aegean city of Sardis in western Asia Minor to Susa, the ancient capital of Persia, also known as Elam (present-day Iran). Fresh horses and new riders were supplied along the road at relay sites, approximately a day's ride apart, which were staffed with grooms and other agents. The Persian postal system thrived and expanded for over 2,000 years. During this time Alexander the Great conquered Persia, dynasties fell, territory was lost, and Elam (Iran) came under Islamic dominance after the death of the Prophet Muhammad. Despite all this upheaval, 930 government postal stations within Iran and Iraq were established at a distance of 7.5 miles apart. In addition, camels and mules were also put into service. The system ceased to operate in 1063 C.E. after the Turks conquered Iran. It was reestablished in the thirteenth century C.E. by the Mameluke Turkish rulers of Egypt and Syria, only to be finally abolished by the Mongols in about 1400 C.E.

The ancient Romans also had a postal system, called *cursus publicus,* which reportedly equaled that of the ancient Persians. It was an almost exclusively governmental service that operated between the Roman emperors and the local governors and administrators. It employed couriers, posthouses, and sailing ships, and enabled all parts of the vast Roman Empire to maintain contact with each other. However, it could not sustain itself after the collapse of the empire in the mid-fifth century C.E. The barbarian rulers who had conquered Rome appropriated some parts of the Roman post, but due to political discord and disorganization, its service became sporadic and inefficient. During his lifetime, Charlemagne, the Carolingian (French) Emperor (742–814 C.E.), attempted to revitalize the system into one that was more organized and permanent. He failed and the system completely collapsed after his death. All vestiges of the *cursus publicus* disappeared from the European continent as a result.

In Mesoamerica, the Incan and the Mayan civilizations employed a foot messenger system, as well as posthouses, similar to their Asian and Roman counterparts. Historians believe that the Incans and Mayans maintained a postal system for at least 1,000 years.

Summary

We need only pick up a newspaper, magazine, or book, turn on the television, radio, or CD player, or pick up a telephone or cellular phone to appreciate the depth of human ingenuity and accomplishment in our quest to communicate with each other. Prehistoric carvings and cave paintings, as well as the development of a written alphabet, were the first steps that humans took on their journey to literacy. While its progression was slow for thousands of years, it inevitably led to the expansion of knowledge and the beginnings of scientific method and discovery. The invention of affordable writing materials, primarily paper, advanced the dispersion of ideas and knowledge en masse. And paper itself has undergone a number of transformations that have enabled it to be used in countless ways in hundreds of industries, not just for writing. Printing, once an almost elite craft, has been largely replaced by computers and word processors capable of producing images and characters of astounding intensity and quality. And lastly, we can communicate almost instantly via e-mail. The personal, stamped, and delivered letter seems almost as ancient as the Persian pony express. When we ask ourselves what separates us from other animals, we must inevitably conclude that it is our extraordinary ability to capitalize on our inherent characteristics—consciousness of self, ingenuity, and the ability to communicate at levels that far exceed all other species.

5

ENGINEERING AND MACHINERY

Background and History

One of the characteristics that sets humans apart from other species on planet Earth is our ability to reason. Taken to another level, it might be interpreted as our ability to understand physics. Obviously, ancient humans did not reason in the abstract, nor did they conceptualize the physical and scientific principles that govern the nature of the universe or the planet on which we live. All that was required for survival, which after all was the most important aspect of early human life, was the retention of knowledge gained by basic trial and error. For example, sticks and stones are inanimate objects without any particular distinction until they are utilized to perform some specific function. Then they become tools that do work or weapons that kill prey or an enemy. Early humans had no theoretical knowledge of gravity, mass, or energy. A rock hurled into the air and directed at a specific target causes damage or injury—if it hits its mark. When a larger rock was used or when it was hurled from a leather slingshot, it caused greater damage and death. Thus, early man "engineered" an object to achieve a more specified goal.

Originally, physics, from the Greek word *physikos* meaning "of nature," described the fundamental beliefs of the early Greek philosophers, such as Democritus, Pythagoras, Heraclitus, and Zeno. They and others rejected many of the myths and legends that their contemporaries relied on to explain the natural world. Instead, they chose to question all aspects of the observable universe and the relationships between humans and all the elements of nature. Sometime after 700 B.C.E. these philosophers began their quest to understand the forces of nature, often arriving at misconceived theories and incorrect conclusions. Nevertheless, their theories and hypotheses were the foundation for what we think of as physics today, a basic physical science encompassing the principles of matter, motion, and energy that is at the very core of engineering, which is the practical application of these principles. What is remarkable, however, is the amount of invention and construction that was achieved prior to 700 B.C.E. Pyramids, stone monuments, water systems, and buildings, many of which still exist, were constructed thousands of years ago by humans who did not possess a formalized set of principles. So, how was this done? One answer lies in mathematics—the language of physics. Many of the ancient philosophers and engineers were primarily mathematicians who first recognized and then demonstrated that a number of natural phenomena could be explained mathematically. The other answer lies in the science of mechanics, which deals not only with machines and how they work but with the *forces* that act upon the most simple to the most sophisticated objects, machines, and devices—both natural and manmade—to give us what is called a "mechanical advantage." In other words, they greatly facilitate the work that is being done.

As it is used in today's vernacular, the term "engineering" derives from the Old French word *engigneor,* which means "contriver," as well as the Latin word *ingenium,* meaning "skill." The ancient builders and craftsmen did not perceive themselves, nor did others, as engineers. They were simply men of intellect, curiosity, and persistence who, most often unknowingly, utilized a combination of mathematics, mechanics, and physical principles to create something of use and benefit. This characterization must also apply to our earliest ancestors who lived thousands of years ago. While it is true that some animals use branches and rocks as tools, at some point in prehistory, early humans (*Homo sapiens sapiens*) began to make their own tools out of these materials for specific purposes. The fashioning of an axlike tool by assembling a head and a haft (handle) may be considered the earliest evidence of engineering. This chapter will deal primarily with examples of those engineering feats and mechanical discoveries that have had the most significant impact on human civilization and our continued intellectual and technological accomplishments.

Simple Machines

A machine is defined as something which enables us to do some form of work, but not until force is applied to it. $W = f \times d$ *(w = work, f = force, d = distance),* which means that when force acts on an object over a distance, work is accomplished. Sticks, stones, even heavy vines, can all be

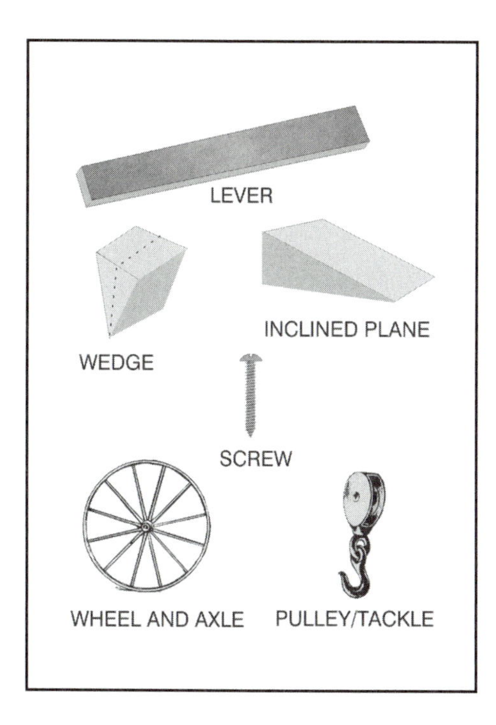

Figure 5.1 Six Simple Machines
The six simple machines—*lever, wedge, inclined plane, wheel and axle, screw,* and *pulley/tackle*—are the basic components of all mechanical devices.

considered as machines—but not until some type of force is utilized that propels them in some fashion. Not surprisingly, historians, engineers, and even teachers cannot agree on the actual number of simple machines that comprise the basic components of all the mechanical devices in the world. Some say two, some four, some five, while others say there are six—the *lever,* the *inclined plane,* the *wedge,* the *screw,* the *wheel and axle,* and the *pulley.* (See Figure 5.1.)The confusion is understandable, since several of these are either combinations of one or more "simple machines" *or* the same design but used for a different purpose.

Lever

This is the most ancient, the simplest, and the most diverse of all the simple machines, and the one which probably had and continues to have the most uses. The lever amplifies physical force when three conditions are present: (1) *force* that is applied to it, commonly called the push or pull; (2) *fulcrum,* the point or the support upon which the lever moves (or rests); and (3) *resistance* or the work to be done. In addition,

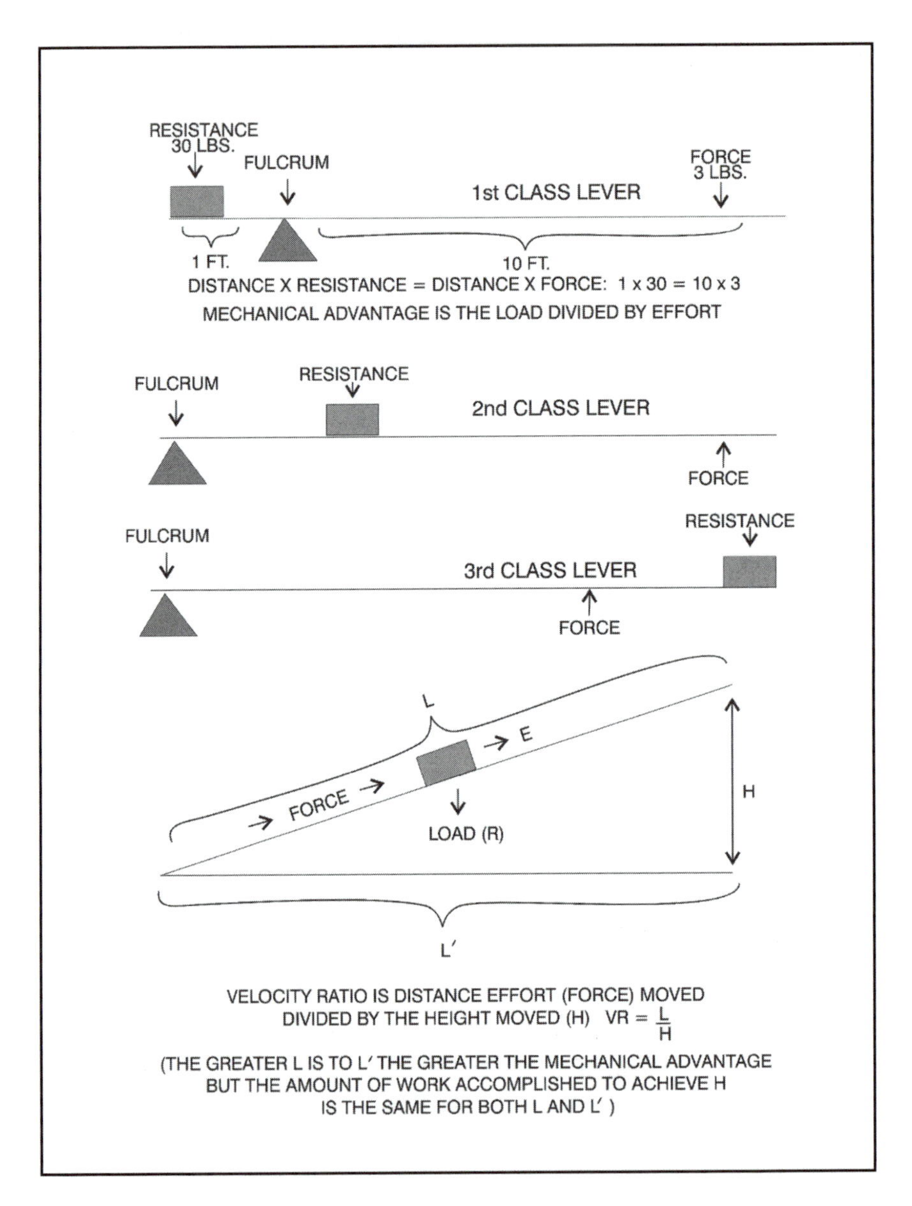

Figure 5.2 Three Types of Levers
There are three types of levers: a) *first-class lever* applies force at one end, resistance at the opposite end, and the fulcrum is somewhere between; b) *second-class lever* has the force and fulcrum at opposite ends, while resistance is somewhere between; c) *third-class lever* means the fulcrum and resistance are at opposite ends, while force is in the middle.

there are three classes of levers. (See Figure 5.2.) A *first-class lever* applies force at one end, resistance at the opposite end, and the fulcrum is somewhere in between. The ancient digging sticks were first-class levers, as are crowbars and scissors. *Second-class levers* (e.g., wheelbarrows and boat oars) arrange the conditions as follows: the force and the fulcrum are at opposite ends, while the resistance is somewhere in between. Lastly, the *third-class lever* means that the fulcrum and resistance are at opposite ends, while the force is in the middle. Examples of a third-class lever are a shovel, a baseball bat, and a fishing pole. (Some scientists believe that both the pulley and the wheel and axle are actually forms of a second-class lever, thus the confusion as two simple machines are eliminated and referred to as levers.) Levers have been used for thousands of years in one form or another. An example of a lever used in Egypt about 7,000 years ago was a balance beam that was used to weigh objects, similar in principle to the weighted scales that were popular until the early part of the twentieth century. However, it was Archimedes in the second century B.C.E. who applied the mathematical principles of the three classes of levers to a number of technological and military challenges.

Wheel and Axle

Historians and anthropologists believe that the wheel was invented in Mesopotamia between 4000 and 3500 B.C.E. They also believe that its first use was as a pottery wheel and/or as part of windlass (pulley system). It is also possible that the wheel was an outgrowth, or later development, of a vehicle called a *sledge.* A sledge, as the name implies, is a sled or a litter that was pushed, pulled, or dragged. Two shafts, usually made from wood, were tied crosswise, and tree material or animal skins were stretched across the shafts, which could support and transport material. At first, humans did the pulling/pushing; later, animals were used for this purpose. The sledge had been in existence for many thousands of years before the wheel was invented, primarily because the Earth's terrain during the last ice age some 40,000 to about 10,000 years ago made the wheel impractical. There was no need for wheeled vehicles in the snow, ice, and subsequent marshy lands that were left after the retreat of the glaciers. Also, some historians believe the wheel may have been a variation of logs or rollers that ancient people used to move objects, although this assertion continues to be disputed by others as lacking evidence. What is not in dispute is the fact that the wheel is a rather late invention, considering its importance and impact on civilization. Nevertheless, once the wheel was equipped with another, usually smaller wheel, called an axle, the invention "took off."

When the wheel and axle are used together, they are considered to be a single simple machine whose primary function is to increase force. (Less often, a wheel and axle can be used to decrease force.) The basic physical principle behind the wheel and axle is that force applied to the wheel will turn the axle. It is believed that the Mesopotamians first used the wheel and axle not for wheeled transportation but for two other purposes. First, it was used to raise water out of wells that were dug in the desert. This device, called a *windlass,* was also employed by the Assyrians to remove dirt and metal from mining areas. And second, the wheel and axle was a pottery wheel. However, archaeologists have unearthed a Mesopotamian vase dating back to before 4000 B.C.E. on which a wheeled chariot was painted, thus raising speculation that the ancient Mesopotamians utilized the wheel and axle simultaneously for several purposes. In any event, once invented, its use quickly spread to other areas in ancient India, China, Egypt, and other places in the Far and Middle East. Wheels were first constructed from wood. At first, one slab of wood was used. Later, it was found that using several separate pieces of wood would produce a stronger, more durable wheel. Still later, probably about 2000 B.C.E., the Mesopotamians improved on the wheel by using spokes, which made it lighter as well as faster.

When we consider the wheel today, our first thoughts would be of transportation vehicles. However, doorknobs, steering wheels, water faucets, and even screwdrivers work on the physical principle of the wheel and axle. (Again, some confusion arises because some scientists believe the wheel and axle is really a lever—as is the screwdriver.) The wheel is also part of another simple machine, the *pulley.*

Pulley

A pulley is simply a wheel around which a flexible rope, belt, or chain is run for the express purpose of pulling it and lifting or moving an object. In mechanical terms, a single pulley, or a combination of pulleys, transmits energy and motion. A *sheave* is a pulley with a grooved rim. (The ancient Assyrian invention, the windlass, which was used to lift water from wells, was a pulley, a wheel and axle, and also a lever. Thus, the pulley, in one form or another, has been around for over six thousand years.) There are three types of pulleys. (See Figure 5.3.) A *fixed pulley* has the pulley (wheel) at the top and the resistance (load) and force (the person doing the pulling) at opposite ends. The raising and lowering of flags on poles is accomplished by using a fixed pulley.

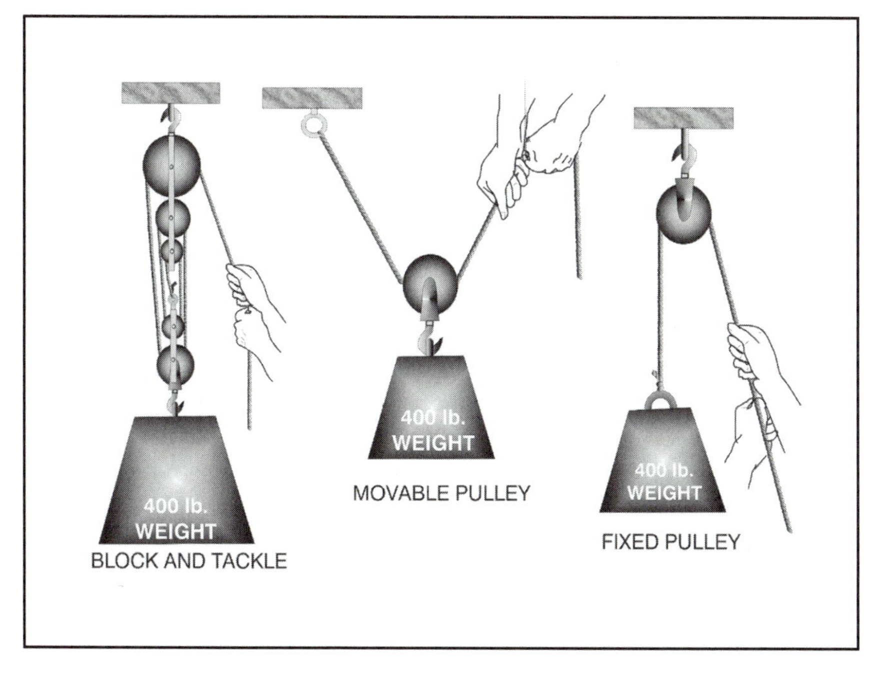

Figure 5.3 Three Types of Pulleys
The three types of pulleys are: a) *fixed pulley,* with the wheel at the top and the resistance and force at opposite ends; b) *movable pulley,* where the object to be moved is attached to the sheave (grooved rim) (pulley is placed between fixed support and the force, while a rope is pulled through the fixed support and around the pulley by the force); and c) *block and tackle (compound) pulley,* a combination of the fixed and movable pulleys.

On a *movable pulley* the object to be moved is attached to the sheave (grooved-rim pulley). The pulley is then placed between a fixed support and the person or force that does the pulling. A rope is pulled through the fixed support and around the pulley by the person doing the work, thus moving both the pulley and the attached object. A *compound pulley,* a combination of a fixed and movable pulley, is also called a *block and tackle.* In this kind of pulley arrangement, the *block* consists of one or more pulleys of decreasing diameters either affixed to each other side by side horizontally, or atop each other vertically, and the *tackle* is the rope or chain that is pulled through the grooved rims of the pulleys. The largest pulley is attached to a fixed support and the object to be moved is attached to the smallest pulley. As force is applied, the pulleys move freely, multiplying the amount of force the person is using and thereby reducing the force needed to move the object. At the same time, the object moves over a shorter distance.

In antiquity, there are two famous examples of the use of pulleys, both involving the famous geometer Archimedes. As a result of a challenge from his friend and cousin, the Greek King Hiero of Syracuse (fl. third century B.C.E.), Archimedes, using the windlass and a system of compound pulleys, reportedly launched single-handedly the heaviest ship in the king's fleet from dry dock into the ocean. Up to this point, the sheer body-force of hundreds of the king's slaves was unable to move the ship from its place of construction into the water. To further add to the challenge, King Hiero reportedly added full freight and passengers to the ship. Undeterred and ultimately successful in his attempt, Archimedes immediately became respected as well as renowned for his "genius." The other pulley system involved a device known as the "Claws of Archimedes." Using machines and weaponry that he had designed, Archimedes supposedly frustrated and delayed the Romans from entering and conquering the city of Syracuse for two-and-a-half years. The Claws of Archimedes, also known as a "ship-shaking machine," was a combination of a lever and pulley system constructed from metal, heavy wooden beams (levers), and hooks (claws) that was built at the edge of the harbor where the invading ships attempted to land. The hooks were attached to the ends of heavy chains that were suspended from long, heavy beams constructed with a specialized joint, called a *carchesion.* The *carchesion* allowed the beams that protruded over the water to move both horizontally and vertically. The pulley moved the suspended hooks or claws into the water where they attached to the ship's hull or bow. When a team of oxen moved the elaborate pulley system, the invading ship was lifted out of the water almost vertically. Tension was then released and the ship crashed back into the water, causing destruction and even death. Nevertheless, the Romans conquered Syracuse in 212 B.C.E. and the entire country of Greece by the second century B.C.E.

Inclined Plane

This simple machine consists merely of any object with a sloping surface that is used to lift a heavier object. A board or plank that is positioned over a curb or a series of steps is an inclined plane. Screws and bolts work on the same principle as the inclined plane, and share its use of mechanical advantage. For instance, if you apply 25 pounds of pressure to a lever in an effort to move a 100-pound bag of grain, you have multiplied the force (through the use of the lever) by four.

The mechanical advantage (MA) of an inclined plane is described as a ratio of the *effort distance* exerted in relation to the *resistance distance.* Since the MA is a ratio for machines, it has no units of measurement. The MA for an inclined plane can be expressed in several ways (e.g., MA = resistance/effort). It is easiest to understand for inclined planes by comparing the length of the sloping plane to the height that the weight is to be lifted. (The angle of the plane can also be used.) This can be expressed as the MA being equal to the length of the plane divided by the height that the load must be lifted. For instance, it requires much effort to lift a 100-pound load 3 feet into the bed of a truck, but it is much easier to slide the 100-pound load up a 12-foot inclined plane into the truck. Therefore, neglecting friction, the MA would require one-fourth the effort (25 pounds) to move the load into the truck bed, compared to the effort required to lift the load 3 feet vertically into the truck—albeit over a longer distance, since the MA is 4 (MA = 12/3). But in both cases, the "work" being done is the same *(W = force* × *distance).* It is just easier to use the inclined plane, as was discovered by the ancients.

Thousands of years ago, the Egyptians employed the principle of the inclined plane to build the pyramids. Earthen ramps that were thousands of feet in length were constructed in order to move the enormous stones to great heights. It is also believed that the enormous stones erected at Stonehenge in Great Britain were moved into place using a system of earthen ramps.

Wedge

The wedge is simply two inclined planes that have been placed together. It is also defined as a "movable inclined plane." As levers had been used by early humans to perform some kind of work, so too were wedges. Prehistoric humans fashioned wedges to break stones and rocks, and to split logs and heavy branches. Ancient Egyptian stonecutters removed large slabs of rock from quarries by using wooden wedges that were soaked with water. The wedge was inserted either into the natural cracks in the rocks, or into manually bored holes in the rock. The wooden wedge was then soaked with water. After a time, the wood swelled and expanded and enlarged the cracks. The process was repeated continuously until a usable portion of the rock broke free. Inclined planes were then used to move the stone. Substituting cork for wood, the ancient Greeks also utilized this method to cut slabs of marble from quarries.

Figure 5.4 Archimedes' Screw
A variant of the inclined plane, the *Archimedean screw* supplies mechanical advantage through circular movement.

Screw

The last simple machine to be invented was the screw, which is a variant of the inclined plane. Historians agree that it was invented in ancient Greece, but are split as to whether it was invented by Archytas of Tarentum (400–350 B.C.E.) or later by Archimedes in the third century B.C.E. A screw is really an inclined plane that is wrapped in a spiral around a shaft. The threads of the screw supply a mechanical advantage through circular movement, just as an inclined plane provides movement in a straight but sloping line. While we generally think of screws as devices to attach or fasten one object to another, the principle of the screw applies to a whole host of other machines, for example, the ancient screw press that squeezed the oil from olives and juice from a variety of fruits. Another famous example from antiquity is Archimedes' screw, an invention that raised water from inside the hulls or cargo holds of ships, as well as raising water from the Nile River in Egypt. The Archimedean screw (water-raising screw) continues to be used in some parts of the world. (See Figure 5.4.) In modern times, the principle of the screw is employed in the construction of water pumps, like the simple home sump pump and the more complex pumps found in wastewater treatment plants.

Building Materials and Construction

At one time during prehistory, when early humans were primarily hunter/gatherers, caves and large, deep overhanging rock ledges were their natural habitats. We can assume the availability of caves was plentiful given the limited human population. They afforded a measure of protection from predatory animals, as well as being a sheltering environment. Once humans ventured into a lifestyle that depended more upon agriculture, the need for shelter outside the cave became a necessity. For example, *mammoth houses* dating back to at least 13,000 B.C.E. were unearthed in Ukraine in the mid-1960s. Constructed entirely from the bones and hides of as many as 100 mammoths, a mammoth house apparently was broken down and reused by succeeding generations of early humans. About 10,000 B.C.E. *mud and wattle houses* began to be constructed in the Middle East. Mud was plastered over a *wattle work,* a skeletonlike structure made from the stalks or reeds of native plants that were woven together. Obviously, this structure was only practical in very dry climates where rainfall was sporadic, but the basic construction principle was utilized and improved over thousands of years and has been found in various forms on both the European and North American continents. (A modern version of mud and wattle, also known as wattle and daub, was the lath-and-plaster method utilized in interior wall construction before plasterboard and Sheetrock.) Egyptian pyramids that are over 4,000 years old and Mesopotamian brick **ziggurats** dating back to about 1100 B.C.E. still stand today. And though many believe them to be symbolic of modern urban development, a*partment buildings* or *insulae* (Latin for "islands"), some as high as four or five stories, were constructed in ancient Rome in the third century B.C.E. What do all of these ancient structures have in common? For the most part and for most of civilization, at least through the nineteenth century, the type of construction chosen was dependent upon the availability of local materials. While prehistoric humans utilized the hides of wild animals for shelter, as well as for clothing and other implements, it was mud, clay, stone, plants, and timber that were the foundation of the ancient construction industry. Some of the earliest techniques for constructing houses and other edifices are still in evidence and in use even today.

Bricks and Tiles

The ancient art of brick making originated in Mesopotamia (present-day Iraq) over 6,000 years ago. This ancient land was then and is now a flat alluvial plain spread between two rivers, the Tigris and the Euphrates. Mud and clay deposits along the banks of these rivers were plentiful, while quarried stone and timber were and still are scarce. The earliest Mesopotamian structures contained walls that consisted of densely packed mud and clay. It is assumed that the earliest bricks were formed by hand, but eventually the Mesopotamians constructed four-sided wooden molds or frames that were packed with mud and clay and then sun-dried to produce uniformly sized bricks. At some time during the evolution of the brick, finely cut straw was added to the mud/clay mixture to add strength and prevent cracking after the bricks were dried. To a great extent, pottery making techniques advanced the further improvement of bricks. Potters learned through a process of trial and error that pottery that was merely baked in the sun would crack and leak, while those that were baked in kilns (ovens) were harder and less likely to chip and leak. In addition, potters also invented a glaze comprised of sand and other available minerals that they spread over the kiln-baked pottery to make it even more strong, glasslike, and water-resistant. This is a process known as *firing*, and about 5,000 years ago Mesopotamian brickmakers utilized the same techniques for their products with the same result—a harder, stronger, and more waterproof brick. The identical technology was used to make clay roof tiles. Before this discovery, thatched roofs made from straw, leaves, and branches were replaced every few years because of moisture and insects. Kiln-dried roofing tiles provided protection and durability, and enabled the architect to design buildings with more aesthetic features rather than the merely practical. Some historians attribute the beauty of ancient Greek architecture to the invention of the fired-clay roof tile.

Egypt, on the other hand, was a country with a slightly different topography, whose soil content was not overly rich in clay. Stone was abundant and was quarried extensively. Most of the public buildings in Egypt were built with stones and mortar. Lesser buildings and houses were constructed with sun-dried bricks primarily because the cost of quarrying and transporting stone was beyond the means of the general population. (Kiln-dried bricks were not used at all because of the lack of adequate clay deposits in the soil.) Today in Egypt, builders still employ the ancient process of mixing chopped straw with mud, pouring it into molds, and allowing it to dry in the hot sun. In the Far East and in Greece bricks were used but not extensively. In China, particularly, timber was readily available and was the preferred material for houses, but bricks were used to build the walls of public structures. Limestone was plentiful in Greece, as were marble quarries, thus stone and marble were the basic building materials of Grecian builders.

The Romans vastly improved the basic concept and took it to higher levels of use. Kiln-dried roofing tiles and bricks had been in existence for thousands of years before the Romans became a dominant force in the then-civilized world. The *insulae* (apartments or tenements) that were a popular form of housing in ancient Rome were, at first, poorly constructed, except for the roofing tiles, which seemed to survive the frequent fires and collapses of these structures. Roman builders salvaged these tiles and used them as building blocks to either greatly reinforce and/or face the walls in which the Roman invention of concrete was used. Thus, the Romans appropriated not merely the process of the kiln-baked roofing tile to make their own bricks, but expanded and varied its original use and the sizes of the bricks to construct edifices that were uniquely Roman. The Roman brick was made in two sizes. The larger of the two was called *bipedalis* and was 60 centimeters square with a thickness of about 7 centimeters (approximately 2 feet by 2 feet by $2^1/_4$ inches). The smaller Roman bricks were shorter, narrower, and rectangular. The typical Roman wall was constructed with intermittent courses (rows) of both sizes of bricks that ran the length of the entire wall, the object being to build a wall that was virtually indestructible. The Romans succeeded. In his treatise *De Architectura* (On Architecture), the famous Roman architect and engineer, Marcus Vitruvius Pollio (fl. 100 B.C.E.), detailed specifics on the preferred methodologies of brick making that had been in existence for centuries, some of which are still valid. Today, bricks are made from natural clays that are mixed with composites of other minerals (e.g., iron or calcium) and are standardized in size to about 20–22 centimeters long, 9–11 centimeters wide, and 5–$7^1/_2$ centimeters deep (8–9 inches by $3^3/_4$–$4^1/_2$ inches by 2–3 inches).

Mortar and Concrete

Mortar is basically any material that bonds other materials to form a structure. The mud that was used in ancient mud and wattle huts could be considered a crude type of mortar. Deposits of gypsum, a sulfate mineral, were common in ancient Egypt. Using gypsum as the basic ingredient, along with lime, Egyptian builders invented a special type of mortar that was compatible with the quarried stone used in their buildings. (Today, we refer to gypsum as *plaster of paris.* It was so named during the Middle Ages because buildings that were constructed in and near the European city of Paris during this time used tremendous amounts of the gypsum mortar. The mineral itself was found in deposits in areas around the Paris basin.) Some historians, however, believe that the Egyptian gypsum mortar was used more as a lubricant for sliding stones rather than as a bonding agent, primarily because the Egyptian stonework was so sophisticated and of such high quality that the stones could be "dry-jointed" rather than set with mortar.

Mortars that used sand were not formulated until much later, most likely around 500 B.C.E. The use of sand as an ingredient in mortar led to two famous Roman discoveries, *pozzolana* and *concrete.* Pozzolana is a mixture of volcanic ash and limestone. At least in the third century B.C.E., and possibly before, Roman builders discovered that deposits of a sandlike substance (which was actually volcanic ash) were present at the site of a prehistoric volcano in the town of Pozzuoli (near present-day Naples but originally called Puteoli in Roman times). When this sand-like substance was mixed with lime, the result was an exceptionally strong mortar. It took decades, but the Romans eventually discovered that pozzolana deposits were not necessarily limited to this one area, but could be found in places nearer to Rome itself. Nor were the deposits necessarily of one color. Depending on the volcanic activity involved, pozzolana deposits varied in color—red, brown, gray, and black. The Romans also refined the lime by burning it in kilns to produce a substance called slaked lime. Once they had perfected the formula for the mortar (two parts pozzolana with one part slaked lime mixed with water), they used it for building walls with a core of rubble. These walls were particularly strong, and again the Romans realized that by actually adding an aggregate directly to the mortar, such as gravel or fragmented rocks, the result was a hardened substance that the Romans called concrete. (The exact date of its invention is unknown, as concrete really evolved from Roman experimentation with various formulas for mortar. It has been in use, however, since the first century C.E.) *Note: Cement,* also known as Portland cement, is the **hydraulic** binding or adhesive agent made from a mixture of pulverized clay and limestone; cement mixed with an aggregate (sand, gravel, broken rocks, or stone) and water forms concrete, the finished, hard, dried material.

Because they recognized that the extraordinary strength of concrete could be utilized in larger structures than merely walls, Roman architects and builders were able to design and build any number of imposing edifices, many of which are still in existence, such as the Pantheon and the amphitheater at Pozzuoli. Modern-day concrete is essentially a variation of the basic Roman formula. Rather than using pozzolana as the Romans did, mixtures of sand and gravel have been substituted. Although less abundant than it was 2,000 years ago, pozzolana in its natural state is still an ingredient for concrete in countries where deposits are accessible. However, a "pozzolana-slag" can now be made artificially in blast furnaces.

Pyramids

The construction of pyramids in ancient Egypt was a consequence of the Egyptians' belief in an afterlife that could only be enjoyed fully if the deceased's body was kept intact. Thus, the practice of mummification and the building of elaborate, virtually impenetrable tombs were begun nearly 5,000 years ago. The first tombs were actually rectangular structures with inward-sloping walls over burial chambers. They were called *mastabas,* derived from the Arabic word for "bench," because they resembled in shape the mud-brick benches of Egyptian peasants. The later design of pyramid-shaped tombs was related to the use of mud-bricks for the early construction of these mastabas. Before the Egyptians used the natural resource of quarried stone for buildings and tombs, mud-bricks were the material of choice. Because mud-bricks were not particularly durable, the Egyptians learned that by sloping mud-brick walls—tapering them from bottom to top—the natural deterioration from the elements, particularly rain, was lessened. Although stone eventually replaced mud-bricks, Egyptian builders continued with this unique style—the pyramid style.

The Egyptian pyramids were built over a period of 2,700 years, beginning in earnest in about 2650 B.C.E. at the beginning of the Third Dynasty and ending with the Ptolemaic period in the first century C.E. The construction of pyramids reached its zenith with the building of imposing structures to honor deceased members of Egyptian royalty. (Egyptian kings were referred to as pharaohs.) Archaeologists have unearthed some 80 pyramids in Egypt. By the time of their discovery, most of these structures were virtually in ruin and their contents plundered. However, there are a number of pyramids that remain intact, and historians have been able to reconstruct the processes that ancient architects and builders applied in the building of these amazing edifices. Pyramids were not simply individual buildings. Rather, they were part of an intricate complex of structures designed to maximize security and prevent tomb robbers from stealing the treasures and artifacts that were carefully selected and placed in these burial chambers. It is also important to remember that all of these pyramids were built using only four simple machines: the lever, the inclined plane, the wedge, and the pulley. In addition, the Egyptians possessed the all-important element of unlimited manpower in the form of slaves as well as "volunteer" laborers. At various times of the year when the Nile flooded and they were unable to cultivate the land, farmers and peasants were conscripted to work at the site of these massive tombs. This challenges a common misconception that the pyramids were built entirely by slave labor. While the peasants and farmers were unpaid for their labors, they were not slaves, although in this case, it may really be a distinction without a difference.

The construction of the first pyramid took place in about 2650 B.C.E. in Saqqarah, the site of an elaborate **necropolis** in the Egyptian city of Memphis. Its architect was Imhotep, the celebrated physician, court magician, and astrologer to Djoser of the Third Dynasty, who reigned from about 2630 to 2611 B.C.E. Characterized by six terraces or levels, it is known today as the Step or Djoser's (also spelled Joser's or Zoser's) Pyramid and was the first building of its kind to be constructed completely using various grades and sizes of limestone. Its unique structure was not conceived by design but rather by dissatisfaction. Originally, Imhotep and Djoser agreed that the design should be in the shape of the traditional rectangular mastaba. But over time, the project became more ambitious and resulted in the final 61-meter (approximately 200-foot) structure with a base that measures approximately 105 by 125 meters (344 feet by 410 feet). Beneath the pyramid, the burial chamber branched out into many corridors designed to hold Djoser's selected treasures. Outside of the Step Pyramid, a walled enclosure encompassed a mortuary temple and living quarters for the priests who were mandated to perform, in perpetuity, rituals that would ensure Djoser's well-being in the afterlife. Though other builders attempted to construct pyramids using the step or terraced design, the steps were eventually filled in and the traditional smooth-sided pyramid became the accepted design in Egypt.

The largest Egyptian pyramid ever built is at Giza. Known as the Great Pyramid of Khufu (also called Cheops), it was built around 2600 B.C.E. Khufu was the second pharaoh of the Fourth Dynasty. When the pyramid was built, it was about 146 meters (479 feet) high on a four-sided base, with each side about 229 meters (751 feet) long, or the length of two-and-a-half football fields. It is estimated that over 2.4 million limestone blocks, weighing between 2.5 and 3.5 tons apiece, were quarried to complete the construction. Over the years it has lost approximately 10 meters (33 feet) of its top due to deterioration. There are several other unusual features of this particular pyramid: namely, the sides of the base are within 7 inches of forming a perfect square, and are oriented to within one-tenth of a degree or less of the true north-south and east-west directions. Another unusual aspect of the construction is that the actual burial chamber, called the king's chamber, was built higher in the structure, rather than in the underground burial vaults, to accommodate Khufu's claustrophobia. He felt that a burial vault beneath the pyramid could be subject to cave-ins on top of his sarcophagus, an unacceptable prospect even in the afterlife. The Great Pyramid contains an elaborate system of airshafts, corridors, galleries, and burial chambers, and remained the tallest structure in the world for nearly 4,300 years. (See Figure 5.5.)

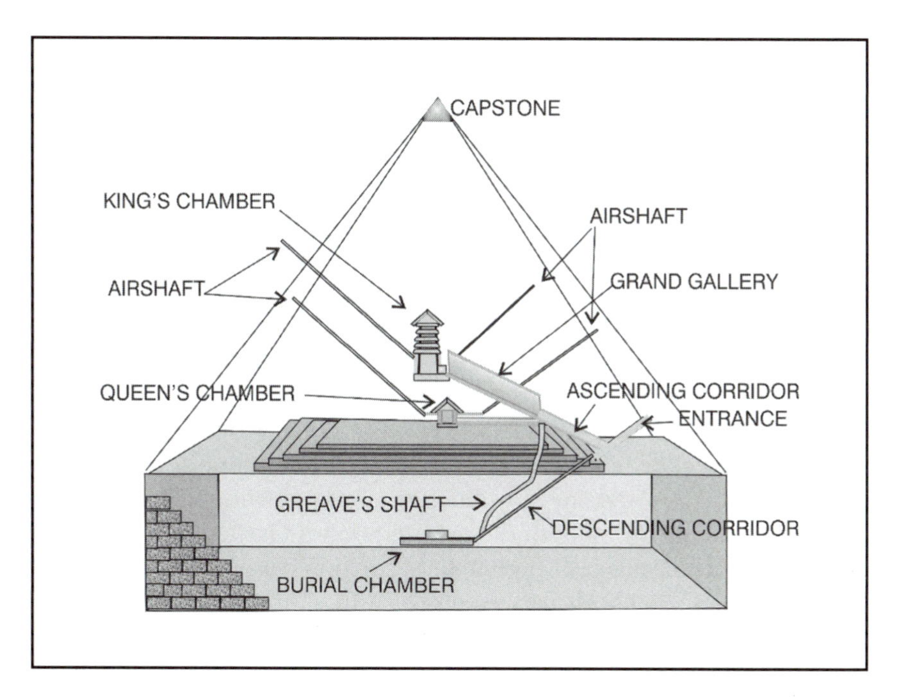

Figure 5.5 Interior of the Great Pyramid of Khufu
Located at Giza, the largest Egyptian pyramid was built around 2600 B.C.E. by the Pharaoh Khufu (sometimes referred to as Cheops). Its interior is elaborately constructed and unusual in that Khufu's burial chamber is situated higher in the structure, rather than in the more traditional underground burial vaults.

In the fifth century B.C.E. Herodotus, the Greek historian, described the process involved in constructing a pyramid. However, this was well over 2,000 years after the actual construction and his information may have been inaccurate in some aspects. There are, however, some generally accepted theories as to how they were constructed, which involved a series of planned and well-focused tasks: (1) The building site was chosen and cleared of sand and gravel. Using a wooden level in the shape of the letter "A," then digging a series of crisscrossed ditches that were filled with water, Egyptian builders ensured the site was completely level. The free-flowing water indicated whether or not the area was level. After the ditches were emptied of water, they were filled with rubble and smoothed over. (2) Architects drew up plans for the exterior as well as interior of the pyramid, many of which contained numerous rooms for the pharaoh's relatives and possessions, and elaborate tunnel systems to confuse tomb robbers. *Caissons,* huge hollow tubelike structures made from stone and brick, were utilized in the building of the tunnel systems in the base of the pyramids. The workers pushed the caisson forward and removed the extruded debris. Other workers then built up the walls surrounding the caisson with stone and brick, thus forming the tunnel. (3) Limestone for the exterior and granite for the interior walls were quarried. The stones were marked and cut. (This is where wooden wedges were employed to split the rock.) Using levers and devices made of palm-fiber ropes (pulleys), quarry workers raised the massive stones onto sledges that were set atop a series of logs to prevent the stones from sinking into the desert sands. The stones were then loaded onto flat-bottomed boats to be transported to the building site. (4) The unloaded blocks of stone were cut into precise sizes at the building site by expert stonemasons. Other workers smoothed off the rough edges, tested the stone with various leveling devices to ensure that it was perfectly level, and chiseled away excess chunks of stone to square the corners of stone. (5) Earthen ramps (inclined planes) were constructed along the sides of the pyramid on which the finished stones were moved to higher and higher levels. Either water or a mixture of gypsum helped to slide the stones more easily. (6) Mortar was not used to secure the stones, as they were cut and fitted so precisely that nothing could be inserted between them—not even the thinnest knife. Egyptian engineers used the concept of a *plumb line,* a sharp pointed stone attached to a cord, to ensure that each layer of stone was placed at the correct angle. The plumb line was lowered until the pointed stone barely touched the ground. The engineer knew that when the line ceased swinging or moving, it was at a right angle to the ground and the stones were set correctly. (The Egyptians did not refer to this as a plumb line, however. That term came later when lead, *plumbum* in Latin, replaced the pointed stones at the end of the cord.) (7) Next came the placement of the *capstone,* the pointed stone with a bottom plug, at the top of the pyramid. (8) The final steps involved polishing the exterior stones, going from top to bottom, and removing the earthen ramps as they went along.

Roads

Before the advent of civilization, ancient humans roamed the land more or less indiscriminately. Over time, crude footpaths were beaten down with the repeated journeys of generations of ancient hunters. Streets only came into being with the growth of settled towns and cities, rather than the mere agglomeration of crude huts and makeshift houses. Roads, on the other hand, were built after the invention of the wheel and the subsequent development of wheeled vehicles. Streets signify community; roads signify commerce. The first streets and roads were merely graded dirt beset by ruts, holes, and mud. When a road became impassable, early humans simply abandoned it and graded another in the vicinity. The ancient Mesopotamians were the inventors of the first paved road over 2,400 years ago, but limited its usage for religious purposes. These roads, referred to as "Processional Ways" and on which warnings were posted (e.g., "The Street on Which May No Enemy Ever Tread" or " Let No Man Lessen It"), were paved with flagstones set in mortar. In some cases, only vehicles that bore the images of the Mesopotamian gods were allowed to travel down these roads. Eventually, paved roads were constructed for other important usages, including the heavily traveled routes between ancient Middle Eastern cities. Paving, however, was extremely limited. The grading of roads was far more widespread, for two reasons. First, it was cheaper to build and repair; second, it was easier on the animals that pulled wagons and carts.

The Chinese began building the first "official" or state-sanctioned roads beginning in the ninth century B.C.E. The project continued until about 200 C.E. when over 20,000 miles of imperial roads had been constructed. China's first emperor, Shih Huang Ti (fl. 215 B.C.E.), mandated that the width of all chariot wheels be standardized at five feet in order to accommodate the width of the roadway. Despite the fact that these roads had as many as nine lanes, most of them were designated for the exclusive use of the emperor and his family, with severe penalties, including death, for those who dared to ride in the "royal lanes." The distinction of the longest single road belongs to the Persians, who built the King's Road (also known as the Royal Road) in about the sixth century B.C.E. It ran some 2,400 kilometers (1,488 miles) from the Persian capital of Susa in Elam (modern-day Iran) to Sardis, a town on the west coast of present-day Turkey. The Greeks, who in history are renowned for their intellectual contributions as well as for their magnificent marble edifices (e.g., the Parthenon and the Acropolis), were not master road builders. Greek streets and roads were muddied, rutted, and often flooded. This was most likely due to the terrain, which was hilly and rocky with many lakes. Though flagstones and mortar, in some form, were available, they were seldom used to pave the streets of ancient Greece. The fact was that projects such as the construction of streets and roads were beyond the financial ability of most ancient civilizations, that is, until the Romans.

Though not the first road builders, the Romans are the most famous, and probably the best, if only because of the sheer mileage and the absolute quality of their construction. By 200 C.E., there were between 80,000 and 92,000 kilometers (50,000–57,000 miles) of well-maintained Roman-built roads in existence. Roads were built on three continents: in Europe, extending from Hadrian's Wall in the northern part of present-day England; in Asia, to the banks of the Euphrates River in the southwestern part of the continent; and in Africa, to the southern part of the Sahara. After the fall of the Roman Empire in the fifth century C.E., the financial burden of the maintenance and upkeep of roads was beyond most governments, and they were not maintained, nor were new ones constructed, until the eighteenth century in Europe. Roman roads were primarily built by and for the Roman armies to allow them to move swiftly from one encampment to another. The Romans also needed adequate roadways to service the commerce and communication of their ever-expanding empire. However, only those that were heavily traveled by garrisons of the Roman army were paved. Provincial or secondary roads were built with gravel, which was easier on the hooves of the horses and oxen that pulled the carts and wagons of merchants and farmers. Unpaved roads were often constructed alongside the hard-paved highways to accommodate the local population.

Roman-built roads (paved or otherwise) have several distinguishing characteristics. First, they were remarkably straight and direct over long distances, deviating only when the terrain was marshy or otherwise difficult. Roman engineers preferred to go over mountains rather than around them. This was done also to accommodate the movement of the armies. Second, the depth of the *paved roadbed* was $1^1/_2$ meters (5 feet). (This is almost twice the depth of modern-day roadbeds.) Using locally available materials, the roadbed contained four to five layers: (1) first was a layer of sand and mortar at the bottom; (2) next came a layer of "squared" stones set in a mortar; (3) then came a layer of gravel set in clay or sometimes concrete; (4) the fourth layer was rolled sand concrete; and (5) finally came the *summa crusta* or *pavimentum*, which was large blocks of hard rock set in concrete.

A third distinguishing characteristic was that Roman engineers had the foresight to camber or slope their roads for drainage purposes, which prevented the deterioration of the roads' foundations. Many of them had curbs that ran the length of the road. The width of Roman roads varied in size, from 5 to 6 meters (16–20 feet) for provincial roads to approximately 10 meters (33–35 feet) for well-traveled main highways. The most famous as well as oldest of the Roman roads is the Appian Way (Via Appia), parts of which are still in existence—and use. Construction on this road began in 312 B.C.E. Originally 212 kilometers (132 miles) long, the Appian Way ran from Rome to the city of Tarracina on the Tyrrhenian Sea, then on to the ancient city of Capua. In about 244 B.C.E. the road was extended another 370 kilometers (230 miles) to the southeastern port city of Tarentum (present-day Taranto) on the Ionian Sea, finally terminating in Brundisium (present-day Brindisi) on the Adriatic Sea. The Appian Way, also called "The Queen of Roads," however, was not a typical Roman road, as it was engineered and built to exacting standards and durability, using lava blocks and concrete that have lasted well over 2,300 years. For most of the 50,000-plus roads, Roman engineers relied on local conditions and resources.

Arches, Domes, and Vaults

It is accurate to state that the arch revolutionized architecture and construction ever since it made its first appearance. However, historians are unsure as to exactly when this occurred or which ancient civilization should be credited with its actual invention. Archaeologists have found artifacts of domed roofs, vaulted ceilings, and arches among the ruins of the ancient Sumerian city of Ur in southern Iraq, at Mycenae in Greece, at Thebes in Egypt, and at Mohenjo-Daro in present-day Pakistan, as well as in China. These artifacts date back thousands of years, and were found on different continents, thus the dilemma surrounding the arch's origins. Some historians believe that the Romans borrowed the concept of the arch from their northern neighbors, the Etruscans. Others are unsure of where the concept originated and whether or if the arch was an independent invention or a shared technology. What is known, however, is that the Romans elevated the arch to technological levels that,

Figure 5.6 Four Steps in the Construction of an Arch
In engineering terms, the *arch* is a curved opening that either supports heavy loads and/or increases the length of a span. The Romans, in particular, used arches in the building of bridges, aqueducts, and public buildings.

theretofore, had not been seen. Even the Greeks, who had built magnificent marble public buildings and temples beginning in the fifth century B.C.E., did not utilize to any measurable extent the principles of the arch until much later, during the dominance of the Roman Empire. Domes and vaults are really extensions or variations of the arch.

In ancient times, an *arch* was a curved opening constructed of stone or brick that either (a) supported extremely heavy weights or loads, and/or (b) increased the length of a span. Roman engineers capitalized on their knowledge that properly built arched structures could carry far more weight than horizontal beams. Thus, they used arches extensively in the building of bridges and aqueducts, as well as in public buildings, such as the Roman Colosseum. The construction of the arch involved several important engineering principles—for example, **vectors**—all of which entailed knowledge of mathematics and physics. First, stable pillars, piers, or other types of supports were built on each side of the arch. Second, during construction a temporary wooden arch-shaped struc-

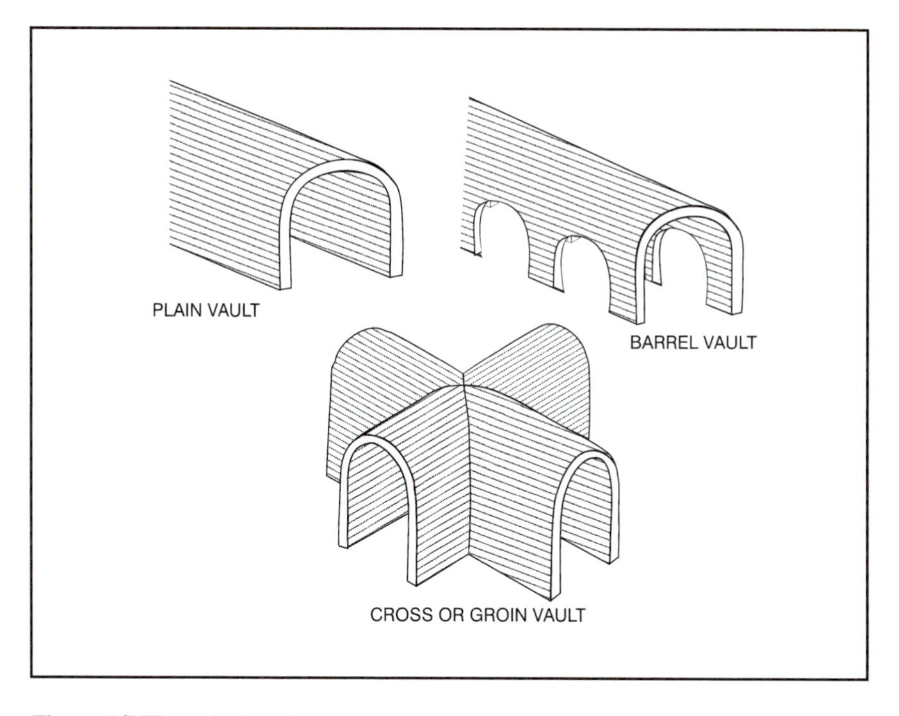

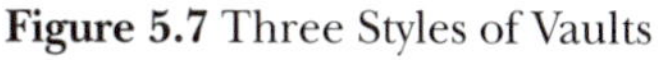

Figure 5.7 Three Styles of Vaults
Vaults are arches that have been linearly extended. In other words, they are arches that are arranged one behind the other to form a roof or ceiling, such as the *plain vault.* The *barrel vault* is a series of arches covering a three-dimensional area. A *cross* or *groin vault* is formed by intersecting two barrel vaults at right angles.

ture, called the centring, was placed inside the arch. This was important to ensure stability. Third, *voussoirs* (wedge-shaped stones or blocks) were set side by side, smaller side downward, in a curved pattern. The central wedge or voussoir is called the keystone. (See Figure 5.6.) The fourth or final step in arch construction was the removal of the centring, leaving carefully shaped and vertically aligned voussoirs on either side of the arch. Arches can be *semicircular* or *segmental,* which is less than one-half of a complete circle, or *ogival* which is pointed with S-shaped sides. If Roman architecture or construction had any constraints, it was the fact that its arches were all semicircular, which restricted their heights to half their spans. Today, arches and archways are commonplace and are used as much for decorative purposes in home and commercial construction as they are for safety and durability.

A *dome* is a three-dimensional arch, hemispherical in shape, used primarily to form roofs or ceilings. Domed buildings require strong supporting walls, since the dome exerts enormous pressure around the perimeter of the structure. Evidence of domed construction, usually small burial chambers, has been found in the Middle East and in India. But the Romans, once again, took the concept to extraordinary heights with the building of the Pantheon, from the Greek word *pantheion* meaning "place for all gods." Construction began in 27 B.C.E. However, it was completely rebuilt sometime in the second century C.E. The dome on this building, the largest in existence until modern times, has a glass skylight oculus, 8.9 meters (29 feet) in diameter at its top, through which light enters the building. The diameter of the dome measures about 43 meters (141 feet), with a height of nearly 22 meters (72 feet) above the base. The Pantheon exists today (as a Roman Catholic Church) almost entirely in its original state. Historians and scientists view this as a remarkable building inasmuch as there are no brick or stone arch supports inside the dome itself (except at the lowest points). Although it is unclear as to the exact methods the Romans employed in building the Pantheon, architects have determined that the dome is buttressed by the enormous brick arches and supports that were built one upon the other inside the 6-meter-thick (20-foot) walls, and that the Romans' use of concrete rendered the building virtually indestructible.

A *vault* is an arch that has been linearly extended, that is, arches that have been arranged one behind the other, to form a roof or a ceiling. (See Figure 5.7.) Barrel-vaulted ceilings have been found in ancient Mesopotamia and Egypt, but here again the Romans improved upon the concept. A barrel vault is basically a series of arches that covers a three-dimensional area. Of necessity, vaulted ceilings or roofs need to be supported by exceptionally strong walls that have few openings or archways. Roman engineers improvised the basic design by intersecting two barrel vaults at right angles, forming a *cross vault,* also known as a *groin vault.* They further discovered that they could build a structure of almost unlimited length or span by using the same cross vault design repeatedly, as in a series. At the same time, the support walls need not be as strong as for a single barrel-vaulted ceiling, because the pressure or thrust is concentrated primarily at the four intersecting corners rather than in the entire length of the wall. However, the stones used in cross vaults had to be cut precisely, a craft that declined with the fall of Rome. The Flavian Amphitheater (better known as the Roman Colosseum) is a series of intricately engineered arches and vaults. It was constructed of stone, brick, and concrete in the first century C.E. Parts of it exist today, although it was extensively damaged by events during the Middle Ages: lightning, earthquakes, and later, vandalism.

Bridges

Bridge building in one form or another has been around since the emergence of the human species, most likely since the first time a log fell across a creek or stream and one of our ancient ancestors realized it could be used to walk across to the other side. Technologically designed and engineered bridges were not built until the appearance of more advanced civilizations, and we must rely on ancient writings to attest to their construction, since many of these structures have long since been destroyed. Ancient bridges were variations of three types of construction: *floating, pillar and beam,* or *suspension.* The Greek historian Herodotus in the fifth century B.C.E. wrote about the oldest stone bridge constructed in the ancient world. And modern-day archaeologists were able to corroborate its existence when they dug up the bases of the bridge's piers. This pillar-and-beam structure, which was also a drawbridge, was built by the Chaldeans of Babylonia in ancient Mesopotamia at least 2,600 years ago and was acclaimed as an extraordinary engineering achievement, almost on a par with the fabled Hanging Gardens of Babylon (which no longer exist). Its span over the Euphrates River was 116 meters (380 feet), and it was built on seven piers that were made from brick, stone, and wood. For centuries, it was the only bridge of its kind. The basic principle of a pillar-and-beam-structured bridge (called a girder in modern-day terminology) is that the support for the road or pathway is concentrated on the foundations at either end of the pathway, thus bearing the entire weight of the bridge itself as well as everything that travels over it. For the most part, ancient bridges were simple pillar-and-beam bridges built from overlapping slabs of flat slate or stones for the pathway and piles of large blocks or stones for the pillars. They are also known as *clapper* bridges, from the Latin word *claperius,* meaning "pile of stones."

Bridges are also a part of warfare, as advancing armies often come upon terrain that would be impassable without the construction of some type of bridge, even if it is only a temporary structure, such as a floating or pontoon bridge. Xerxes the Great (ca. 519–465 B.C.E.), the Persian king who invaded ancient Greece in 480 B.C.E., ordered the building of a massive floating bridge over the Hellespont, the ancient name for the Strait of Dardanelles. Two bridges were actually constructed, but the first was destroyed during a storm. The second bridge was built using cables made of papyrus, planks, and massive amounts of brush and dirt. (These ancient engineers applied the principle of density to the construction of this floating bridge—that is, they used materials that were less dense than water, a process that probably involved trial and error.) The bridge construction was successful, allowing over 150,000 soldiers, and at least that many noncombatants, to pass safely over the waterway. It did not ensure Xerxes' victory however, as both his army and navy were defeated in a series of famous battles. Engineers who marched along with the Roman armies under the command of the famous general and statesman, Gaius Julius Caesar (100–44 B.C.E.), built pontoon bridges over the Saône River (in one day) in Gaul (present-day France), and the Rhine River in Germany (in 10 days), during military campaigns. Unlike the Persian, Xerxes, Caesar was victorious.

The concept of the suspension bridge is as ancient as hooking vines across a river's gorge. Again, historians must rely on ancient writings to attest to its history since the actual bridges no longer exist or have undergone numerous reconstructions, rendering the dates of their actual origins impossible to ascertain. However, it is very likely that suspension bridges were built independently by many civilizations and on many continents. In simple terms, a suspension bridge derives its support from abutments on either side of the bridge that serve as anchors, as well as from the parallel cables on either side of the bridge's pathway that transfer the weight to the anchors on either side. Heavy vines, hemp, and bamboo were used extensively in suspension bridges' early construction. Historians generally credit the Chinese with inventing the true suspension bridge, based on two things: the An-Lan Bridge, which can be found in the province of Szechuan and presumably dates back to the third century B.C.E., and their use of iron cabling in the building of suspension bridges dating to the first century C.E. The An-Lan Bridge, made of wood and rope, is a catenary bridge; that is, its pathway of chain-like planks follows the curves of the cabling rather than merely hanging flat. When originally built, it contained five spans and was nearly 4 meters wide (13 feet). After centuries of rebuilding and reinforcement, its walkway today is only $2^3/_4$ meters (9 feet) wide, but it has been extended to eight spans, for a total of 320 meters (1,050 feet). Like the Chinese, bridge builders on the Indian subcontinent were skilled in iron working. And also like their neighbors, they used iron cables in their suspension bridges, but not until several centuries after the Chinese.

In ancient times, however, it was the Romans who were the prodigious bridge builders. While they may not have brought anything of note to the advancement of pure science, they did master the principles of applied science in all their engineering projects. In bridge building, Roman engineers relied on the *cofferdam,* as well as the concept of the arch. In ancient times, a cofferdam was a watertight enclosure made from timber that was built at the site in a river where the pillars or piers

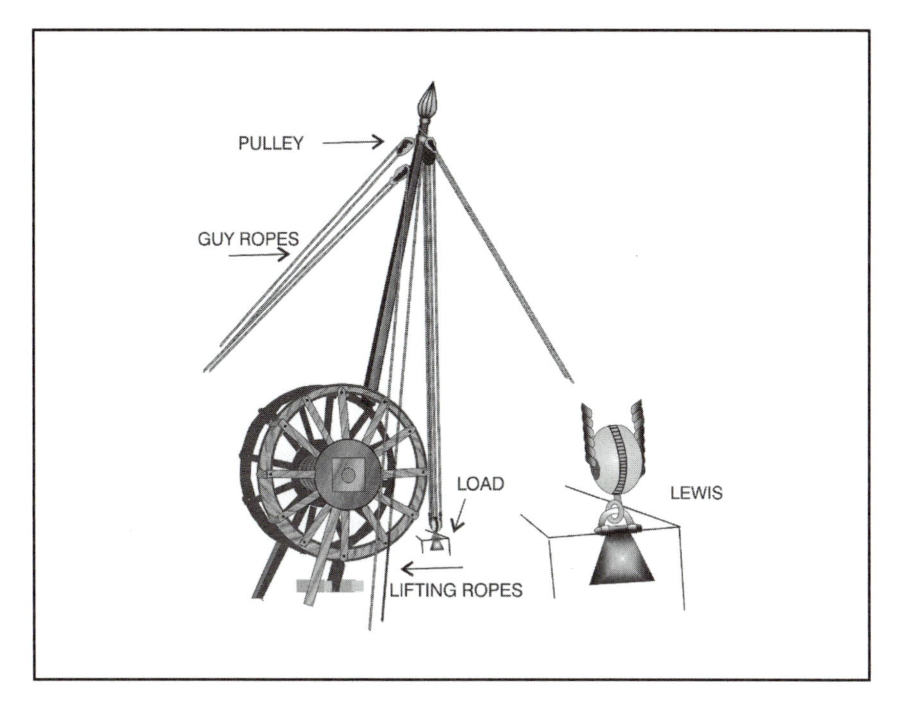

Figure 5.8 Crane, Pulley, and Lewis
A *crane* is defined as a machine that hoists and moves heavy objects through the use of cables attached to a movable rope or cable. Laborers pulling on the lifting ropes that were part of a *pulley* system could maneuver and lift the huge stones or blocks of cement used in the construction of large edifices. The stone to be lifted was secured by the *lewis,* a device for attaching the crane's rope to the stone or cement block.

of a bridge were to be constructed. Water was pumped from inside the cofferdam to expose the riverbed, thus enabling workers to construct the base of the pillars from massive stones and blocks, and later concrete. For large bridges and aqueducts that required deep foundations, the Romans built double cofferdams between which clay was solidly packed, and an Archimedean screw (waterscrew) was used to pump out the water. For smaller projects that required less manpower and engineering expertise, simple water buckets were used to lift the water out of the cofferdam. A key step, however, was to allow the riverbed or lakebed to dry completely before constructing the foundation. The construction of a bridge or aqueduct entailed massive amounts of materials and manpower, as well as a number of machines, including cranes to raise heavy stones or blocks, treadmills that powered enormous pulleys, and a wedge-shaped iron tenon designed to fit into a dovetail-shaped hole in a piece of block or stone. After securing the tenon in place, usually with a metal key to which a rope was attached, the stone or block was hoisted into place. Afterwards the key was removed. Today, this ancient machine is called a *lewis.* (See Figure 5.8.)

The Romans used the concept of the arch in all their brick and stone bridges. While a properly built arch provides incredible stability, it also posed some engineering challenges. The massive pillars that were constructed to buttress the wide spans that were built over powerful rivers actually blocked large sections of the river itself. These pillars acted as dams that affected water levels on either side of the bridge. As a consequence, water rushed under the bridges at tremendous speeds that, in time, undermined the foundation of the massive pillars, sometimes to the point of collapse. The solution was either to build more massive pillars, most likely with tons of concrete, but again with the effect of drastically affecting the water flow—or using wider arches, thus reducing the number and size of the pillars in the water. This particular solution posed more problems, since the Romans only built semicircular arches that restricted their height to half the span, meaning the wider the span of the arch, the higher the bridge would have to be built. If the river's banks were low and flat, the bridge would be higher in the middle than at either end, necessitating steep grades or extremely long approaches. Oftentimes steps would have to be built, making the bridge inaccessible to wheeled traffic. The problem of the semicircular arch in the building of bridges remained until the time of the Renaissance, when architects and engineers redesigned and reengineered arches into semi-ellipses, thus increasing the size of the span without increasing its height. Throughout the European continent, it is possible to view many of these ancient structures, in whole or in part. Among them are the Ponte Grasso Bridge near Urbino, Italy, and the Pons Aemilius in Rome, both of which were built in the last century B.C.E., and the Augustan bridge over the River Nera at Narni, Italy. Many architects believe the Lacer's Bridge over the River Tagus at Alcántara in Spain is the finest bridge that the Romans built. This particular bridge was built entirely without mortar. Each of its six spans is nearly 50 meters (165 feet) high, and its length is 200 meters (655 feet).

Water Systems

Water is a basic requirement of life. Some life forms require larger amounts of water than others, but all plants and animals need some amount of this liquid to sustain the processes involved in their survival as a species. In particular, all land mammals, whether they have gone extinct or are still thriving, have searched for a sustainable source and supply of fresh water from the time of their earliest emergence as viable life forms. Unlike other land mammals that first must seek and then go to the source of water, humans have the unique ability to devise methods of bringing the water to them and their communities. The first settled communities of early humans were most certainly located near rivers or lakes where the water supply, as well as edible vegetation and access to prey animals that also used the river, was readily available. As humans' curiosity grew along with their population, and they journeyed farther away from rivers and lakes, it was necessary to find sources of water, to access the supply, and finally to deliver it to the community.

Wells were the first manmade invention to solve the problem of access, and we can surmise that the first well that was dug by human hands was probably done so serendipitously—digging a hole for some other purpose on a piece of ground with a high water table, thus releasing a steady supply of hidden water. The discovery of groundwater, however, was not a solution to the problems of source and access. Rather, it was just the beginning of a conflict that still rages today, namely, water rights. For instance, the Sumerians and Babylonians who lived in Mesopotamia 5,000 years ago recognized the conflicts associated with the possession of water sources and the delivery of that precious commodity. These desert civilizations relied on the water of the Euphrates River for drinking and irrigation, and the local governments of the towns that sprung up along the banks of the river and beyond fought constantly over water rights, particularly after a series of canals and dams were built. (Even though Mesopotamia was located between two rivers, the Euphrates and the Tigris, its inhabitants relied almost exclusively on the Euphrates because the Tigris was a more powerful and swift-running river, making it more dangerous and thus less accessible.)

Water was so important that many of the laws that were written at the time and included in the famous *Law Code of Hammurabi* dealt with the intricacies of the canal system: maintenance and repair, as well as water rights. Ancient civilizations invented canals, dams, pumps, aqueducts, sewers, and water-driven energy sources. Variations of these same engineering feats are utilized today in every country, region, and city on our planet, albeit with improvements and safeguards. And while these basic concepts that ancient people first exploited still serve us well, we continue to be challenged by the same issues that these ancient civilizations encountered, only to be compounded by the demands of dynamic population and climate changes.

Canals and Dams

Most historians believe that settled farming originated in the Middle East about 10,000 years ago. At that time, the region contained lush vegetation and fertile river valleys. Nevertheless, the demands of an ever-increasing population over thousands of years required that a more consistent food supply be grown, and this required land, labor, and water. Land and labor were plentiful, but the two rivers were essentially the only major sources of water. The dilemma was how to channel the river's water to irrigate the fields of crops. We do not have an exact date when the first canal was dug. Some historians credit the Egyptians with digging the first irrigation canals about 5,000 years ago. However, we do know that sometime during the third millennium B.C.E., one of the Mesopotamian kings ordered the building of a large canal between the Tigris and Euphrates rivers along the forty-sixth meridian. Evidence of this ancient irrigation canal, about 62 kilometers (38 miles) long, can still be seen in the line of lakes, streams, and marshes that exist in this region today. Sometime after the digging of this canal, an even longer one, some 125 kilometers (78 miles) long and 122 meters (400 feet) wide, was dug parallel to the Tigris from Baghdad (at the time, a tiny town) to what is the site of present-day al-Amārah.

The building and maintenance of these ancient canals were daunting and laborious feats and presented a number of challenges to Mesopotamian farmers. Lack of timber and stone in the area meant that the mud banks were constantly compromised. The canal itself had to be slightly higher than the land to be irrigated and with a slight slope to enable the water to run freely. If the slope was too great, the water ran too fast and undermined the banks. If it ran too slowly, weeds grew and sediment blocked the canal's channels. In addition, goats, native to this region, further undermined the banks by climbing up and down almost at will. Dredging was done on a constant basis. To solve the problem of a deficiency of timber and stone to reinforce banks, Mesopotamian farmers used cane reeds that were either woven into mats or tied into bundles as buttresses for the mud banks. Because of the ever-present need for cane reeds, Mesopotamian communities maintained a "municipal marsh." This was a deliberately built swamp in which cane reeds were cultivated and harvested. Also, the water from these ancient canals flowed into basins and, as such, they did not have gates or sluices. Thus, gaps in the embankments were dug to release the water, and then closed up to block the water—all done by laborers. Remarkably, because of this labor-

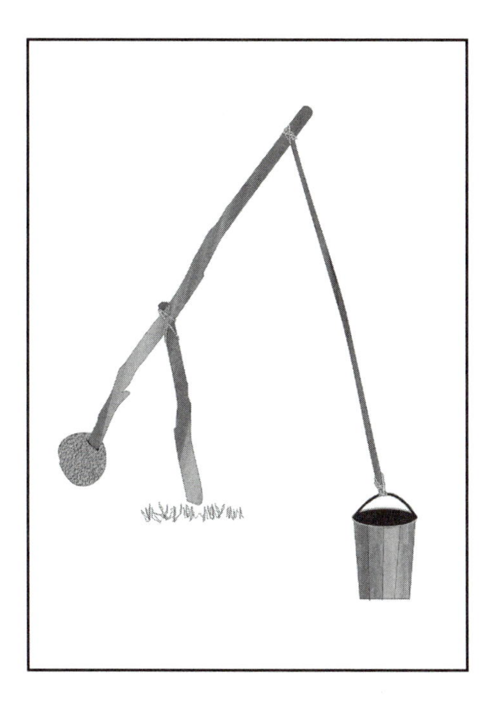

Figure 5.9 The *Shadûf*
One of the most ancient of inventions, the *shadûf* (also called the *swape*) was invented in Egypt by farmers or peasants who manually lifted water from the Nile River or canals with the device and then deposited the water directly onto their fields.

intensive system, canals lasted about 1,000 years or so. New canals were usually constructed parallel to abandoned systems. The canal system flourished in this area (present-day Iraq) for nearly 4,500 years. Its demise came at the hands of Mongol invaders who destroyed the irrigation system and decimated the population.

While the Mesopotamians used their canal system primarily for irrigation, with travel and commerce being secondary, the Egyptians constructed canals that could be utilized as channels for dams, and for transport, as well as for irrigation. And like the Mesopotamians, the Egyptian pharaohs and their territorial governors were consumed with building and maintaining a canal system. Each autumn, the Nile River flooded after the summer rains. In order to take better advantage of this yearly occurrence, Egyptian engineers built a series of canals, the purpose of which was to direct the water to distant tracts of land around which dykes and basins had been built. When water filled the basins, which were actually fields of crops, the dykes were closed off and the water remained until the ground was completely soaked. The excess water was then released into the canals. If the fields to be flooded were too high, Egyptian farmers or peasants manually lifted the water from the canal and/or from the Nile and moved it directly to the fields. They used an apparatus called a *swape* or *shadûf*—a long pole balanced on a wooden beam with a weight at one end and a bucket on the other. (See Figure 5.9.) Present-day Egyptian farmers continue to use this ancient invention for light irrigation. The swape was also popular with the Mesopotamians.

The construction of the pyramids was a factor in the building of the first dams in Egypt. Engineers channeled the Nile River into canals and built dams to store water for the workers at the stone quarries, who cut and moved the massive stones used in the building of the pyramids. The oldest dam in Egypt, constructed in about 2500 B.C.E., was located about 13 kilometers (8 miles) south of present-day Cairo across the Wadi Garrâwi in the eastern Egyptian desert. The masonry dam is 90 meters (295 feet) thick and 125 meters (410 feet) long. Sometime between 1500 and 1300 B.C.E., the Egyptians built a larger dam across the Orontes River in what is present-day Syria. The dam is three-fourths of a kilometer (approximately one-half mile) long and its building created the Lake of Homç. The lake and dam are still in existence and are still being used for irrigation purposes.

In about 600 B.C.E. the Egyptians began construction on a ship canal connecting the Red Sea and the Mediterranean. The canal was to run east and west between a point on the Nile River (present-day Zagazig) to Lake Timsâh, then turn south to follow the route of the present-day Suez Canal, and finally to reach the Red Sea. Superstition on the part of the pharaoh led him to abandon the project, which was completed by the Persian conqueror King Darius I (550–486 B.C.E.) almost a century later. Although a few succeeding Egyptian rulers improved the canal, even adding gates and locks, it fell into disrepair over the centuries. It closed for good sometime during the eighth century C.E. and did not reopen until the Suez Canal was completed in 1869. The Egyptians, like the Mesopotamians, suffered an ignominious fate at the hands of foreign invaders. When the Islamic Arabs conquered Egypt in the seventh century C.E., they abandoned the building and maintenance of the entire Egyptian canal system, resulting in massive starvation.

The Chinese also built extensive canal systems that were used for flood control, irrigation, and commerce. The first Chinese canal system was constructed in about the fifth century B.C.E. between the Yellow

River and the Pien and Ssu rivers. It ran for approximately 162 kilometers (100 miles). A 150-kilometer (93-mile) canal between the city of Ch'ang-an and the Yellow River was built in 133 B.C.E. However, the most famous of the Chinese transport canals was the "Grand Canal," which was begun in 70 C.E. It eventually reached 1,000 kilometers (620 miles) upon its completion in 610 C.E. All transport canal systems are problematic in that there are unequal levels of terrain. Consequently, navigation becomes difficult. The Chinese solved this problem by inventing a network of slipways (sloping inclines which enabled the boats to be dragged over the unequal levels). Sluice gates, which diverted excess water and/or held it back into a basin, were also constructed.

Not surprisingly, Roman engineers took full advantage of the many rivers on the European continent and built canals and dams that were mainly used for transport and flood control. Europe's climate provided sufficient rainfall for crops, and canal irrigation systems were less important than in the desert regions of the Near and Middle East. A famous canal built in 480 B.C.E. was dug exclusively for purposes of warfare. Xerxes, the Persian king who employed the construction of a pontoon bridge in his invasion of Greece, ordered, during the same campaign, the construction of a canal approximately one kilometer (0.6 miles) long and wide enough to accommodate two warships sailing abreast of each other. Remnants of this canal, known as the Athos Canal, remained until the early part of the twentieth century.

Aqueducts

The word *aquaeductus* or aqueduct derives from two Latin words, *aqua* meaning "water" and *ducere* meaning "to lead." Generally, when the term aqueduct is used, it brings to mind the Roman-built system of 11 aqueducts that traversed some 578 kilometers (358 miles) in Europe and North Africa, some of which still exists. However, the concept was not a Roman invention, but one that the ancient civilizations of Persia, Egypt, Mesopotamia, Greece, and India devised hundreds of years before. These water systems were called *qanats* by these ancient Middle Eastern civilizations. The Romans did not begin construction of their aqueducts until the fourth century B.C.E. A *qanat* or an aqueduct is a manmade passage that carries water from one point to another and is basically a variation or innovation of the principle of the canal. Whereas canals are dug in the ground, *qanats* or aqueducts were designed to bypass sections of terrain where canals could not be dug. Early *qanats* were built of stone, brick, and lead pipes.

In about 691 B.C.E. the Assyrian king Sennacherib (704–681 B.C.E.) ordered the building of a water system that would bring water from the Atrush or Gomel River in the hills of Bavian to the ancient capital of Nineveh, which was a distance of about 80 kilometers (50 miles). Stones for the tunneling project were quarried at Bavian. In places the *qanat* was approximately 20 meters (66 feet) wide. In one location over a valley stream, a 24-meter (79-foot) bridge with five pointed **corbelled** arches was built. The tunnel or *qanat* was constructed of masonry bricks lined with stone and sealed with *bitumen,* a naturally occurring byproduct of petroleum. Bitumen deposits were common in the Middle East thousands of years ago. Bitumen pooled and dried in the hot sun, or it floated to the surface of the waters of the Dead Sea after seeping from the seafloor and hardening. Ancient Mesopotamian builders mixed the bitumen with cut-up reeds, straw, sand, and limestone, and then used it as a waterproofing mortar. In 522 B.C.E., the Greeks built an 1,100-meter (3,608-foot) tunnel to supply water from one side of Mount Castro on the Isle of Samos to the other. Construction on this tunnel, which in places cut through hills that were 300 meters (984 feet) high, began at both ends simultaneously. Remarkably, because of superior surveying techniques, they met precisely in the middle. The construction of some famous modern-day tunnel systems also employed the same technique—starting at both ends simultaneously. Construction crews working on the 1939–40 Pennsylvania Turnpike project, which cut several tunnels through the mountainous areas of western Pennsylvania, met near the center, just a few inches off. The same was true for the Lincoln and Holland tunnels beneath the Hudson River in New York and New Jersey.

But like so much else in ancient history, the Romans appropriated the concept of the *qanat* and advanced it to a greater technological level. The construction of the Roman aqueduct system took place over a 500-year period, beginning in 312 B.C.E. and ending in 226 C.E. The first Roman aqueduct was built, in part, for defensive purposes. In 312 B.C.E. the Romans were engaged in the second Samnite War. Fearful that the enemy could poison the waters of the Tiber River, which was Rome's only water source at the time, Roman engineers embarked on the construction of the *Aqua Appia.* Rome's expanding population was another reason for the building of the aqueduct that was 17 kilometers (10.5 miles) long. It is important to note that an aqueduct is not a bridge that carries water. However, because of the terrain, it usually becomes necessary to integrate a bridge or a series of bridges into the aqueduct's design. Because of the expertise that Roman engineers had demonstrated in building bridges throughout the empire, the problem of constructing aqueducts over inhospitable terrain was somewhat diminished. The use of arched construction and concrete enabled Roman builders to erect aqueducts that still stand today. An example is the two-level aqueduct in Segovia, Spain, that was completed sometime in the second century C.E. and is still in working order. It is approximately 728 meters (2,388 feet) in length and about 30 meters (98 feet) high. In actuality, only about 19 kilometers (12 miles) of Roman-built aqueducts contain the famous arched bridges over valleys and rivers, out of a total of some 161 kilometers (100 miles).

The Romans employed three different kinds of materials in the building of the conduits that carried the water: masonry (stone, bricks, and mortar), lead or bronze piping, and terra-cotta (earthen) piping, although in the earliest construction, wood and leather were used as well. By far, masonry conduits were the most widely constructed, probably because of the availability of the material. Roman engineers built their aqueducts on an incline, so that gravity would ensure the flow rather than a pressure system. Early construction was done in large part underground, again to forestall the possibility of enemy contamination of the water supply. Later construction employed the use of covered arched brick and stone structures that protected the water from evaporating under the sun's rays. (This type of construction was done prior to metal piping design and construction.) Much of what we know today about the construction of these edifices was found in *De architectura* (On Architecture), written by Marcus Vitruvius Pollio sometime in the first century C.E. Though the treatise gives little detail about the actual building methods, it does contain abundant information relative to the design and specifications, as well as the date of construction, of many of the most notable Roman aqueducts.

The construction of the aqueduct required the skill of the *librator* or surveyor, since many kilometers of the aqueduct would be cut through unfamiliar and unmapped sloping hillsides. It was up to him to determine the most appropriate water channel and approximate the height of its slope. In addition to construction challenges, maintenance and rebuilding were constant as the slope of the hill exerted tremendous pressure on the downhill wall of the aqueduct, causing leaks and collapse if the walls were constructed poorly. While the construction entailed the labors of thousands of men, the surveyor relied on a number of ancient leveling instruments: the *dioptra,* an upright distance sight level with limited capability; the *leveling staff*, which was used with the *dioptra* to indicate the horizontal line of sight; and the *chorabates*, a long, narrow tablelike device with a water channel that indicated whether the actual construction had been leveled or sloped correctly.

Aqueducts were vastly important to the life of the Roman Empire. Rome itself had nine aqueducts supplying the city's water. When the water entered the city, it was channeled into a series of distribution tanks. Archaeologists, using the *Aqua Marcia* as a model, estimate that at the end of the first century C.E., one million liters (220,000 gallons) of water flowed into Rome hourly. These systems supplied the fresh water for the public baths and fountains, from which most of the residents drew their water. At the time, the population of Rome was approximately one million residents. For those who could afford it, connecting pipes from the aqueduct were installed directly into their homes. For those who did so without permission and without paying, stiff penalties were imposed. Nevertheless, many "illegal" piping systems existed in Rome. The Romans, however, made no provision to store water, a shortsighted policy that caused problems during times of drought. Water ran continuously into the baths and fountains in public buildings and private homes with no thought of conservation or waste. However, the Romans did use the excess water to flush out the city's sewers. When Rome fell in the fifth century C.E., the aqueducts rapidly deteriorated and for the most part were never used again. A number of the Roman aqueducts, however, remain in existence. One of the most famous is the Pont du Gard, near Nimes in southern France. Today, the largest aqueduct system is located in the state of California. New York City's water supply is provided by three different aqueduct systems.

Plumbing (Sewers, Toilets, Pipes, Baths)

Personal hygiene and waste disposal were of no, or little, concern to our most ancient ancestors. Surviving, after all, was the main preoccupation for hundreds of thousands of years. It is most likely that ancient humans first addressed these problems when they began to group into urban cities or centers with denser populations. There the problem was more immediate and visible. Ancient civilizations dealt with water delivery and sewage disposal with foresight and ingenuity and devised a number of methods for sanitation that can be considered far advanced of those found in some present-day, underdeveloped Third-World countries. The term "plumbing" originates with the Latin word for lead, *plumbum*, which was the metal that eventually replaced wood, terra-cotta, and stone piping or conduits in ancient water systems. As in modern times, ancient plumbers were skilled artisans who worked on all aspects of the water system—designing, building, maintaining, and repairing. Artifacts and remnants of ancient pipes, sewers, lavatories, and baths, some of which date back 5,000 years or so, can be found in India, China, Egypt, and the Middle East, as well as on the European continent.

There is evidence that 5,000 years ago Neolithic tribes living on present-day Orkney Island, which is just off the coast of Scotland, constructed drains lined with slabs of stone that ran underground from their huts to nearby cliffs and then out to the sea. Their huts also contained crude toilets that were connected to this most ancient *sewer system.* In about 2500 B.C.E. in Mohenjo-Daro (present-day Pakistan), engineers built a highly advanced brickwork sewer system, where water from each house flowed into a main drain and then into a cesspool that could be cleaned periodically. These two examples are somewhat unusual in that it is presumed all members of the population were "connected" to these main facilities via toilets located in their houses. In reality, in most countries it was only royalty or the very wealthy who had access to such a system. In the first millennium C.E. the Mesopotamian kings had elaborate sewers connected to their palaces, all of which had personal bathrooms and toilets. The piping and drains were built from clay mixed with straw, sealed with bitumen for waterproofing. Later, pipes were alloyed from such metals as bronze, copper, tin, and antimony, and later lead. The peasants, on the other hand, simply deposited their waste onto the road in front of their homes. When it built up to unacceptable levels, a layer of clay was laid on top, which necessitated the building of steps into the houses. This process was repeated until the houses became unreachable or unlivable.

Another example of royal privilege, as well as engineering adeptness, has been found on the island of Crete in the Mediterranean. Sometime between 3000 and 1500 B.C.E., ancient Minoan builders constructed a highly advanced system of underground sewers and drains in the Minoan Palace of Knossos on Crete that rival those that are built today. (Minoans were an advanced Bronze Age people who lived on the island.) The drainage system consisted of a series of terra-cotta channels that fed into a main sewer, lined with stone. The water system also supplied cold water to fountains and both hot and cold water to faucets in the palace. The palace contained a prototype of the flush toilet with a wooden seat and a reservoir of water. Engineers took full advantage of the island's weather, which saw frequent and violent thunderstorms. The water would run down the channels with such force that it would clean out the sewers. Some parts of the system still exist and help to carry off excess rainwater. The ancient Greeks constructed a municipal sewer system as early as the fourth century C.E., to which many of the inhabitants' homes were connected. Greek engineers even erected ventilating shafts on the sewers to help alleviate the strong odors that were a constant problem. Waste collected in cesspools that were emptied periodically.

The historic city of Jerusalem and a number of surrounding towns are the sites of advanced sewer and drainage systems, as well as underground wells and cisterns, dating back to at least 1200 B.C.E. Improvements to the system, in the forms of bathhouses and fountains, were made in Jerusalem when the Romans vanquished the region. When they retreated in 73 C.E., the Romans destroyed it all, leaving the city in disarray. Ancient people, particularly those living on islands or near coastal areas, were unaware of pollution and contamination of fresh water. For inland populations, effluent was usually discharged into the nearest river, which eventually made its way into the open sea, but not before rivers themselves became flowing cesspools.

Initially, the *Cloaca Maxima,* the ancient sewer of Rome, was built by the Etruscans in the sixth century B.C.E. The Romans began to refurbish it in the third century B.C.E., using semicircular vaulting construction. Up until 33 B.C.E., when it was enclosed upon the orders of Emperor Augustus, it was merely an open drain. (At this time, there were about one million people living in Rome, and the problem of odor was overwhelming. Hence, the emperor's order to close it up.) Even so, the waste simply poured into the Tiber River. The *toilet* was, most likely, simultaneously and independently invented by various civilizations and at differing times in history. (The flush toilet came much later, in the nineteenth century, and credit for its invention is also clouded with controversy.) Royalty in Mesopotamia, Egypt, Greece, India, and China all had personal bathrooms with private toilets or privies that were connected to the palaces' sewer systems. The ultimate disposal of the waste was left up to either slaves or what today would be considered highly paid laborers.

The material used for ancient sewer systems was usually dependent upon the availability of local materials. Thus, the earliest water systems were constructed using terra-cotta or earthen bricks, as well as stone and mortar, for their conduits and *piping.* In Mesopotamia, where bitumen deposits were commonplace, engineers lined the pipes and drains with a waterproofing mortar made from this mineral, as well as bricks of asphalt, which is a mixture of bitumen, sand, and gravel. Almost 1,500 years before the elaborate water systems of Mesopotamia were built, the Egyptians were using copper, bronze, and brass in their mortuary pyramids. The earliest evidence of the use of metal piping dates back to about 2500 B.C.E., when approximately 396 meters (1,299 feet) of copper piping, along with lead stoppers, brass rings, and drain pipes, was installed in the pyramid of King Sahura at Abusir. Metal piping in water systems became fairly standard by the end of the first millennium B.C.E. In about 200 B.C.E., the Greeks installed a highly advanced system of water pipes in the city of Pergamum in Turkey. The water, piped in channels most probably made out of lead piping, originated from a spring about 366 meters (1200 feet) above sea level, crossed two valleys, and was then piped up again to a point that was only about 40 meters (131 feet) below the height of the originating spring. Lead pipes were made by a method of folding sheets of lead into a cylindrical shape and then soldering the seams, as well as the lead collars that joined the pipes together. This system was remarkable for its time because of the engineering prowess involved in building and maintaining the natural water pressure in a system, the ultimate destination of which was up a hill—not down.

After a time, lead-lined pipes replaced stone and terra-cotta conduits in the Roman aqueducts that brought water to public fountains and private homes. In addition, most of the population of Rome used cooking pots made of lead and lead alloys. Since lead is slightly soluble in water, it was thought that much of the population was poisoned, to some degree, by lead. There is scant evidence that mass lead poisoning existed, and speculation that lead may have helped accelerate the decline of the Roman Empire has been discounted.

One of the most popular and famous (or infamous) innovations associated with ancient Rome is the public *bathhouse.* For most, except royalty and the wealthy, personal bathing was limited to a basin or urn of water poured over one's head, or a dip in a lake or river. Nonetheless, most ancient civilizations believed that personal cleanliness was an important consideration. (It was only after the fall of Rome that personal and public hygiene was neglected.) By 1500 B.C.E., long before the Romans began to build these famous structures, the Greeks were already capable of supplying hot and cold running water furnished by their system of aqueducts. However, they felt that bathing with hot water was unmanly, and thus the male population, for the most part, used cold water for personal cleanliness. (This may be the origin of the term "Spartan" when describing behavior that is self-disciplined or austere. Sparta was a city of ancient Greece whose inhabitants were famous for exhibiting such behaviors.) Wealthy Greek citizens did have the luxury of private bathrooms and earthenware tubs—the water temperature being a personal option. The Greeks were also famous for building **gymnasiums,** all of which contained hot and cold water baths as well as toilets. The construction of these arenas for sports and exercise took place in all major Greek cities and followed the creation of the Olympic games in 776 B.C.E. The games themselves were an offshoot of the desire for personal fitness and hygiene.

The public bathhouses that the Romans built, virtually in every major city they ruled and on every continent, were designed as centers for entertainment, gossip, and contact (both business and personal). The Romans, in general, were not shy and not embarrassed by public exposure of private functions, including toileting. In fact, most of the Roman baths were built with toilets, sometimes as many as 12 side by side (no doors), and conversation was encouraged during what most people would consider a private event. The huge bath areas were fed with hot and cold water but soon became unhealthy pools that were malodorous, stagnant, and bacteria-filled. In general, they were emptied and refilled once per day. Hot water was supplied by a furnace system that consisted of heating hollow clay bricks that were placed under the floor of the pool or tub. Bathhouses were constructed with vaulted ceilings and skylights and were elaborately decorated with marble statues. The marble tubs or pools were designed with steps and benches for ease of entering and exiting. A lead piping system both fed fresh water into the bathhouse and flushed away the wastewater from the tubs and sewers underground. At about the time of the fall of Rome, there were 11 public bathhouses in operation, although not all the population was permitted to use the facilities.

As with all municipal projects built during the height of the Roman Empire, bathhouses fell into disrepair and disrepute. The barbarian hordes that conquered Rome in the mid-fifth century C.E. had little concern for sanitation, either personal or public. Interestingly, Christianity itself may have played a part in the regression of both public and personal cleanliness. Early Christians rejected anything that was vaguely "Roman" (e.g., bathing, toilets, fresh water) and also discouraged personal vanity. In other words, it was unseemly to be clean or to have or partake of any worldly possessions.

Machinery

Ancient engineers, builders, and inventors incorporated the concept of the "prime mover" into the development of a number of devices that harnessed the energy of water. In physics, a *prime mover* is defined as the initial force that transforms energy into work. When humans first harnessed an animal to move an object, they "invented" the prime mover—that is, something other than themselves to perform a task. Ancient humans invented a number of labor-saving devices that could be considered prime movers, all of which incorporate one or more principles of the four simple machines.

For example, the *waterwheel* works on the principles of the lever and the wheel and axle. Its exact date and place of origin are unknown, but we do know that it was in use in Greece by at least the first century B.C.E., hence the name *Greek Mill* is often applied. It was merely a set of wooden paddles affixed to a wheel that was placed horizontally in a fast-moving river or stream. The force of the water moved the paddles. As the waterwheel turned, it moved a shaft that was connected to a geared wheel that was connected to another device, a millstone that also revolved. Waterwheels were first utilized to grind grain (although the process was painstakingly slow), to "full" cloth (increasing the weight of the cloth by beating and shrinking it), and to raise water levels. The Chinese also designed a horizontal waterwheel, which had an intricate system of gears. In the first century C.E. their waterwheel powered a blast furnace that manufactured metal plowshares. The Romans adapted the horizontal waterwheel by turning it vertically. Not surprisingly, this is called the *Roman Mill* or *Vitruvian Mill,* named after Marcus Vitruvius, believed to be its inventor. There are two variations of the vertical waterwheel, the *undershot* and the *overshot.* In an undershot waterwheel, fast-moving water presses against the paddles in the water, thus turning the wheel. The overshot wheel incorporates the inclined plane in its operation as follows. Water pours over the top of the wheel via a ramp (chute). The wheel is turned in a forward direction by the weight of the water on the paddles. An undershot waterwheel sits in a moving stream of water (millrace) that turns the wheel backward as the current strikes the paddles. The overshot waterwheel is more efficient, since falling water weighs about eight pounds per gallon, making the wheels move faster and with more power. Vitruvius also designed another application of the vertical wheel that used a system of gears. By far, the overshot wheel was more efficient. However, the use of vertical waterwheels was never widely implemented in Rome, primarily because with the availability of slave labor, the Romans saw no need for widespread mechanized systems. When Rome fell and manpower was not quite as abundant, the use of waterwheels became increasingly popular.

Another variant of the vertical waterwheel is the *chain pump,* invented in China sometime between 100 B.C.E. and the first century C.E. Its design consisted of a series of square paddles connected in a chainlike pattern to a movable shaft that was attached to a wheel. The first pumps, powered by foot treadles, helped Chinese farmers in irrigating their fields, particularly rice fields. Using this device, water could be raised as much as 4.5 meters (15 feet) from rivers and ditches and deposited into fields of rice that needed to be flooded with water. The difference between a waterwheel and a chain pump is the source of energy. Whereas a waterwheel is powered by water, the chain pump depends upon human power. Centuries later, the chain pump principle would be used to move sand and earth, as well as water, much like a modern-day conveyor belt. Its invention transformed agriculture because water could be brought to otherwise dry areas without having to dig elaborate canal systems. The chain pump also transformed civil engineering projects because water could be drained from previously unsuitable building sites, and it could bring supplies of fresh drinking water into population areas that were not connected to piped water systems or aqueducts. Conveyor belts continue to be used in various manufacturing industries and warehouses.

Summary

It is often said that there are no new ideas—only new applications or improvements. This may be hyperbole or exaggeration, but there is some validity to the statement. It is easy to take for granted the ease with which people avail themselves of modern-day conveniences and technology without appreciating the history and turmoil that preceded them. Today, the integrity of the construction of a stadium or high-rise building is seldom questioned, nor is the ability of the architects and builders or the tools and methods that are utilized in these massive endeavors. Few of us realize that the basis of all engineering derives from four simple machines: the lever, the wheel and axle, the pulley, and the inclined plane. Modern-day construction equipment, such as cranes and elevators, work on the basic principles that Archimedes first applied to his study of plane geometry and its practical applications.

Local interest groups often fight to preserve historic structures and buildings while other "lesser" hotels or office buildings, usually built within the last 40 years, are demolished within seconds by dynamite charges. So it is easy to take for granted the fact that the step pyramid at

Saqqarah in Egypt still stands after 4,500 years. Hoover Dam, the Golden Gate Bridge, and the Superdome in New Orleans are all modern-day architectural achievements, and are but a few examples of engineering achievements built upon the inventions, innovations, and technological expertise of others who lived thousands of years in the past. Their roots are in the ancient Persian and Roman aqueducts, the manmade Chinese suspension bridge of the third century B.C.E., and the Roman Colosseum, which seated 50,000-plus people and could be flooded with water for mock sea battles. Modern-day farmers continue to build canals equipped with pumps to irrigate their fields. Cities and rural areas are supplied with fresh water from aqueduct and water systems much like those that were built several thousand years ago. The exact range of engineering and mechanical inventions and discoveries that have been developed in the last 6,000 to 7,000 years is nothing short of astounding, as well as being a testament to the intellectual and physical skills of humanity.

6

MATHEMATICS

Background and History

Notwithstanding all that we have discovered and learned about human evolution and ingenuity, each year brings new insights and discoveries that suggest that our primitive ancestors may have demonstrated some forms of human behavior much earlier than was first believed. For instance, as reported by Michael Balter in the January 11, 2002 issue of *Science,* archaeologists in 1999 and 2000 unearthed from soil deposits in Blombos Cave in South Africa two artifacts that are some 40,000 years older than the cave paintings in Chauvet, France. These two chunks of red ochre are engraved with geometric crosshatches. As yet, there is no universal agreement among scientists as to what this discovery represents. One archaeologist believes it to be an "intentionally incised, abstract geometric design." Another believes it could simply be "doodling." What is not in dispute is the age of the two bits of ochre: 77,000 years. Many in the scientific community believe that calendars, based on lunar cycles, existed during the Stone Age, more than 30,000 years ago. These discoveries are all suggestive of the ability of prehistoric humans to project complex forms and to understand natural manifestations, such as the cycles of the moon. Our most primitive ancestors recognized, albeit in practical rather than abstract terms, the concepts of "more than one," progression, and time; and along with their ability to communicate by speech, prehistoric humans were able to count, recognize problems and cause-and-effect relationships, and ultimately arrive at explanations.

Just as we have no precise knowledge as to when early humans developed an identifiable spoken language, we do not know when they first exhibited the ability to transform empirical observations into reasoned judgments. We can safely assume that primitive humans saw the symmetry that existed in nature and in themselves. For example, they recognized that they had an arm and a leg on either side of their body with the same number of digits on each limb, as well as two eyes, two ears, one nose, etcetera, and that the animals they both feared and hunted manifested similar physical proportionality. During the Neolithic Period, some 10,000 years ago, early humans displayed an awareness of spatial relationships as evidenced by geometric patterns on pottery and metalwork. There are several elements to consider in the development of mathematics, as follows.

Numbering

The actual verbal assignment of numbers to describe the obvious is probably less important than the recognition of their existence. In other words, early humans knew that hunting was more successful when done in multiples of more than one, that a larger animal provided more hide (protection) and more nourishment (meat) than a smaller one, and that a longer spear was more deadly than one that was shorter. Thus, by necessity, early humans "did the math" and counted and measured everything in their world. Evidence for this can be traced back to the Paleolithic Era and the advent of the crudest of stone tools. A tally stick made from a wolf's bone believed to date back about 750,000 years was found in Moravia (present-day Czech Republic) in 1937. This bone is marked with 55 sharply carved notches. The first 25 of these notches are arranged in groups of fives and then followed by a single notch that is twice as long as the others in the groups and appears to designate an end to the series. The next notch is twice as long and begins a new group that goes up to 30, indicating that Paleolithic humans grasped the concept of bundling or batching (Modern set theory).

Early humans counted the periodicities of the moon to calculate time, particularly for planting and harvesting crops. They measured the size and shape of implements and tools based on purpose and effectiveness—longer spears for hunting, shorter scrapers for tanning hides. Their math, or arithmetic, was rooted in experience as well as instinct. They did not possess measuring or counting devices, such as a ruler or an abacus; they had no formalized numbering system. These inventions are perhaps a bit older than 5,000 years, yet humans have been counting and measuring for far longer. Agriculture was the bridge to civilization that led to the development of formalized writing (language) and arithmetic (mathematics). When humans abandoned the hunter/gatherer lifestyle for one that was, for a long time, harsher and more unpredictable, settled farming, interdependence, and communication became necessary for survival within these primitive communities. In the Middle and Near East, particularly, these settlements eventually became towns, then cities and urban areas where the agricultural bounty was charted mathematically for the first time. Arithmetic or mathematical symbols coincided with the formulation of cuneiform writing by the Sumerians in Mesopotamia sometime around 3500 B.C.E.

Defining Mathematics

There are several ways to define mathematics. One is the study of how many or how few, how big or how small, how long or how short, or how far or how near—that is, in the quantities, magnitudes, and relationships between objects or symbols. Another more technical definition is that mathematics is a system built on principles related to numbers and spatial relationships. Still another is that it is the body of knowledge based on specific axioms and assumptions that may or may not be proven in the real world. Today, mathematics is considered the language of science inasmuch as it calculates, formulates, recapitulates, and communicates consistently and impartially, regardless of where on earth or in the universe a mathematical problem is posed and/or solved. Various civilizations developed independent and distinct arithmetic or numbering systems based on varying factors. For example, the Sumerians used a sexagesimal system (base 60), while the early Egyptians and the Chinese both employed a decimal system (base 10). Modern-day digital computers used a base 2 numbering system.

The important point is that ancient mathematics or arithmetic, while developing independently among all ancient civilizations, was founded on universal precepts. In other words, these ancient peoples marked time and religious rituals, thus the need for calendars. They were curious about the heavens and the forces of nature, thus their interest in the periodicity of heavenly bodies. Taxes had to be collected to both maintain prosperity and subsidize growth, thus monetary systems were established. These were all mathematical problems. Another is the concept of pi (π), which has been known for thousands of years: the ratio of the circumference to the diameter of a circle is constant. Over 4,000 years ago, the ancient Babylonians and Egyptians each assigned a value to π that was close to that which is accepted today.

Over 5,000 years ago men with vision, curiosity, perception, and persistence laid the foundation for all branches of modern mathematics. Moreover, it was done without benefit of sophisticated intellectual resources or instrumentation. While a detailed account of the complexities of each of the above civilizations' mathematical systems is beyond the scope of this book, this chapter will examine the development of and contributions to mathematics that were made by them.

The Mathematics of Mesopotamia (Sumerian, Akkadian, and Babylonian)

Although primitive humans grasped the concept of counting in their daily lives (e.g., notches on a piece of bone), true mathematics only came after a formalized accounting record represented by a numbering system was developed by the Sumerians. An advanced civilization that flourished around 3500 B.C.E., the Sumerians were successful farmers and engineers who had built a series of irrigation canals to channel water for agriculture as well as for human consumption in their cities. The Sumerians also developed legal, administrative, and postal systems. Thus, it is assumed these profitable enterprises necessitated the development of an accounting system to record the details of daily and long-term activities. It is believed that the Sumerians were the first to develop a formalized system of writing, cuneiform. Records that date back approximately 5,000 years illustrate that they recorded crops that were harvested and then stored in and distributed from the religious temples by Sumerian priests. This charting of goods and/or crops most likely coincided with the invention of pictographic cuneiform writing—text as well as a numbering system—that was well established by about 2500 B.C.E.

The Sumerians employed a sexagesimal numbering system. The earliest Sumerian numbers were written on clay tablets in the following manner: (1) Numbers up to 10 were recorded by the appropriate number of slanted markings. (2) Tens and multiples of 10 were recorded by vertical markings. (3) Alongside these markings (which were really based on a decimal system) was a marking that represented the number 60. (4) Smaller reeds were used to record the markings of units and tens; larger reeds were used to record units of 60 (slanted) and units of 600 (vertical). When cuneiform script became popular in about 2500 B.C.E., the wedge-shaped stylus replaced reeds. For example, a single vertical marking illustrated any power of 60—1, 60, 3,600, etcetera—while two markings made at an angle that formed an arrowhead symbol illustrated the numbers 10, 600, 36,000, and so forth. Sometime around 2300 B.C.E., the Akkadians, fierce but less culturally advanced, invaded and ruled Sumeria until they themselves were overthrown two centuries later. During their rule, the Akkadians developed a crude form of arithmetic that employed addition, subtraction, multiplication, and division of numbers. (Arithmetic, from the Greek word *arithmos,* meaning "number," is the most elemental branch of mathematics, dealing with numerical computation, measuring, and basic number theory.)

By about 2000 B.C.E. both the Sumerians and Akkadians, as distinct civilizations, had disappeared from Mesopotamia and were replaced by the Semitic Babylonians. Babylonian mathematics was also a sexagesimal system, most probably appropriated from the Sumerians. However, it was more advanced in that it was a positional system; that is, the position of the symbol determined its place value in the number that is represented. Ancient Babylonian mathematics had some peculiarities and inconsistencies. For example, it contained characteristics of a decimal system. Unlike the nine numbers plus zero of our current Arabic decimal system, the Babylonians used only two symbols to record all of their numbers; thus the position of each of these two symbols was all-important. (In Figure 6.1, see numbers 11 and 70, which illustrate the importance of placement. See also number 59, which is represented by

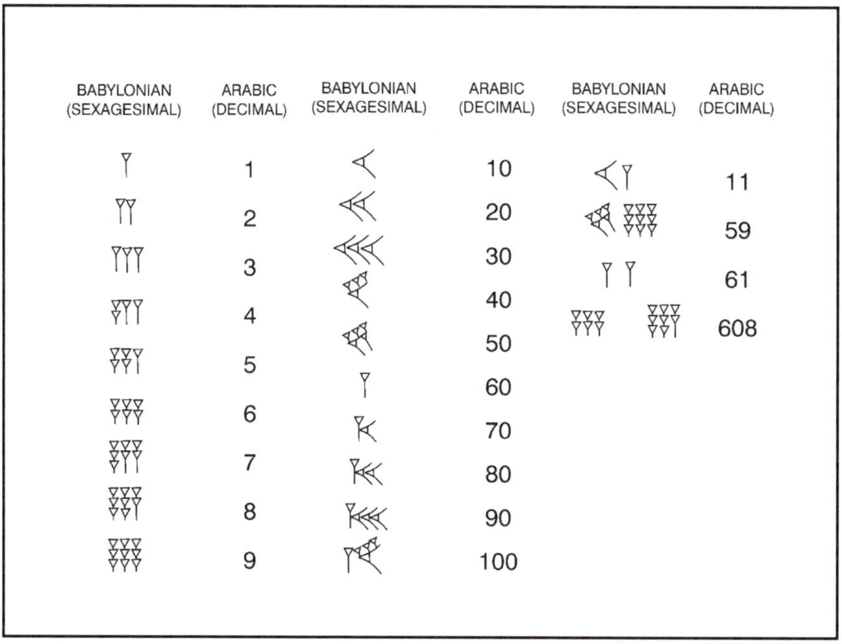

BABYLONIAN (SEXAGESIMAL)	ARABIC (DECIMAL)	BABYLONIAN (SEXAGESIMAL)	ARABIC (DECIMAL)	BABYLONIAN (SEXAGESIMAL)	ARABIC (DECIMAL)
	1		10		11
	2		20		59
	3		30		61
	4		40		608
	5		50		
	6		60		
	7		70		
	8		80		
	9		90		
			100		

Figure 6.1 Babylonian and Arabic Numerals
Babylonian mathematics was a sexagesimal system (base 60), with elements of a decimal (base 10) system as well. Placement of the numerical symbols was important, as they had only two symbols to represent all numbers. See the numbers 11 and 70, which are illustrative of the importance of placement.

the symbol for the unit 50 and the symbol for the number 9. This is suggestive of a base 10 or decimal system.) Also, the symbol for the numbers 1 and 60 are identical; therefore, writing the numbers 2 and 61 could be confusing. This was solved, to some degree, by having the symbols touch each other when recording the number 2; when representing the number 61, the symbols would be separated. Another major problem was that there was no symbol for zero; thus, when writing a number that contained what we call zero (for example, 608), the Babylonians simply left a blank space. This system was fraught was errors, particularly if the scribe, who may have had limited space on a clay tablet, omitted the blank space, either by necessity or through simple error. In about 700 B.C.E., the Babylonians did devise a symbol, a hook-like marking, to indicate "nothing." Apparently, they did recognize the problem of nothing or zero in mathematical recordings. (A more detailed account on the history of zero follows later in this chapter.)

For centuries, historians have pondered why the Babylonians (and before them the Sumerians) chose a sexagesimal system. To some degree, we still follow this ancient system, since we continue to use 60 seconds in a minute, 60 minutes in an hour, as well as 360 degrees in a circle (6 × 60). There is no reliable answer, but several theories have been expounded while research into this mystery is still ongoing. The Alexandrian mathematician Theon (fl. fourth century C.E.) proposed that 60 was chosen because it was the smallest number divisible by 1, 2, 3, 4, and 5. The twentieth-century German mathematician, Otto Neugebauer, suggests that the sexagesimal system is based on Sumerian/Babylonian weights and measures, which divided commodities into thirds, thereby using basic fractions (although the reverse can be argued, that because of the base 60 system, it was logical to fractionalize into thirds). By far the most popular theory is the one that is based on celestial events. For example, to the Babylonians, who were unaware of the rotation of the Earth, the length of time it took for the Sun to apparently move completely around the heavens relative to the stars was about one degree per day or 365 days per year. Thus, the sum of 360 is thought to have derived from this phenomenon. Also, the Babylonians named and identified the zodiac, which is made up of seven planets, the paths of which pass through a set of stars that they divided into 12 constellations. Thus, all these numbers were either multiples or fractions of the number 60.

Regardless of the premise for the base 60 system, the Babylonians made several major contributions to the science of astrology/astronomy and the discipline of applied mathematics. They were the first to use mathematical calculations to predict the future position of certain planets, thus the first to develop mathematical astronomy. They were also the first to calculate that the square of the hypotenuse of a right triangle is equal to the sum of the squares of the other two sides of the triangle. (In a right triangle, the side opposite the right angle is called the *hypotenuse.*) Evidence for this discovery is recorded on clay tablets that date back to the Babylonian era, which is more than 1,000 years prior to the lifetime of the Greek philosopher/mathematician Pythagoras of Samos, for whom this theorem was named (the Pythagorean theorem).

The Mathematics of Egypt

Unlike the civilizations of Mesopotamia, who benefited from contact with other tribes who journeyed on the Middle Eastern caravan routes, Egypt was an isolated country that had few neighbors in the surrounding desert regions. This isolation, as well as the fact that it was, geographically, an easily defended country, allowed the Egyptians to advance and prosper in relative peace for centuries. The regularity of the dry and rainy seasons, along with the yearly flooding of the Nile River, provided fertile land and bountiful crops that supported a population of successful farmers, engineers and builders, mathematicians, and astronomers. About 3000 B.C.E., at the beginning of the Old Kingdom, two separate nations united under one ruler to form a single Egyptian civilization. In time, a complex administrative system was formed to collect the taxes that supported the Egyptian armies and other civil projects and entities. By about 3100 B.C.E. the Egyptians had already developed a system of hieroglyphic writing for both words and numbers. At first, the hieroglyphs were carved in relief into stone. Later, with the discovery and use of papyrus, reed pens, and ink, an easier and faster hieratic script developed for writing and mathematics.

Much of what we know about Egyptian mathematics was contained in two papyrus documents, the Rhind Papyrus and the Moscow Papyrus. The Rhind Papyrus was named after a Scottish dealer of antiquities, Alexander Henry Rhind (1833–1863), who purchased it in Luxor in 1858. Ahmes the Moonborn (ca. 1680 –1620 B.C.E.) in 1650 B.C.E. copied the document, the basis for which was material that was known some 200 years earlier, in 1850 B.C.E. It has become more popular to refer to it as the Ahmes Papyrus (rather than the Rhind) in honor of the Egyptian scribe. The Moscow Papyrus, so named because it is housed in the Museum of Fine Arts in Moscow, is often referred to as the Golenischev Papyrus in honor of its purchaser, V. S. Golenischev, who died in 1947. The Egyptian author or scribe of the Moscow Papyrus is unknown, but it is believed that the document contains material that also dates back to circa 1850 B.C.E. There are 87 mathematical problems written on the Ahmes Papyrus; the Moscow Papyrus has 25 problems.

Egyptian mathematics concerned itself primarily with the practical elements of basic arithmetic that could be used in daily commerce, construction, and civil administration. Ahmes, the scribe of the Rhind Papyrus, also explained how to multiply and divide fractions, and how to calculate the area of a circle, square, and triangle, as well as the volume of solids. This mathematical ability was helpful to merchants and traders, government tax collectors, surveyors and builders of the pyramids, and the priests who tracked the levels of the Nile River through the alleged actions of the Nile god, Hapi. Egyptian priests invented vertical stone gauges, called *nilometers,* built into the banks of the river, that recorded the water level. A level of 7.6 meters (approximately 25 feet) was optimum. Higher levels caused flooding, while lower levels meant poor crop production and harvest.

Hieroglyphics

There are several important points to consider about Egyptian mathematics. First, it was based on a decimal (base 10) system that evolved from hieroglyphs into hieratic script after the invention of papyrus. In the earlier phase of their base 10 system, separate hieroglyphic symbols represented the numbers 1, 10, 100, 1,000, 10,000, and 1,000,000. The system was somewhat ungainly in that numerous symbols were needed to write relatively small numbers. For example, 15 symbols were required to write the number 69, and 19 symbols for the number 478. (See Figure 6.2.) Secondly, it was not a positional system; thus, the symbols could be placed in any order. Third, since zero did not exist in their mathematical vernacular, a blank space represented a lack of a number. Fourth, hieroglyphs were used exclusively when they were carved in relief (that is, in stone) even after the invention of hieratic script.

Hieratics

Hieratic script numerals are quite different, although they are still based on a decimal system. Rather than simply learning six or seven hieroglyphic numerical symbols, the Egyptians were forced to memorize 36 distinct hieratic symbols that went from the numbers 1 through 9,000 (in multiples of 10). (See Figure 6.3.) These mathematical symbols were used when written on papyrus. However, fewer symbols were required to transcribe numbers. For example, rather than 15 hiero-

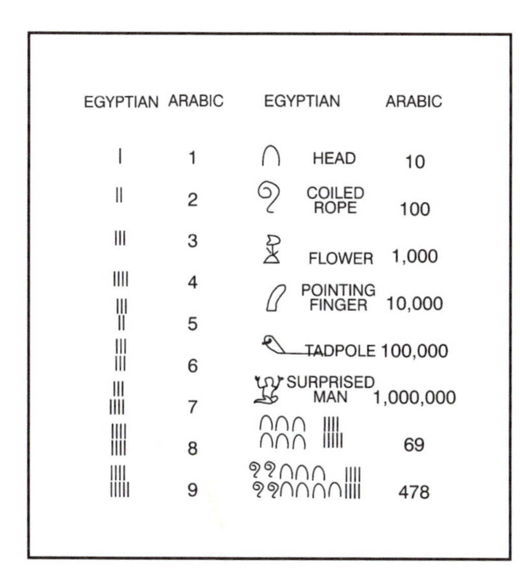

Figure 6.2 Egyptian Hieroglyphics and Arabic Numerals Egyptian mathematics was a decimal (base 10) system that incorporated symbols, called *hieroglyphs*, into the representation of numerals. Placement was unimportant, and zero was nonexistent.

glyphs for the number 69, only two hieratic symbols were used; and for the number 478, three hieratic symbols, as compared to 19 hieroglyphs. As with hieroglyphs, the position of the symbols was of no consequence; they could be written in any order. The character of hieratic mathematical symbols also evolved over the centuries, while retaining a similar style.

Fractions

Egyptian mathematics poses some problems of its own. Since it is not a positional system with place values, simple arithmetic calculations are complicated. The Egyptians could perform simple addition, but multiplication and division were difficult if not impossible. To overcome this deficiency, their forms of multiplication and division were essentially variants of addition and subtraction, as are all systems of arithmetic. The Egyptians were somewhat proficient in the use of fractions, probably because trade and commerce necessitated that they deal in "parts." However, their use of fractions was limited to "one part of," and subsequently they devised a symbol that represented that concept. (See Figure 6.4.)

Figure 6.3 Egyptian Hieratic and Arabic Numerals
Hieratic script numerals were also decimal based. Thirty-six distinct *hieratic* symbols represented the numbers 1 through 9,000, multiples of 10. Thus, there were fewer symbols to record. See the numbers 69 and 478. Placement of the hieratic symbols was also inconsequential. They could be written in any order.

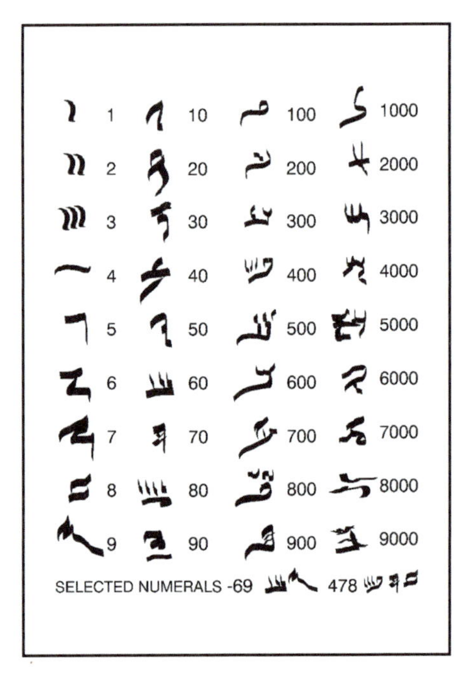

Figure 6.4 Egyptian Fractions
Egyptian fractions contained their hieroglyphs as well as a symbol, similar in shape to the elliptical orb of an eye, to represent a fraction of the recorded number. However, it was limited to "one part of" that number.

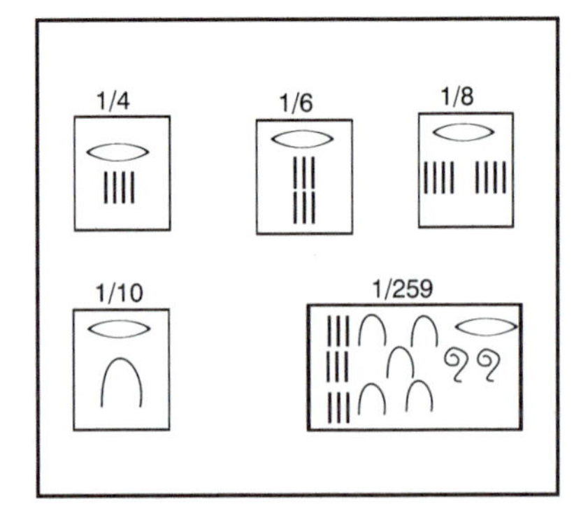

Weights and Measures

The Egyptians invented a standardized measurement system that used the body of the reigning pharaoh. For instance, the *cubit* was based on the length of the pharaoh's forearm—from the tip of the middle finger to the elbow. The size of his palm and the width of his four fingers held together were also used as a measurement. They quickly learned that each succeeding pharaoh's body size was different, and eventually created standardized and official measuring sticks. They did retain the body-part names (for example, 7 palms = 1 *cubit* or 46 centimeters [18 inches]). The Egyptians also used standardized stone weights. Geometry played an important part in Egyptian mathematics, although they did not refer to it as such. For example, surveyors who needed to measure the size of irregularly shaped crop fields simply divided the field into triangles. They then calculated the area of each triangle and added the sums together to determine the total area of the field. Thus, they recognized the concept of a right angle. And while basic algebra was another concept recognized by the Egyptians, as evidenced by problems contained in the Rhind Papyrus, the character of Egyptian mathematics is based on practicality rather than the abstract.

The Mathematics of China

The ancient Chinese dynasties had several things in common with the ancient Egyptians of the Old Kingdom. Like the Egyptians, the Chinese were an isolated civilization with little contact from foreign visitors. Second, they were a practical people who used mathematics for astronomical/astrological calculations (the lunar month), surveying and building (measuring distances and quantities of goods), and recording chemical formulations (alchemy). Lastly, their system of mathematics was based on the decimal system, although there is pictographic archaeological evidence, dating back to about 1500 B.C.E., illustrating the early use of a sexagesimal system. This may or may not be suggestive of a Babylonian influence, but the Chinese obviously abandoned this system in favor of the decimal system, evident in an inscription from about 1300 B.C.E., where the number 547 is written "five hundreds plus four decades plus seven days." The basis of Chinese arithmetic is contained in the text called *Nine Chapters on the Mathematical Art.* It is believed that this document, which is a purely mathematical work, was originally written during the Zhou Dynasty (ca. 1000 B.C.E.) and rewritten during the Han Dynasty (ca. 200 B.C.E.–220 C.E.). The most influential of all Chinese mathematical texts, the *Nine Chapters* addressed such problems as the area of triangles, trapeziums, and circles; the use of fractions; and how to calculate the volumes of cones and pyramids. It also contained more mundane advice regarding the apportionment of agricultural produce and the construction of irrigation and transportation canals. The Chinese mistakenly believed the value of π to be 3, later revising the value to equal the square root of 10. (However, it was not until the second century C.E. that Chinese mathematicians began using the correct value for π, 3.14159.) The book also contained problems of proportions and percentages, as well as volumes using square and cube roots. *Nine Chapters* contained problems leading to a system of linear equations (algebra) using negative numbers that appeared for the first time in history. Another text, called the *Zhou bei* or *Chou pei,* contains some mathematical material, in particular an examination of the formula that later became known as the Pythagorean theorem. The *Zhou bei,* believed to be written at about the same time as *Nine Chapters,* contains material that is contemporary to that found in the *Nine Chapters.*

Numbers

Ancient Chinese numbers or numerals were uncomplicated inasmuch as in the beginning, they were represented by an arrangement of bamboo sticks, called counting rods. Later, they were written as a system of vertical line and horizontal lines with place values. (See Figure 6.5.) On the other hand, it was a complicated system since only two characters—a vertical and a horizontal line—represented, in one form or another, all numbers. The numbers 1 through 9 were written as vertical lines that represented units, while a horizontal line meant an entity of 5 units. For example, the number 3 would be written as ||| while the number 7 would be written as 𝍦. For higher numbers, 10 to 90, horizontal lines represented 10 units each, with a vertical line representing 50 units; for example, the number 20 was written as = and the number 70 was written as 𝍮. Somewhat complicated, however, was the fact that hundreds were also written as vertical lines. For example, the number 1 was written as |. However, when the single vertical line was followed by two other characters, it represented hundreds; for example, the number 123 would be written as | =|||. Sums that represented thousands were written in the same way as tens; ten thousands and millions were written in the same way as units, but hundred thousands were written in the same way as tens. What differentiated the numbers was that adjoining place values would be written with a different set of symbols. For example, the number 7,722 would be written as 𝍮 𝍦 = ||. A blank space rep-

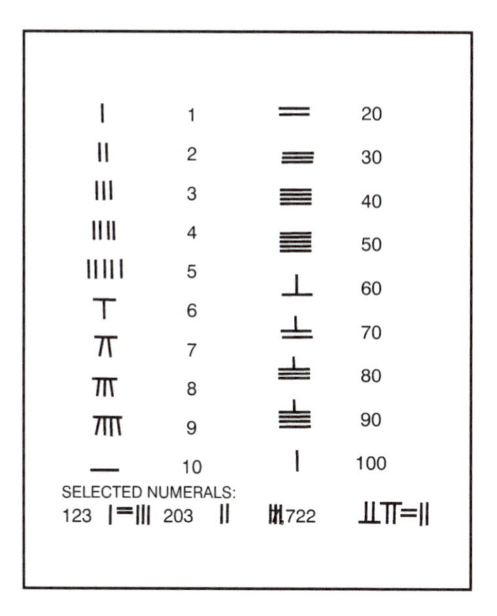

Figure 6.5 Ancient Chinese and Arabic Numerals
Ancient Chinese numerals were written as a series of vertical and horizontal lines. While it appeared uncomplicated, in reality, it was not. For example, a single vertical line could represent the number 1 or 100. While the number 50 is written with five horizontal lines, the number 60 is a single horizontal line (for the number 10) and a single vertical line that represents the number 50. In this sense, position was all-important. Zero was nonexistent. A blank space mean "nothing."

resented zero or nothing; the number 203 would have been written as || |||. The blank space was problematical in that the talent of the Chinese scribe could influence the reader's ability to accurately interpret a number. In other words, a number could be easily misread, but the problem would not be solved in China until the latter half of the first millennium C.E. Later Chinese numerals would evolve into pictographic symbols rather than vertical and horizontal lines.

Negative Numbers and Fractions

The ancient Chinese system of mathematics can be differentiated from that of other civilizations in several areas. For example, by 200 B.C.E. the Chinese understood and used the concept of negative numbers. By contrast, although European mathematicians in the latter part of the third century C.E. acknowledged the existence of negative numbers, they did not accept their validity nor would they use them until the mid-sixteenth century C.E. In China, negative numbers on counting boards were represented as black rods, while positive numbers were red in color; or if colored rods were unavailable, diagonal or slanted rods represented negative numbers as opposed to the vertical rods for the positive. The use of fractions, and later decimal fractions, was another area where the ancient Chinese were somewhat unique. They were not confounded by "leftover numbers," as were other civilizations that ignored or avoided them altogether. The Chinese were a practical people who used mathematics in their everyday lives—for example, to apportion food inventories and to distribute monies. Their calendar year was based on a cycle of 365 and one-fourth days. Fractions are mentioned in the ancient text *Nine Chapters* with regard to the extraction of square roots resulting in numbers that were not integral; that is, something was left over—a fraction. On the other hand, decimal fractions, or "leftover numbers" represented by decimal places, were not developed until sometime after 200 C.E. and would not be widespread for several more centuries. For instance, while the value of π was known to be 3.1415927, it was always written with words. It would not be until the end of the eighth century C.E. that π would be expressed numerically using decimal fractions.

Equations

The ancient Chinese were also the first to employ equations to represent a specific form. In other words, they used algebra in geometry. This mathematical process was documented in *Nine Chapters,* the famous Chinese mathematical text. The problem posed was to find the cube root of 1,860,867. The answer given in *Nine Chapters* was 123—which is correct. The famous nineteenth-century mathematician, William G. Horner (1786–1837), devised a process of standard extraction, known as "Horner's method." He was, however, unaware at the time he published his findings in 1819 that his methods were similar to that used by the ancient Chinese more than 2,000 years earlier. However, the ancient Chinese chose not to exploit this on a wider scale. While they used equations to describe the areas and peripheral shapes of various structures, they did not take the concept to the next level by developing analytical geometry. Analytical geometry uses algebra to describe geometrically shaped objects in numerical terms. One reason may be that the Chinese saw no reason to measure conical forms, that is, ellipses and parabolas.

The Mathematics of Ancient Greece

Unlike the civilizations of Mesopotamia, Egypt, and China, little is known about the origins of early Greek mathematics. The reasons for this are twofold: war and the use of papyrus. First, the history of the

Mediterranean region was dominated by war and the shifting of populations, which culminated in the disappearance of several Aegean civilizations, among them the Minoans and Mycenaeans. This upheaval resulted in increased trade and coined currencies, as well as the development of a more facile alphabet that enabled greater portions of the population to participate in commerce as well as private interests. Thus, a new kind of society emerged, one that was wealthier, with more time for leisure and philosophical pursuits. The second reason why there is no written evidence of early Greek mathematics is the Greeks' use of papyrus as a writing material. Because it is made of plant fiber, papyrus, which was invented in the dry desert regions of Egypt, deteriorated rapidly in the humid climate of the Mediterranean countries.

The oldest complete surviving Greek mathematical text is Euclid's *Elements,* which has been continuously copied since it was first written around 300 B.C.E., presumably because it was a superior work that rendered all previous mathematical treatises obsolete. *Elements* is a compilation of older Greek mathematical theories of which there is no actual written record. It is based on the concept of *rationalism* that began in approximately 600 B.C.E. In other words, the Greeks were not simply content to find out *how* to solve a problem but they needed to know *why* as well. This was the beginning of the process of scientific thought. The early Greek philosophers/mathematicians used geometry, which is at the center of ancient Greek mathematics, to examine and understand the universe and humanity's place in that universe. The word "geometry" is derived from two Greek words, *geō,* which means "Earth," and *metria,* meaning "to measure." Thus, the ancient Greeks used mathematics to find logic, order, and rationality within the world. Not surprisingly, almost all of the great astronomers and philosophers of this Hellenistic period were first and foremost mathematicians, who sought answers to complex problems rather than concern themselves with the mundane arithmetic of the merchant class. Their continuing study of mathematics, primarily geometry, enabled them to make important discoveries in astronomy, which is intricately entwined with mathematics, as well as important contributions in the field of philosophy. There are literally hundreds of ancient Greek mathematicians, all of whom made contributions of varying importance to the collective body of work in mathematics. The most influential of these individuals are the following:

Thales of Miletus

He is often called the "father of Greek mathematics," despite the fact that none of his writings survive. His achievements, however, were recounted by other respected mathematicians and philosophers, among them Herodotus and Aristotle. Most historians accept such attribution with skepticism if it cannot otherwise be validated. Nonetheless, Thales is believed to have discovered five geometry theorems: (1) A circle is bisected by its diameter. (2) Angles in a triangle opposite two sides of equal length are equal. (3) Opposite angles formed by intersecting straight lines are equal. (4) The angle inscribed within a semicircle is a right angle. (5) A triangle is established if its base and the two angles at the base are known.

Pythagoras of Samos

The founder of the famous Pythagorean Order whose motto was stated as, "All things are numbers," Pythagoras believed that all events and things can be reduced to mathematical relationships. Today, he is most famous for developing the Pythagorean Theorem, which states that the square of the hypotenuse of a right triangle is equal to the sum of the squares of the other two sides of the triangle. However, the original concept of this theorem is at least 1,000 years older than Pythagoras, dating back to the Babylonians, who first conceived that any three-sided figure with sides containing the ratio of 3:4:5 would form a 90-degree (right) angle. Proof for the theorem ($a^2 + b^2 = c^2$) was derived by the Pythagoreans, with credit for the proof given to Pythagoras. He was the leader of this academic "cult" of mathematicians who believed their work was sacred and should be kept secret. The other important discovery that is attributed to the Pythagoreans that supposedly changed Greek mathematics was the principle of incommensurability—that is, having no common measurement—of the diagonal of a square with its side. This discovery showed the existence of irrational numbers, which are defined as any real numbers that cannot be expressed as the quotient of two integers. For instance, no number exists among integers and fractions that will solve the problem of determining the square root of 2. It is important to note that the secrecy of the Pythagorean Order is one reason it is difficult to determine which writings were actually Pythagoras's and which were done by his fellow cultists.

Hippocrates of Chios (fl. ca. 440 B.C.E.)

An Ionian philosopher, Hippocrates of Chios's contribution was the discovery of a *lune.* (See Figure 6.6.) Webster's defines this as "a section of a sphere enclosed between two semicircles with their common end points at opposite poles." A lune (plural, *luna* or *lunulae*) is the crescent-shaped area created by the overlapping of two circles. The importance

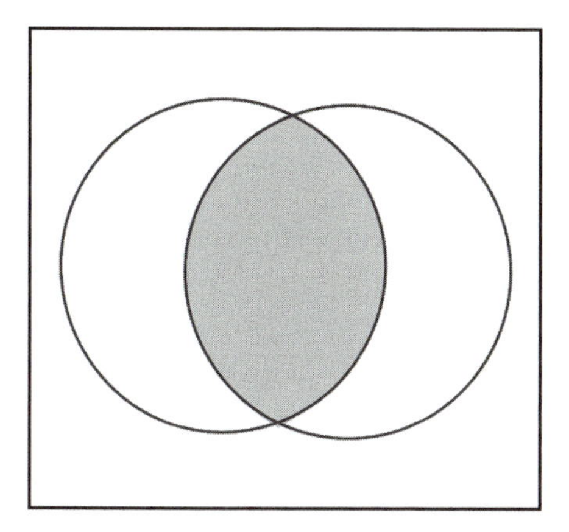

Figure 6.6 Representation of a Lune
Hippocrates of Chios discovered the *lune,* which is a crescent-shaped area created by two overlapping circles.

of the discovery was its relationship to a central problem of ancient Greek mathematics: the quadrature of the circle. In other words, Greek mathematicians tried to develop the formula that would determine the dimensions of a square that is equal in area to a specific circle. The quadrature problem was important inasmuch as it could not be solved geometrically, thus influencing other mathematicians, among them Apollonius, to explore new areas of mathematics in an effort to devise a solution. (Hippocrates of Chios should not be confused with Hippocrates of Cos, the physician.)

Euclid (ca. 330–260 B.C.E.)

A Greek mathematician and the author of the famous geometry text *Stoicheia,* in English known as *Elements,* Euclid lived in Alexandria in Egypt during the reign of Ptolemy I. Reportedly, when the Egyptian pharaoh, who did not want to plow through the 13 volumes of *Elements,* asked him to devise a quicker method to learn mathematics, Euclid's famous reply was, "There is no royal road to geometry," meaning that no matter what position one held in life, one had to follow the same difficult road in order to master mathematics. Euclid's **postulate** stated that all **theorems** must be stated as **deductions** arrived at as self-evident propositions or **axioms,** for which a person can use only propositions already proved by other axioms. This is known as Euclid's **paradigm** for all bodies of knowledge, which led to his great achievement in the field of plane geometry, a series of postulates and notions. (Today, the word "axiom" has replaced the term "postulate" in the current mathematical vernacular.)

Euclid's mathematical axioms can be stated as follows: (1) Given two points, there is one straight line that joins them. (2) A straight line segment can be prolonged indefinitely. (3) A circle can be constructed when a point for its center and a distance for its radius are given. (4) All right angles are equal. (5) If a straight line falling on two straight lines makes the interior angles on the same side less than two right angles, the two straight lines, if produced indefinitely, meet on that side on which the angles are less than the two right angles. (Source: *Encyclopedia Britannica* CD 2002.)

In the 13 books of *Elements,* Euclid brought together his predecessors' many statements and discoveries related to geometry into a logical, systematic form of mathematics in which algebraic reasoning is interpreted entirely in geometric terms. It is believed that *Elements* is the second-most copied and scrutinized text in history. (The Bible is the first.) Books 1 through 4 cover plane geometry; Book 5 deals with Eudoxus of Cnidus's theory of proportions; Book 6 analyzes the similarity of plane figures and the Pythagorean theorem; Books 7 through 9 deal with number theory; Book 10, considered the most difficult, addresses the problems of quadratic irrationals and quadratic roots posed by Thaetetus of Athens; Books 11 through 13 cover solid geometry, the volumes of prisms, pyramids, and parallelepipeds, and a discussion of Plato's theory of five regular bodies. The importance of *Elements* lies in its historical record of mathematical theories prior to Euclid's lifetime and the inclusion of Euclid's own contributions; it is also the basis for the advanced studies of his successors in the field of mathematics. Apollonius of Perga, who developed a theory on conic sections, and Archimedes of Syracuse both relied on *Elements* in their work on mathematical theory. Historically, 9 of the 13 books of *Elements* were taught verbatim in most elementary geometry classes, and the Euclidean principles continue to be an important part of the curriculum of modern-day geometry.

Archimedes of Syracuse

Unlike many of the forefathers of Greek mathematics and his contemporaries, who merely theorized about the mathematical aspects of the world, Archimedes endeavored to apply his mathematical genius to practical purposes.

Thus, he is considered by most to be the greatest mathematician of his time. The famous Greek philosopher and biographer Plutarch (ca. 46–ca. 120 C.E.) wrote that Archimedes' pragmatic approach was considerably at odds with the existing philosophy of his time. In other words, anything done for use and profit was somehow "ignoble." Whether Archimedes believed this to be true or not is subject to speculation, since other biographers of Archimedes mention no such conflict. What is known, however, is that he was an accomplished theoretical and applied mathematician who developed experiments to test his ideas. He then expressed their results mathematically. A few of his most important mathematical discoveries follow. (For his *theory of "perfect exhaustion,"* see the section on the history of π< below.)

Theory for the Volume of Spheres states that the volume of a sphere is two-thirds the volume of a cylinder that circumscribes (surrounds) the sphere. Historically, measuring the volume of a sphere was difficult, while measuring the volume of a cylinder was easy. Therefore, if one knew the volume of a cylinder that surrounded a sphere, its volume could be determined.

Archimedes' Concept of Relative Density and Specific Gravity states that the compactness (amount of matter) of an object is related to the ratio of its weight divided by its volume. Archimedes used his concepts of buoyancy and displacement to measure the relationship between the weight and volume of an object. Reportedly, he came to this discovery after being asked by King Hiero of Syracuse to ascertain whether a crown that the king had commissioned was pure gold or whether silver had been substituted by the goldsmith. As the story goes, Archimedes pondered the question while lowering himself into the water at a public bath. As he sank deeper into the tub, water spilled over the sides. Grasping the significance of this, he jumped from the tub and ran naked down the streets, shouting, "Eureka! Eureka!" ("I have found it! I have found it!") He then filled to the brim a bucket of water, lowered the crown into it, and caught and measured the water that had overflowed to measure the volume of the crown. He did the same with weights of gold and silver equal to the crown's weight. Gold has a greater density than silver, thus the ball of gold was smaller and less water spilled over the edge. Once he measured the volumes of the water representing the volumes of the gold, the silver, and the crown, he divided the figures obtained for the volume of each item into their weight and calculated a ratio representing their comparative densities. He then determined how much of the crown was gold and how much was silver. (Supposedly, the crown was not pure gold and the goldsmith was executed!)

The importance of the principle led to the expression of density as the weight (mass) of an object divided by its volume ($d = m/v$).

Apollonius of Perga (ca. 262–ca. 190 B.C.E.)

Apollonius was called "The Great Geometer" by his peers. He was a prolific writer, but only two of his works survive in written form: *Conics* and *On Proportional Section.* His other works are known only through the references made to them by other ancient writers. *Conics* is an eight-volume treatise that explains the characteristics of the curves that are created when a cone is divided by a straight line. Only the first seven volumes of *Conics* survive, with the eighth volume being lost. Books 1 through 4 survive in the original Greek and deal with the fundamental principles of conics, most of which were addressed by other mathematicians, namely Euclid, Aristaeus (fl. ca. 320 B.C.E.), and Menaechmus (fl. ca. 350 B.C.E.). Books 5 through 6, which survive only in the Arabic translation, are texts of his original theories. In mathematics, conics is defined as the study of plane curves (circle, ellipse, hyperbola, parabola) that are created by the intersection of a right circular cone and a plane. (See Figure 6.7.) His discoveries dealing with the measure-

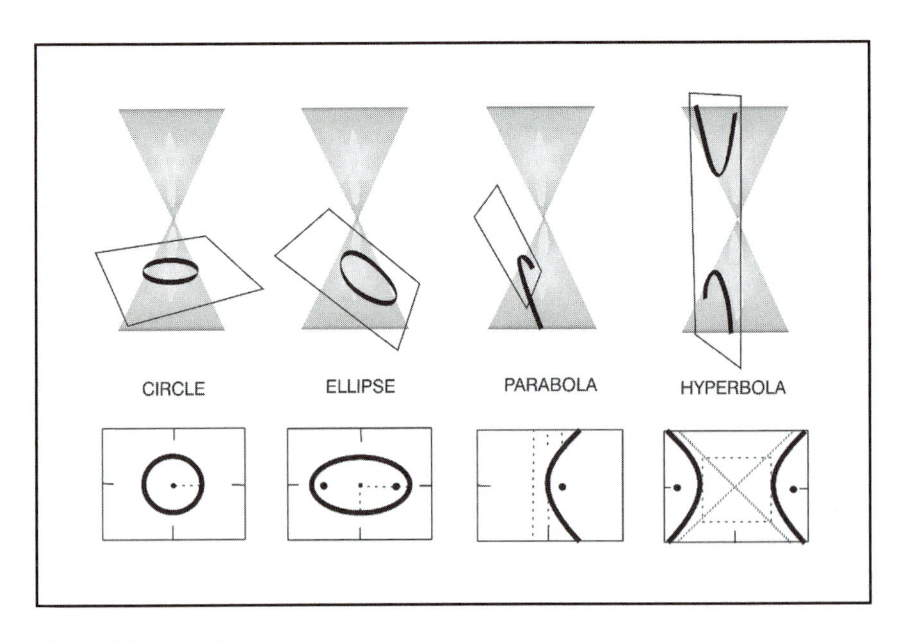

Figure 6.7 Conic Sections
In mathematics, the study of *conics* is defined as the study of plane curves, that is, the *circle,* the *ellipse,* the *parabola,* and the *hyperbola.*

ment of conic sections were important because they assisted succeeding mathematicians and astronomers in developing valid theories regarding planetary orbits.

Apollonius's theories also influenced Hypatia of Alexandria, the daughter of Theon, the last of the influential Alexandrian mathematicians. Hypatia was the only Greek woman of antiquity known to have worked in the field of mathematics. She headed her own school in Alexandria in Egypt and expanded on the theories of other famous mathematicians, as well as those of Apollonius. Her writings were lost after she was murdered in 415 C.E. by the monks of Saint Cyril, presumably because of her school's resistance to the rise of Christianity and her involvement with Orestes, Alexandria's pagan magistrate.

Hipparchus of Nicaea

This Greek mathematician and astronomer is known as the "father of trigonometry." "Trigonometry" (*plane* and *spherical*) comes from the Greek words *trigōnon,* meaning "triangle," and *metria,* meaning "to measure." It is that branch of mathematics that deals with the specific functions of angles and their application to geometric calculations. *Plane trigonometry* deals with angles and distances in one plane. *Spherical trigonometry* deals with similar problems in more than one plane of three-dimensional space. In other words, a great circle on a sphere is the intersection of the sphere with a plane through the center. It evolved from the need to compute angles and distances, particularly for astrological mapmaking. The early Greek mathematicians/astronomers worked primarily with spherical trigonometry (although plane trigonometry was inherent in the application of spherical trigonometry) inasmuch as they viewed the universe—stars, Sun, planets, Moon—as celestial circular disks that progressed in a great circle. Thus, spherical trigonometry was basic to the ancient Greeks' study of astrology/astronomy—as it is today. Using this system, Hipparchus made two important astrological/astronomical discoveries, namely, the precession of the equinoxes and the length of the year (accurate to within 6.5 minutes). Over 200 years later, Hipparchus's work greatly influenced Ptolemy of Alexandria in his study of the arcs and chords of great circles. Ptolemy's famous treatise, *The Almagest,* covers spherical trigonometry in great depth.

The Mathematics of the Indus Civilizations

The contributions made by Indian mathematicians are less well known and, for the most part, have received less recognition. Dating back over 4,500 years, the Indus civilization, alternately called the Harappan civilization, was quite advanced. Its two major cities, Harappa in Punjab and Mohenjo-Daro in Sind, were located near the Indus River in what today is Pakistan. Other villages and smaller towns were spread over the region. The population was literate, religious, and skilled in urban planning and many practical aspects of construction and engineering. In mathematics, ancient Hindu texts, including the famous *ūlvasūtras,* parts of which date back to 500 B.C.E., reflect that the people utilized the principles of geometry in the construction of religious edifices, particularly altars. They used a decimal system for numbering, and created a consistent method of weights and measures. The Harappan civilization thrived until about 1750 B.C.E., when it is believed that a combination of climatic conditions, disease, and the likely invasion of a northern tribe, the Aryans, forced a migration from the northern regions to other areas on the Indian subcontinent.

Religion played an important part in the development of Indian mathematics, inasmuch as the Aryans followed the teachings of the *Rig-Veda* (also called the *Vedas*), a collection of religious hymns and sacrificial rituals that was written in Sanskrit. When the Vedic religion fell out of favor, it was replaced by other religions. One of these, Jainism, appeared about 600 B.C.E. In about 150 B.C.E. the Jains developed a number of mathematical systems, including number theory; arithmetical computations; geometry; fractions; simple, cubic, and quartic equations; theories about infinity; as well as the early concept of logarithms. It is presumed that astrology/astronomy, which was intertwined with religious interests, was the driving force behind these mathematical discoveries. Also, the concept of using a set of 10 symbols, each of which has a place value and an absolute value, originated on the Indian subcontinent. And from these numerals, the universally accepted Arabic numerals evolved. Their origins are believed to be a set of Brahmi numerals first drawn in about the third century B.C.E. (Brahmi is derived from the Sanskrit word *brahman,* meaning "the essential divine reality in Hinduism." It is also the name of the highest "caste" in India, which identifies the order of social groups.) There is disagreement among historians as to how this evolution occurred, but there is written evidence that their gradual development took place over a period of about 1,000 years. (See Figure 6.8.) Another contribution made by Indian mathematicians is the development of the base 10 positional system. These mathematical developments traveled the caravan routes westward from

C. 1ST CENTURY C.E.	C. 4TH CENTURY C.E.	C. 11TH CENTURY C.E.	
—	—	[illegible]	1
=	=	[illegible]	2
≡	≡	[illegible]	3
[illegible]	[illegible]	[illegible]	4
[illegible]	[illegible]	[illegible]	5
[illegible]	[illegible]	[illegible]	6
[illegible]	[illegible]	[illegible]	7
[illegible]	[illegible]	[illegible]	8
[illegible]	[illegible]	[illegible]	9

Figure 6.8 Evolution of Arabic Numerals
Arabic numerals is a bit of a misnomer, as their origins are in the mathematics of the Indus civilization (Brahmi numerals) from about the third century B.C.E. However, since they traveled westward from India onto the European continent, via trade routes and conquest, they were named after the group (Arabs) who introduced them to the wider world.

India to the Arabic countries and eventually to the European continent with the conquering armies of Islam. Some historians believe that the naming of Arabic numerals is another slight to Indian mathematicians. Rather than naming the numerals after the country of origin, they were named after the group who introduced them to a wider world.

Roman Numerals

The Romans, by nature, were not theorists or scientists. Rather, they applied the discoveries of others to useful purposes, often improving or refining the invention or method. The Romans made no mathematical discoveries save for the invention of base 10 Roman numerals, which are still used today. Their history is unclear, including how the letters were selected. Many historians believe that it is based on the practical use of counting with ten digits and the variations that can be made with the fingers. Numbers one through three are easy to understand by holding up one, two, or three fingers, but the letters V, X, C, and M remain a puzzle. In addition, the term "Roman numerals" is really a misnomer, since the system uses seven letters—not numbers—to represent sums up to 4,999. They are:

I = 1

V = 5

X = 10

L = 50

C = 100

D = 500

M = 1,000

When calculating sums larger than 5,000, the use of a horizontal bar above the set of numerals indicates the number is to be multiplied by 1,000. For example,

$\bar{V}$ = 5,000

$\bar{X}$ = 10,000

The currently used Arabic numerals were firmly in place by 1000 C.E., but Roman numerals continued to be used in bookkeeping and accounting systems in post-Renaissance Europe. There are a number of deficiencies (or inconsistencies) in Roman numerals, although they are more striking or attractive symbols for formal designations—for example, Queen Elizabeth II, World War I, XXVI Olympic Games, Superbowl XXXIV, Pentium II, subsection II.4.v. (This is probably the reason they continue to be used.) Lowercase letters are often placed in the front matter of published texts. The letter "u" was often substituted for "v," particularly in medieval texts. Representing large numbers was cumbersome. For example, the number 4,999 is written as MMMMCMXCIX. There was no symbol for zero. As with other numbering systems, zero was represented by a blank space. Arithmetical computations with Roman numerals were cumbersome, except for simple addition and subtraction. Thus, the Romans used an abacus for computing higher values, and limited the numerals to recording the actual sums. However, the Roman abacus was nearly as awkward in its construction as Roman numerals. It is presumed that the incommodious nature of all aspects of "Roman mathematics" led to its demise once the Hindu-Arabic numbering system became more widely known.

Abacus

It is unclear what civilization is actually responsible for inventing the abacus, a crude counting device that is essentially the precursor of the adding machine and modern-day computer. The abacus can be used for addition, subtraction, and the related functions of multiplication and division. It works with any base numbering system, and neither pen nor paper are required. Some historians believe it was the Chinese who invented the abacus, while others believe it was the Romans. The confusion probably arises from the fact that the abacus was not a discrete invention. Rather, it was an evolution of improvements and refinements that took place over thousands of years, finally becoming the finely engineered and exclusive counting tool used by many ancient civilizations. And like so many other ancient inventions, the abacus may have been invented contemporaneously by more than one civilization. One version of the history of the abacus begins with the Akkadians, an early Mesopotamian tribe, who are often credited with its invention. The term "abacus," however, was not used until much later in history, most probably by the Greeks, who called it an *abakos.* The word is believed to be either Semitic in origin (*abaq,* meaning "dust") or Phoenician (*abak,* meaning "sand"). This seems logical, as the first abacus was simply sand spread over a flat board on which crude writings or markings were recorded. It was not meant to be permanent but rather to be used over and over. The ancient Egyptians also used sand abacuses (or abaci) for counting, but by about 500 B.C.E. they had constructed a counting frame made with beads and wire. Archimedes, the brilliant Greek mathematician, is said to have been kneeling on the ground working out a problem on a sand abacus when he was killed by a Roman soldier, presumably for ignoring the soldier's warning.

Eventually, these sand boxes were replaced by *counting boards,* which were reusable. The oldest extant counting board was discovered in 1899 on the Greek island of Salamis. It is believed to date back to the Babylonians over 2,300 years ago. Made of marble, it is carved with two sets of 11 vertical lines that represent 10 columns, with a blank space between them. A horizontal line crosses each set of vertical lines, and Greek symbols are placed at the top and bottom. Small stones were then moved around to facilitate the calculating. The Romans constructed counting boards made of wood that contained painted lines, naturally called the *lined abacus.* The Romans improved the design somewhat by replacing the painted lines with incised grooves. Small metal balls rather than stones or pebbles moved around on this *grooved abacus.*

Commerce was a factor in the further evolution of the abacus as it traveled the trade routes via the Silk Road sometime around the second century B.C.E. The Chinese made further refinements and constructed an abacus on a wooden frame, with vertical metal rods and wooden beads. The earliest reference to China's use of an abacus is documented in a book written during the Han Dynasty in about 190 C.E. Today, the abacus with which we are most familiar is the *rod abacus,* also called the Oriental calculator. It is set on a frame (usually wood) with vertical metal rods that run through seven beads. The beads are separated by a horizontal beam, with two beads above the beam and five beads below. However, the rod abacus was not developed until about the thirteenth century C.E.

The abacus was accepted throughout the civilized world, from the Dark and Middle Ages through the end of the Renaissance, although the design of the European abacus was slightly different than that used in the Orient. When the Hindu-Arabic decimal numerical system was introduced and adopted in Europe and the Middle East, the computation of numbers became easier and the abacus became less prevalent, except in the Orient. Even after the introduction of Hindu-Arabic numbers, China, Japan, and some other Asian countries chose to retain their own numeric symbols, a practice which is still true today. Chinese and Japanese numeric characters are unwieldy, thus the popularity of the abacus for calculating purposes continues, particularly in China. In the 1980s China created an electronic abacus that supposedly integrated, as well as simultaneously used, the basic principles of the Oriental methods of addition and subtraction and Western technology for multiplication and division.

Pi (π) or Squaring the Circle or "Perfect Exhaustion"

The concept of π has been known for thousands of years: the ratio of the circumference to the diameter of a circle is a universal constant. In other words, it is the same for all circles, no matter what the size, or where located in the universe. Why this mathematical **conundrum** intrigued early humans is a conundrum itself. It may have been the fact that early humans placed a great deal of importance on the circle itself. After all, the most visible celestial bodies, the Moon and the Sun, are circles, and the primary role they played in the daily lives of Earth and its inhabitants was readily evident. Early humans also realized straight lines do not exist in nature and recognized curved lines in the shape of rocks, plants, animals, and other objects. As humans evolved and civilizations grew, we can also assume that practical considerations influenced their quest for the solution, since π is needed in order to calculate the area of a circle and the volume of a sphere. What we do know is that the search to find the value of π began nearly 4,000 years ago in Egypt and Mesopotamia. By about 2000 B.C.E. people recognized and roughly calculated the relationship of a circle's measurement in the sense that the larger the circle's diameter, the greater is its circumference. The ancients, however, did not refer to this number as "pi." This came several millennia later, as did the assignment of the symbol "π." (In 1647 William Oughtred, an English mathematician, used the abbreviation [d/Pi] to express the ratio of the diameter of a circle to its circumference. In 1706 a Welsh mathematician, William Jones, wrote that 3.14159 andc. = Pi. As a consequence, the term was adopted. It was Leonhard Euler, the famous Swiss mathematician, who assigned the symbol "π" in 1737. Since then, it has become the standard notation.)

The earliest record that assigns a value to π is in the Rhind Papyrus, a document written by Ahmes the Moonborn in 1650 B.C.E. but containing mathematical data that were at least two centuries older. The Egyptians' value for π was 3.162. The Babylonians also tackled the problem, arriving at 3.125. Passages in the Old Testament of the Bible refer to the concept of π, assigning it a value of 3. And in India, ancient mathematicians used the square root of 10 to determine its value. However, in ancient times, the most important mathematicians to work on this problem were Archimedes of Greece and the Chinese mathematicians Liu Hui, Tsu Ch'ung-Chih, and Tsu Keng-Chih.

By the time of the great Greek mathematicians, it was understood that this ratio was consistent for all circles, since they measured and compared the diameters and perimeters of various circles. Using his knowledge of the geometry of many-sided plane figures, such as squares and multiple polygons, Archimedes proposed his *theory of perfect exhaustion.* He demonstrated this by drawing a circle and inscribing several polygons on both the inside and outside of the circle. At first, he used polygons with just a few sides. Later, he used multiple polygons with as many as 96 or more sides. The theory is called *perfect exhaustion* because, theoretically, a polygon with an infinite number of sides could be used that would approximate a circle's circumference. Using geometry and fractions, Archimedes measured the inside polygons and compared them with the measurement of the outside polygons. He concluded that the polygons touching the circle on its outside circumference (perimeter) were slightly larger than π and that the polygons touching the inside of the rim of the circle were slightly smaller than π. Therefore, π must be a value somewhere between these two measurements. His value for pi was 3.14163, which he calculated as the figure between the inner and outer polygons ($3_{10/71} < \pi < 3_{1/7}$). His figure for π was developed using Euclidean plane geometry, which has physical limitations for this purpose. Ptolemy, the great astronomer and mathematician, worked on the value of π, and in 150 C.E. came up with the figure 3.14166. Mathematicians in China also struggled with the concept of π and over the centuries several of them made significant strides in calculating the exact ratio.

Pi is an irrational number, and as such, it can be run off to an infinite number of decimal places. After Archimedes and Ptolemy, the Chinese continued to find a more accurate value for π. One way was to increase the number of sides in the polygons that were placed in the circles so that their areas would more closely approximate the area of the circle. Archimedes' polygon had 96 sides. In the third century C.E., Liu Hui (fl. 260 C.E.) used polygons that had as few as 192 and as many as 3,072 sides to arrive at a value of 3.14159 for π. In the fifth century C.E., Tsu Ch'ung-Chih (430–501 C.E.) and his son, Tsu Keng-Chih, worked out an even more accurate computation of π to 10 decimal places: 3.1415929203. Although all the original material for these ancient computations was lost, they were all recorded in a number of historical documents. An interesting note is that the circle in which these many-sided polygons were inscribed was about 10 feet in diameter. The value of π determined by the Tsu family was verified in the fourteenth century C.E. by another Chinese mathematician who used a polygon with 16,384 sides. Today, using supercomputers, mathematicians have run off π to 6.4 billion decimal places. However, the fact remains that the first five decimal places for π—worked out on sand, clay and wax tablets, papyrus, and only much later paper—have remained constant since the fifth century C.E.

Zero

In modern times, zero appears to be a self-evident mathematical maxim. In actuality, it was a concept with which generations of ancient mathematicians wrestled until the time of the Greek mathematicians/astronomers. Zero has two functions in mathematics. First, it is used as a symbol for an empty place in our decimal place-value numbering system—for example, in the number 407. Second, it is a number itself, 0. Also, there are variants in both these functions, in particular the use of negative numbers in algebra and the application of higher mathematics. The origin of zero, both as a symbol and a number, remains unclear because there is no clear-cut, decisive documentary evidence tracing its discovery to one ancient civilization over another, namely, the Greeks, the Chinese, or the Indians.

The ancient Mesopotamians, Egyptians, and Chinese used mathematics to solve practical problems. In general, they were not theorists. Even though their numbering systems were place-value systems, they were not perplexed by the concept of "nothing." They simply left a blank space to indicate something that was missing or left out. On the other hand, the Greeks, probably the most famous mathematicians of antiquity, were theorists who did not use a place-value numbering system. To them, this posed no particular dilemma since they were essentially geometers who worked with numbers as dimensions rather than values. (Out of necessity, Greek merchants used a more conventional numbering system.) The one exception in Greek mathematics was in its application in astronomy. Greek astronomers used the symbol "0" in their calculations. Historians disagree as to why "0" came to signify zero. Some suggest that the first letter of the Greek word *ouden,* meaning "nothing," is omicron, hence, the letter 0. Others believe it stands for *obols,* coins of little value that were used as counters on sand counting boards. When the coin was removed from the sand board to indicate an empty or blank space, a circle-like depression was left. Thus, "0" came to signify "nothing." In his book *The Almagest,* Ptolemy, the great mathematician/astronomer, used the symbol "0" to indicate the end of a number, as well as the space between digits. However, its use was limited until it appeared in Indian mathematics sometime after 500 C.E.

The Indians used a place-value numbering system where a dot, and later the symbol "0," indicated an empty place. Historians, again, disagree whether this was an independent invention or whether it was borrowed from Greek astronomers. What is certain is that around 650 C.E. zero, as a number, was used in Indian mathematics, although the use of "0" as a symbol for zero would not be recorded until 876 C.E. in India. In the Far East (Cambodia and Sumatra), the symbol "0" can be traced back to 683 C.E., giving some credence to the fact that the Chinese may have actually been the first to use the symbol "0." However, it would not be until the thirteenth century C.E. that the symbol "0" would be recorded by Chinese mathematicians. The actual word "zero" is derived from *zefiro,* the Italian form of the Arabic word *ifr,* meaning "empty."

Summary

Calculators and computers have in many ways taken the fear out of mathematics. Problems that were once thought unsolvable are routinely deciphered, and most of us are not awestruck by these mechanical machinations. The discovery of galaxies and star systems and planets would not be possible without the application of algebra, geometry, and trigonometry. The Hubble Telescope could not have been built without the knowledge and value of π. Commerce, industry, the stock market, transportation, architecture, engineering, and physics all have their foundations in mathematics. But mathematics is more than numbers. It encompasses abstract theories and possibilities. When Einstein grappled with his energy/space/time concept of mass, the result of which was $e = mc^2$, his challenge could be compared to that of Archimedes' concept of relative density and specific gravity. Both used, for their time, ingenious methods to measure a mathematical puzzle, albeit Einstein's problem was more of a challenge than Archimedes' need to find the amount of gold in a monarch's crown. The mathematical discoveries and inventions made by the people of ancient Mesopotamia, Egypt, Greece, China, and India have not been relegated to the dustbin of history. Succeeding generations of mathematicians, astronomers, and others in commerce and financial fields have improved and expanded them to provide the world with inventions such as the automobile, airplane, radio, television, instruments for space exploration, adding machine, calculator, personal computer, satellite and missile technology, space travel, and the discovery of black holes. It is not inconceivable that scientists, either in the present day or certainly in the future, may be able to explain the origins of the universe using a part of mathematics that was first developed 4,000 or more years ago. While many scientists do not consider the ability to count as a foundation for mathematics, it nevertheless is basic to the science.

7

MEDICINE AND HEALTH

Background and History

Modern medicine is dominated by technological, diagnostic, and pharmacological achievements. Remarkably, most of these advancements were beyond the imagination and aspirations of physician-researchers as recently as the mid-twentieth century. Penicillin is a classic example—a discovery that was first made in 1928 by Alexander Fleming (1881–1955) but not fully recognized or appreciated until the 1940s, when British scientists made further tests and developed its potential for use as an antibiotic agent. Up until 1944, Allied servicemen and servicewomen injured during World War II were treated, at times unsuccessfully, with sulfa and other anti-infective drugs. The mortality rate from infection dropped significantly once penicillin was available and used on a widespread basis. Today, the use of antibiotics is commonplace to the point of their being overprescribed. Another significant advancement in medicine was the development of computer- and laser-guided technology, such as Magnetic Resonance Imaging (MRI) and Positron Emission Tomography (PET) scans that have enabled physicians to more effectively diagnose and treat scores of diseases and injuries. The first modern computer was not developed until 1952; lasers were introduced in the late 1950s.

While the level and scope of medical advancements that have been made in the last 50 years may be the most important, the history of medicine, which began many thousands of years ago, is replete with discoveries, experiments, and inventions that were made with one purpose in mind—to alleviate someone's pain or suffering. That, in itself, sets us apart from other mammals, who when ill or injured will crawl off to "lick their own wounds." Humans will intervene in the care or treatment of their fellow humans. Evidence of this can be traced back some 200,000 years. In a cave in southern France, anthropologist Erik Trinkaus of Washington University in St. Louis found a fossilized jawbone that contains broken roots of teeth as well as significant abscesses. However, the jawbone also shows indications of bone growth that had refilled the sockets of lost teeth. The significance is that the man lived long enough for this to occur—but only because someone had taken care of him and helped him either prechew his food or grind it to a pulp. This was no small feat, since his diet consisted mainly of wild goats and horses that, according to Trinkaus, had "the consistency of old shoe leather."

By examining and studying the fossilized bones that have been discovered over the centuries, *paleopathologists* believe that our most ancient ancestors suffered from the same afflictions as modern humans, among them arthritis, tuberculosis, bone fractures, and dental problems. There are even indications of deliberate amputations, as depicted in ancient cave paintings, although many scientists believe these amputations, primarily of fingers, were performed as part of some ritual rather than for medical purposes. Obviously, the study of *paleomedicine* is handicapped by the absence of conventional written records. Paleopathologists and archaeologists, therefore, must rely on fossilized artifacts and pictographs to reconstruct how the earliest humans practiced the art of medicine.

Primitive Medicine

Knowledge of the manner in which ancient humans survived in the brutal pre-Neolithic Period, though somewhat speculative and anecdotal, is nevertheless genuine, for they obviously not only survived but continued to procreate and populate the earth. Most anthropologists and historians believe that prehistoric humans designated one person, either man or woman, in each clan or tribe as a healer. That person was responsible not only for treating the illness or injury but also for somehow preventing the "evil spirits" from causing the sickness. While early humans attributed maladies to the whims of unknown external forces, they did use the available products of their environment along with incantations and charms to treat them. Tree limbs were fashioned into splints to immobilize and treat fractures, although most likely without actually setting the bone. Various plant and herb concoctions found to be effective for medicinal purposes were administered in the treatment of wounds, infections, sores, and intestinal diseases. Several plants used in modern pharmacology bear a prehistoric imprint. Among these are *foxglove* (digitalis), *willow bark* (aspirin), *papaver plant,* also known as the poppy (morphine), *mahuang* (ephedrine), and *cinchona tree bark* (quinine). Exactly how ancient humans learned to use herbal remedies is also speculative, but *ethnobotanists* and *zoopharmacologists* posit that the instinctive behavior of ill or injured animals who ate the leaves of certain plants, shrubs, and trees may have set the example for early humans. Trial and error was most likely the other factor.

There is one procedure, however, that is indicative of prehistoric medical treatment or intervention, namely *trephining* (also known as *trepanning*). This so-called treatment entails boring a hole in the skull, presumably to release evil spirits, and/or to decrease cranial pressure. Trephined skulls have been discovered in numerous Neolithic sites on both the European and South American continents. (See Figure 7.1.) Paleopathologists and archaeologists believe that the ancients employed this procedure to alleviate the suffering of those inflicted with incurable headaches (migraines), epilepsy, insanity, or other neurological disorders. Trephining, as well as the practice of medicine itself, began as part of rituals founded in beliefs in the supernatural and magic, which were espoused first by primitive clans and tribes, then

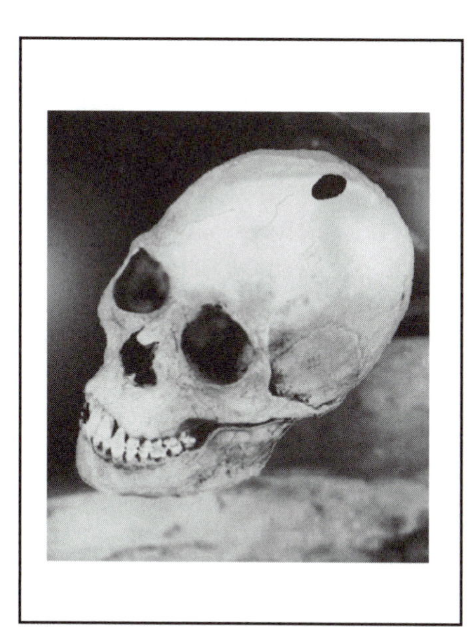

Figure 7.1 A Trephined Skull
One of the oldest "medical" practices, *trephining,* which involved boring a hole in the skull of the afflicted, released the evil spirits that caused headaches, seizures, insanity–or any other unexplained ailment related to the head.

later by the priest-physicians of the first civilizations. Science was an unknown, but primitive humans needed to understand the world around them and the events that transpired; which often led to the illness and death of their fellow clan members. Trephining was believed to be a foolproof way of releasing unknown and evil spirits from the body. The ancient Mesopotamians, Chinese, Romans, Greeks, early European tribes, and Arabs performed trephination for hundreds of years, even as each of these civilizations began to develop more sophisticated and successful medical treatments that remain valid even today. Trephination continues to be practiced in certain areas on the African continent.

History of Disease

A history of medicine would be incomplete without briefly detailing the history of disease, primarily the "diseases of civilization." Our most primitive ancestors, the hominids, who lived 4.5 to 5 million years ago, were hunter/gatherers who lived in small groups. Apart from death by trauma and advanced age, their mortality was determined by their contact with other wild mammals and the zoonotic infections and diseases carried by them. These early humans were infected with lice, worms, trichinosis, tetanus, schistosomiasis, salmonella, and treponema (the bacterial spirochete that causes yaws and syphilis), among many others, which they caught from handling wild animals, eating their flesh, and coming in contact with their wastes. However, these afflictions were primarily limited to individual members of the group rather than the whole. The mobility of the group also meant that they sampled a variety of foods that supplied them with an adequate array of nutrients necessary for the maintenance of good health, which in turn was necessary for survival of the species. Evolution and the eventual extinction of various species of hominids culminated in the appearance of our species of *Homo sapiens sapiens* about 40,000 years ago. Their descendants, who lived during the Neolithic Period 10,000 years ago, developed the concepts of settled farming and animal domestication. They also lived with their by-products, namely, the catastrophic diseases of civilization that have continued to plague the world since that time.

The hunter/gatherer lifestyle, dominant for most of humankind's existence, limited, but did not eliminate, exposure to disease. Presumably, each hominid species continued to thrive and multiply until some unknown force or forces intruded and extinction resulted. The surviving human species' abandonment of a nomadic lifestyle brought with it unimagined consequences as agriculture and animal domestication exposed humans to new and various types of pathogens. Rather than living in small groups, ancient humans lived in communities in close proximity to each other and their animals. Fleas, ticks, lice, dysentery, mosquitoes, hookworms, typhus, chronic diarrheal diseases, and tuberculosis, as well as other bacterial diseases, such as measles and smallpox, were rampant among our earliest ancestors. On the other hand, a relatively stable food supply resulted in an increase in population that exceeded the mortality from these new and unchecked scourges. The quality of life improved as agriculture and animal domestication also provided more leisure time with the division of labor, resulting in the development of tools, the arts, and so forth. It cannot be known with any degree of certainty how civilization eventually overtook the brutal and unsanitary conditions of the very first farmers. But we do know that the earliest civilizations of Mesopotamia, Egypt, India, and China showed remarkable awareness of the sources of many diseases (if not the microorganisms that were the true origins) and developed treatments and systems that, for their time, were often quite effective. Each civilization made independent medical discoveries, yet many were quite similar in character—for example, trephination. They also developed various hygienic and sanitary systems to deal with human waste and water delivery. Civilization has made remarkable leaps since the advent of the Neolithic Period. Populations increased steadily, along with the number of discoveries and inventions that not only vastly improved the conditions of the daily lives of these ancient peoples, but added longevity to them. Around 500 B.C.E. the new deadly pathogens—smallpox, diphtheria, leprosy, influenza, mumps—began to impact negatively on the growth of populations as trade and commerce quickly spread disease from one region to another. Immunities that may have been built up among one tribe or population center were absent in those newly infected by travelers. Notwithstanding the fact that the ancient world dealt with diseases and illnesses that disabled and killed many, it did not suffer the catastrophic plagues that beset the European continent after the fall of Rome in the fifth century B.C.E. and that reached their height during the Dark Ages.

Medicine in Mesopotamia

Mesopotamia, the "Cradle of Civilization," was home to a number of tribes—the Sumerians, the Akkadians, the Assyrians, and the Babylonians—all of whom viewed their daily lives, including bodily functions and malfunctions, in astrological terms. They were superstitious and considered a wide variety of events to be signs of impending death, illness, or recovery. In addition, the Mesopotamians believed that spirits or demons were responsible for certain illnesses or conditions that were unleashed upon humans as punishment for either individual or collective sins. For example, *Nergal* was the god of death, *Nasutar* was the plague demon, *Axaxuzu* was responsible for jaundice, while *Asukku* caused consumption. Historians have a fairly accurate record of the medical practices of these ancient civilizations from 40 cuneiform clay tablets, known as the *Treatise of Medical Diagnosis and Prognosis,* that date from about 1600 B.C.E., as well as from another clay tablet that is believed to be over 4,000 years old. These tablets record in great detail the symptoms of the afflicted patient, usually by the affected body part, and the expected outcome. Hundreds of other clay tablets containing varying amounts of medical data have been unearthed, including several accurate representations of the liver. The Mesopotamians believed the liver to be the seat of the soul (life). To them it was the largest organ in the human body, through which the will of the deities could be determined.

The First Physicians

Mesopotamian physicians or healers were actually priests who were responsible to the various spirits. The *asu* or water-knower practiced a form of empirical medicine. He prescribed herbal remedies, treated wounds with washing and bandaging, and treated fractures with plasters. Eye problems, for example, were treated with sliced onion, which seems to be a reasonable approach. Exposure to the onion produces tears in the patient's eyes; tears contain a substance called *lysozyme,* a naturally occurring antibacterial agent. The *asu* used a wide range of "drugs" made from fruits, flowers, tree bark, the roots of trees and plants, as well as minerals such as copper, iron, and aluminum, and even animal excreta, which were formed into pills, powders, and suppositories. Interestingly, the bitter chemical that is found in the bark of willow trees was a favorite prescription of the *asus.* Today, it is the primary ingredient of aspirin. The popular phrase "pouring salt in the wound" originates in Mesopotamia, as salt was a cleansing *antiseptic,* while the compound *saltpeter* was used to close a wound. Another type of physician or healer was known as the *ashipu,* a sorcerer or conjurer. The *ashipu* diagnosed the illness by conjuring what particular demon was responsible for the patient's incapacitation, and then attempted to drive away the offending spirit. He was also the first "physician" to refer the patient to another medical practitioner, either the *asu* or a surgeon who was a layperson—not a priest—and thus responsible to the state government for his actions rather than a spirit, god, or demon. Presumably, it was the surgeon who performed trephination on patients suffering from insanity, headaches, epilepsy, or other unexplained diseases of the head.

The First Medical Laws

One of the important contributions to medicine made by the Mesopotamians (more precisely the Babylonians) was the development of a medical code that addressed the practitioner's accountability, both to the patient and to the state, as well as a set of fees for each procedure. These medical concepts are contained in the *Law Code of Hammurabi,* a collection of legal pronouncements made by King Hammurabi (ca. 1792–1750 B.C.E.). These precepts are interesting in the scope of their fairness, as well as in their inherent unreasonableness. For example, Hammurabi's law held surgeons responsible for errors made with "the use of a knife." The surgeon, however, received 10 shekels (the currency of the time) for performing a successful operation on a nobleman. (The rate for a common citizen was five shekels, while a slave brought two shekels.) If the operation was deemed unsuccessful—that is, the patient (a citizen or nobleman) died—either one or both of the surgeon's hands were cut off. If the slave died, the surgeon merely had to repay the owner for the cost of the slave. Many historians believe, however, that these rather harsh punishments were seldom carried out. The Mesopotamians were the first civilization to study the human body, albeit through the dissection of animal cadavers rather than human. In addition, their system of sewers and lavatories is indicative of a civilization concerned with public hygiene. Their medical practices and pharmacology were probably not unique, in that other civilizations (e.g., the Egyptians) may have developed some of the same techniques independently and contemporaneously.

Medicine in Palestine (Israel)

The ancient Hebrews (Israelites), who flourished from about 1200 to 600 B.C.E., were responsible for preventive medicine that resulted, in many respects, from their experience as captives of the Assyrians and Babylonians of Mesopotamia. The ancient Hebrews lived in the Near East region of Palestine, which was named after an ancient tribe of mariners, the Peleset, who settled in this southern portion of Canaan near Gaza. The region was rife with conflict for centuries and one tribe was often absorbed by another, more victorious one. Historically, the Hebrews settled into this land of Palestine, and their history and laws were recorded beginning in about the eighth century B.C.E. in the Old

Testament, primarily in the first five books, also known as the *Pentateuch.* The entire body of Jewish law and learning, including sacred scripture, is contained in the *Torah.*

The Beginning of Preventive Medicine

While certain aspects of Hebrew medicine ascribed causation of disease to evil spirits, it is most notable for its emphasis on prevention and hygiene. The priest-physicians of the ancient Israelites passed laws prohibiting the contamination of public wells, as well as prohibiting the eating of pork, primarily because of the very real possibility of parasitic infections. Feminine hygiene was addressed in the provision of ritual baths for women of childbearing age. Males also participated in ritual baths, but this was to cleanse them of guilt (real or unknown) for having contact with something or someone considered to be unclean. Those afflicted with venereal diseases, leprosy, and other contagions that would spread to the population at large were isolated. This was the beginning of the concept of medical quarantine. Homes contaminated by contagious diseases were scrupulously disinfected. Leprosy was probably not as widespread as the Old Testament suggests. Rather, other skin conditions, such as psoriasis, were also treated and viewed as the same as leprosy. Many scholars believe the only surgery performed by the ancient Hebrews was circumcision, a practice that began as a prophylactic but evolved into religious ritual. However, the Talmud, which is an interpretative text of the Old Testament, describes surgeries involving anal fistulas as well as what came to be known as Caesarean section. (It should be noted that the story of Julius Caesar's mother giving birth to him by this procedure, whereby it was so named, is pure fiction. The naming of this procedure did, however, involve Roman law. See the section, "Medicine in the Roman Empire.") In a sense, the ancient Hebrews "discovered" a set of higher standards for the practice of medicine, standards that emphasized common sense based on empirical evidence rather than mere reliance on the capriciousness of the deities.

Medicine in Egypt

The civilizations of Mesopotamia and Egypt were similar in their practice of medicine in that they were both theocratic (assigning specific deities or demons the responsibility for either causing or curing ailments and diseases). Thoth, the Egyptian god of the sciences, whom the Greeks called Hermes Trismegistus, was the most powerful in the medical arts, as he systemized all illnesses that were first revealed by the other gods. Medical historians and scientists have learned a tremendous amount about the nature of Egyptian medicine through two sources. The first was the discovery of medical papyri, the two most important being the Edwin Smith Surgical Papyrus and the Ebers Papyrus. The second source was the unearthing of over 36,000 mummified bodies. As a result, numerous archaeologists and anthropologists were able to examine both their tissue and skeletal remains. The Smith Papyrus was named after Edwin Smith, the American who purchased it in 1862 from an Egyptian named Mustafa Agha.

The Medical Papyri

The Smith Papyrus dates back to about 1600 B.C.E. but is believed to be a copy of an older document written between 3300 and 2360 B.C.E. It contains instructions for the examination, diagnosis, and treatment of some 48 medical conditions that are listed by body part, from the head down to the feet. The priest-physician, also called the initiate, pronounced that either the condition was treatable and curable or hopeless and therefore untreatable. Visible wounds and maladies were treated with a combination of surgical techniques and medicines. Treatment of illnesses with causes that were less apparent consisted of charms, potions, and incantations aimed at breaking the evil spell of the offending god (for example, Sekhmet). The priest-physician used stitches to close wounds, performed skin grafts that involved removing skin from one part of the body and placing it on another, set fractures and splinted them with birch limbs covered with bandages, lanced boils with copper instruments, and applied compresses containing a mixture of grease and honey. All their operations were of an external nature, since they were prohibited from dissecting any organ of the body. Thus, their knowledge of physiology was extremely limited and surgery on the internal portions of the human body was unheard of. There is some dispute among historians, however, concerning the ancient practice of trephination. Some believe it was the one invasive surgery that was permitted, while others dispute that this can be proven definitively, although copper instruments for the treatment of cranial injuries have been discovered.

The Ebers Papyrus is more of a medical reference book, although it is considerably longer than the Smith Papyrus. It, too, was purchased by Edwin Smith in 1862 but sold in 1872 to George Ebers, the Egyptologist, after whom it was finally named. It contains incantations for various ailments, including those for removing bandages. The pathology of skin conditions, worm-related diseases (schistosomiasis), arthritis and rheumatism, and even a diabetes-like disease were described. Treatments varied from magic spells to potions, amulets, and even exorcisms. The papyrus lists more than 500 herbs, plants, and minerals from which 876 formulas were created. One of them was the chewing of castor berries for intestinal problems. (Castor oil is still an ingredient in modern medicine to treat upset stomachs.) A mixture of poppy pods (opium, a calmative, is the derivative) and "fly dirt" was given to infants suffering from colic. A mixture of hot wine and salt was the prescription for earaches. (Today, boric acid powder and alcohol are often the ingredients in ear drops.) The Egyptians perfected an oral contraceptive made from acacia and honey, which when mixed together become lactic acid, an ingredient in some modern-day contraceptive formulas. Cleanliness and hygiene were emphasized as counteragents to disease. While a valid concept, its roots were religious rather than clinical in origin. The most famous priest-physician in ancient Egypt was Imhotep, who was also an architect and builder of some of the most famous pyramids. The Egyptians worshipped him as a god. It is believed that the Ebers Papyrus contains potions and formulas created by Imhotep. Other medical papyri have been found that deal with various aspects of medicine, from gynecology to veterinary science, but the Smith and Ebers papyri are considered the most significant and influential.

Mummification

This ancient ritual resulted from the Egyptians' religious belief that a body could accompany its spirit to another world—if it was preserved forever, an objective that was quite successful, given the number of amazingly well-preserved Egyptian mummies now in the possession of museums the world over. The process of mummification took approximately 70 days, during which time spells and incantations were performed as the organs were removed and, along with the body cavity, washed with wine, a natural preservative and germicide. The body and its organs were packed with hydrous sodium carbonate (natron) for 60 days, after which they were sealed with resin and wax. The body was wrapped in linen while the organs were kept in urns called Canopic jars. In actuality, this elaborate process was not entirely responsible for mummification. Rather, it was the dry, hot climate of Egypt itself that contributed to preservation of the body. Mummification did advance the medical knowledge of Egyptian priest-physicians. Embalmers were a select group limited to this task only. Even though dissection was prohibited, so that any surgical examination of diseased organs was forbidden, empirical studies enabled them to understand the structure of human internal organs. The benefit of mummification came thousands of years later, when modern-day archaeologists and anthropologists discovered that these ancient Egyptians suffered from a variety of the same ills that afflict humans today, among them heart disease, arthritis, dermatological conditions, gynecological maladies, and cancer.

The Character of Egyptian Medicine

Aside from a belief in the supernatural, Egyptian medicine was characterized by other components. First, it was primarily the wealthy and elite who could afford medical care, as was true in almost all early civilizations. Second, Egyptian medicine was divided into three categories: (1) *Physicians,* who were specialists dealing with one area of the body, such as the "eye physician" or the "stomach-bowel physician." These specialists were highly respected but relegated to a hierarchical order by the pharaoh. (2) *Exorcists,* who administered potions and recited incantations over statues of certain deities believed to have influence over the demon that was afflicting the patient. (3) "The priests of Sekhmet," the *surgeon-specialists* who while attempting to find a physical cause for the malady nevertheless ascribed illness to a spiritual causation. And while Egyptian priest-physicians recognized that the heart was at the center of circulation in the human body, they believed, erroneously, that breathing was responsible for circulation; thus, to them respiration was of primary importance in the body. Taking the pulse of an individual was reserved for only the most esteemed priest-physicians, usually the "priests of Sekhmet."

The First Dentists

Egyptians were probably the first to practice dentistry. Among them was Hesi Re, known as the "Chief of Toothers and Physicians," who lived about 2600 B.C.E. Evidence for this was found in two archaeological finds: one of a jawbone believed to date back to 1800 B.C.E., in which a hole was found under one of the molars, presumably to drain an abscessed tooth; and the other a skull containing a set of false teeth, dating back to approximately 2500 B.C.E. So, while Egyptian physicians may have been limited by their belief in supernatural or religious causation for illness, they did make significant contributions to the field of medicine, primarily in the pharmacological aspects, as well as in dentistry.

Medicine of the Indus Civilizations

Vedic Period

The origins of Indian medicine can be traced back approximately 3,500 years, to when it is believed that the books of the *Rig-Veda* were written. The *Vedas,* a series of four sacred books written in Sanskrit in hymn or verse form, include the *Ayurveda* (or *Veda* of long life), which addresses medicine in its earliest form. As with all archaic civilizations, medicine was based on the supernatural, and disease and illness were believed to be the products of sinful behavior or demonic possessions. The *Vedic* or *Vedantic* period of Indian medicine lasted until about 800 B.C.E. Writings concerning this period are filled with descriptions of illnesses that in all probability are malaria, cholera, tuberculosis, diabetes, and plague.

Brahmanic Medicine

The Brahmanical period of Indian medicine followed the Vedantic and was marked by the writings and discoveries of two Hindu physicians, Charaka and Sushruta. Brahmanic medicine was named after the social order that was enforced for all inhabitants of the Indian subcontinent after they were conquered sometime during the second millennium B.C.E. by the Aryans of the north, who imposed a strict social order among the population. One of the social divisions was the *Brahman* (priests). The *Dasas,* or dark-skinned inhabitants, who were vanquished, were forbidden from assimilating into the population and became *shudras,* or the lowest of the social order. *Kshatriyas,* the highest social division, were warriors; *vaishyas* were farmers. Physicians were placed below the *Brahman* priests. The word "caste," Portuguese in derivation, did not appear until about the sixteenth century C.E. It is, however, still used in modern times, and *Brahmans* have replaced *kshatriyas* on the top of the social order or caste system.

Historians believe that Charaka wrote his eight medical treatises either at the beginning of the first millennium C.E. or possibly as late as the second century C.E. Sushruta wrote his books sometime between 400 and 500 C.E. Although there is no consensus on the dates, historians agree the material written by both men, which is, in essence, the foundation of Indian medicine as we know it, contained aspects of Vedic medicine. Indian medicine is characterized by five components. First, a belief in the supernatural ascribes illness to a possession of the soul (Karma) by demons as a result of a sin or sins committed in a previous life. The antidote was the chanting of prayers, as well as physical interventions. Second, the practice of medicine tended to be subdivided or partitioned by concepts, making it difficult to determine whether these concepts were discovered by Indians or taken from other civilizations. The last three components—surgery, diagnostics, and herbal prescriptions and preparations—were quite advanced.

Surgery in India

Indian physicians, skilled in surgical techniques, were, ironically, deficient in the knowledge of anatomy. This deficiency stemmed from religious laws that prohibited any surgery with a knife on a deceased individual. Thus, any anatomical examinations were strictly observational (usually after the body was immersed for a period of time in water and the decomposing tissue could then be subsequently removed). Indian physicians performed operations using some 121 surgical instruments that are described in detail in the *Sushruta Samhita.* These surgeries included removal of neck tumors and anal fistulas, tonsillectomies, lancing of abscesses, suturing, and amputations. The one surgical area in which Indian physicians excelled was in reconstructive surgery, primarily rhinoplasty, the reshaping of the nose. In ancient India, adultery was punishable by cutting off the offending individual's nose. Using the leaf from a plant as a pattern, the surgeon cut skin away from the patient's cheek. The skin was then applied to the stump of the nose and sutured onto it. Hollow reeds were inserted into the nostril areas for breathing. Thus, the surgery was done more for necessity than for aesthetics.

Diagnostics

The fourth component of Brahmanic medicine deals with diagnostics. The ancient Indians believed diseases or disturbances in the body were caused by an imbalance among the humors—namely the spirit, bile, phlegm (mucous)—including the moral humors. Nevertheless, Indian physicians knew that malarial diseases were the result of mosquito bites and plaguelike illnesses were attributed correctly to contact with rats. Indian physicians assumed a hands-on posture with the patient by examining the abdomen, skin, and eyes. Tasting the patient's urine was part of the diagnostic procedure for detecting diabetes; Indian physicians recognized that a sweet taste to the urine was indicative of this serious condition. They also recognized that spitting up blood was a symptom of tuberculosis (consumption). Given today's level of medical sophistication, these signs of disease may appear to be self-evident. But it is necessary to view this in the context of a primitive culture of 2,500 years past. Pain, age, and temperament of the patient were considered, as well as the perceived prognosis of treatment. Those conditions that were considered incurable or hopeless were not treated, and Indian medical treatises specifically warned against all attempts to intervene in such cases. Preventing disease and illness were as important as treatment. Thus, personal and community hygiene were emphasized, as well as a vegetarian diet and abstinence from alcohol.

Herbalism

The fifth and last component of Indian medicine was the use of herbal remedies (pharmacopoeia). Aside from religious incantations, numerous herbals were prescribed for purgatives, enemas, and emetics. Hemp, cannabis, and henbane, a poisonous plant, were among the first plants used for anesthesia as well as for antivenins. Rauwolfia, a tropical tree native to southeast Asia, was given as a tranquilizer. Belladonna, another poisonous Eurasian plant, was administered to colicky infants and those suffering from asthmalike symptoms and indigestion. (Belladonna continued to be prescribed by physicians for these very same symptoms as late as the latter half of the twentieth century, until the development of more advanced pharmaceuticals.) And for some 2,000 years, Indian folk healers have used the resin from the Commiphora mukul tree, commonly known as the guggul tree, to treat various ailments, including obesity. Recently, researchers have found that the resin, called guggulsterone, has the potential for regulating unhealthy levels of cholesterol in the human body.

Religion and war impacted negatively on the continued advancement of Indian medicine. Buddhism, which took hold in the latter part of the sixth century C.E., forbade the study of anatomy, a proscription that hampered even further the limited knowledge of Indian doctors of the actual functioning of the human body. And the Islamic invasion of the Indian subcontinent in the eighth century C.E. discouraged new investigations or research into the pathological origins of disease. Despite these two significant events, Indian medicine was noteworthy for its emphasis on the undisputed principles of hygiene and diet. Ayurvedic medicine continues to be practiced by many alternative or holistic specialists who believe that the methods espoused by the physicians of the early Indus civilizations are preferable to more traditional Western medical principles.

Medicine in China

Many of the principles of Chinese medicine, some of which are nearly 5,000 years old, remain unchanged even as they are practiced in the modern world. These principles date back to the beginning of the dynastic periods of ancient China, and three of their most ancient emperors are credited with inventing specific aspects of Chinese medicine. They are Fu-Hsi (fl. 2900 B.C.E.), believed to have established the philosophy of the *yin* and *yang;* Shen Nung (fl. 2838 B.C.E.), the inventor of *acupuncture* and *pharmacopoeia;* and Huang-Ti (fl. 2600 B.C.E.), the reputed author of *Neiching,* an internal medicine treatise. There are some similarities between Chinese medicine and that which was practiced by other ancient civilizations. For example, religious laws forbade the dissection of the human body. Therefore, the Chinese were deficient in their knowledge of the anatomy of the human body. And like the Mesopotamians, Egyptians, and Indians, the Chinese believed in and used herbal remedies (pharmacopoeia) to treat a variety of ailments, from skin conditions to pain. Unlike their ancient contemporaries of Mesopotamia and Egypt, the Chinese did not ascribe illness to demons or deities, but rather to an imbalance in the forces of nature. (The Chinese, however, did believe in deities with positive powers or influences—for example, the god of the Sun, also called the god of medicine.) The principles of yin and yang, acupuncture, and moxibustion are the three characteristics of Chinese medicine that, along with herbalism, make up the foundation of the Chinese approach to the pathology of disease.

Yin and Yang

The Emperor Fu-Hsi is generally believed to have originated the philosophy of the yin and yang. (Some references credit the Emperor Shen Nung with its introduction.) Nevertheless, the roots of the yin and yang are in the ancient Taoist religion, whose god, Pan Ku, created the world and established order over confusion based on two opposite poles, the yin and yang. The yang principle was positive, light, active, and male, characterized by the sky, the light, strength, warmth, dryness, and the left side of the body. Conversely, the yin was negative, dark, passive, and female, characterized by the Moon and earth, weakness, dampness, cold, and the right side. Illness was attributed to an imbalance between these two opposing poles, which caused a disruption in the life force of the body, known as the *ch'i.* When the yin and yang ceased to flow entirely, death ensued. It was the physician who was charged with restoring the balance. Treating this imbalance also took into consideration the blood, which the Chinese believed to be a vital fluid, since it circulated throughout the body. In addition to the balance between the yin and yang, it was also necessary to maintain harmony among the five elements (wood, fire, earth, metal, and water), which were related to five planets, five seasons, five colors, five sounds, five directions, and the five organs in the human body. Thus, the Chinese physician had a seemingly daunting task. Treatments to reestablish balance included surgery, herbal remedies, moxibustion (moxa), and acupuncture.

Acupuncture

Acupuncture, an ancient physiotherapy practice dating back at least 4,500 years, was an outgrowth of the philosophical principles of the yin and yang. The practice was a way to treat the imbalance between these two opposing poles that resulted in the obstruction of the life force in the body. The ancient Chinese believed the human body was inwardly constructed of "canal-like tubes" or vessels that carried the body's two vital substances—blood and air—and that these vessels became obstructed when the yin and yang were out of balance, resulting in disease or pain. The insertion of long, thin needles into specific points on the body, which lay on a series of invisible lines or meridians associated with specific and major bodily organs, unblocked these vessels, called the *chin, loh,* or *sun,* and allowed the body's energy (the *ch'i*) to flow once again. Prior to about 800 B.C.E., stone needles, called *bians,* were used. They were replaced by bone needles, then by various metals, such as bronze, copper, iron, gold, and silver. The needles varied in length from 2.5 to 28 centimeters (1 inch to 10 inches) and were inserted into some 365 acupuncture sites along 12 meridians. (See Figure 7.2.) Acupuncture was, and still is, used to treat pain and any number of ailments, ranging from hepatitis to nosebleeds to depression. A skilled acupuncturist, then and now, can insert the needles without inflicting pain at the site. Widely practiced in China even today, reportedly acupuncture can be an effective anesthetic even during major surgeries. There are several theories that explain the increasing popularity and supposed success of this ancient procedure. One speculates that the insertion of the needle stimulates the production of the body's own natural painkillers or opiates, namely endorphins or enkephalins. Another theory states that the procedure acts on impulse transmission to the central nervous system, blocking selected neurological sensors to effectively prevent the transmission of pain from various parts of the body. For instance, inserting a needle into a specified site on the foot may in fact be focused on an abdominal complaint or pain. Others

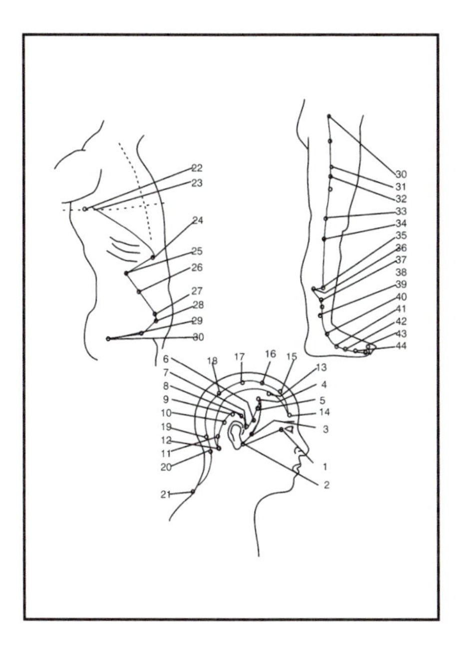

Figure 7.2 Meridian Lines and Acupuncture Points for the Gallbladder
Dating back at least 4,500 years, the ancient physiotherapy practice of *acupuncture* treats the imbalance between the two opposing poles within the human body, that is, the *yin* and *yang*. Using long, thin needles that are inserted into specific points on the body, the acupuncturist attempts to alleviate pain associated with arthritis, migraines, as well as other internal disorders, such as gallbladder disease and stomach ulcers.

believe that acupuncture may engender a powerful placebo effect, particularly in the area of pain management. However, its effectiveness in actually curing disease has yet to be proven, at least by the conventions of Western medical researchers and practitioners. During the latter part of the twentieth century it became more widely accepted as an alternative form of medicine, particularly in the United States, for the treatment of arthritis, migraines, eye diseases, and stomach ulcers, as well as for pain management.

Moxa or Moxibustion

This was a surgical procedure that involved heat rather than the insertion of needles along specified sites on the body. The practitioners of moxa treated chronically ill patients suffering from long-lasting illnesses, such as epilepsy, strokes, blindness, insanity, and even back pain. For example, burning cones of dried herbs and leaves, such as mugwort and wormwood, were placed directly on certain sites on the patient's body. The burning would raise a painful blister into which the ashes of the burnt plant material were rubbed, presumably acting as a local counterirritant to the more generalized disturbance. For obvious reasons, this practice was not as popular as acupuncture, since it caused great pain and disfigurement in the form of scarring. Nevertheless, moxibustion found its way into the Western world and continued to be practiced in Europe during the Renaissance.

Herbalism

The ancient Chinese were culturally averse to the spilling of blood. For this reason, surgery was limited and alternative medical treatments, such as acupuncture, moxa, and herbal remedies, were the procedures of first resort. The knowledge of Chinese herbalists was highly developed and passed on from generation to generation. Over 4,500 years ago, the Emperor Shen Nung wrote an herbal treatise entitled *Pen T'sao Ching,* meaning "herbal," in which 365 herbs, prescriptions, and poisons are described. Many of these ancient herbal remedies, minerals, and treatments continue to be used in modern medicine. Among them are opium (pain relief), rhubarb (laxative), rauwolfia (tranquilizer), kaolin (diarrhea), ephedrine (respiratory problems), iron (anemia), sodium sulfate (purgative), rock alum (styptic), and ginseng (vigor and sexual prowess). In addition to the administration of valid herbal medicaments, Chinese physicians also prescribed potions and elixirs that had more metaphysical than medical elements. These included "dragon bones," which were actually the remains of fossilized animals. Chinese herbalism reached its zenith in the mid-sixteenth century when a 52-volume work, entitled *Pen T'sao Kang Mu,* meaning "great herbal," was published by Li-Shi-Chen.

Two other important aspects of ancient Chinese medicine involved their diagnostic procedures and their concept of the circulatory system. Unlike modern doctors, ancient Chinese physicians did not believe the medical history of their patient was important. Rather, they relied on two areas: first, the examination of the tongue, which according to them had a range of 37 different shades, all of which were indicative of some ailment or condition; and second, the examination or observation of the pulse. Depending upon which ancient volume is read, the Chinese believed there were either 28 or 52 different types of pulse, each with its own particular prognosis. All this was done within the belief system of the yin and yang. The Chinese were also the first to identify that blood circulated, continuously and in one direction, within the internal vessels of the human body, and that when a person dies, the circulation ceases. These statements were made in about 100 C.E., some 1,500 years before William Harvey developed his theory on the circulation of blood. The Chinese are also credited with the first experiments involving endocrinology, using sex and pituitary hormones from human urine for medicinal purposes, and being the first to recognize that certain diseases were caused by dietary deficiencies.

Chinese medical practices did advance to other parts of the Far East, namely Japan, by about the fourth century C.E. While not much is known about Japanese medicine prior to the fourth century, there is some evidence that their ideas of medicine were based on the supernatural, with a limited knowledge of anatomy. However, once it was introduced to them, the Japanese readily adopted the Chinese approach and abandoned the then-existent primitive medical practices of most Japanese islanders. Chinese-style medicine continued to flourish in Japan until the Portuguese arrived in the middle of the sixteenth century C.E., at which time the Japanese adopted the Western approach to medicine with as much enthusiasm as they had earlier with the Chinese. Chinese medicine, on the other hand, languished, primarily because of cultural influences. Respect for the wisdom of their elders hampered further research and investigation into practices and procedures that in other cultures would have been questioned in light of new knowledge and insights. As a result, Chinese medicine stagnated until about the nineteenth century, when Western medicine finally reached Chinese shores.

Medicine in Ancient Greece

Classical Western medicine has roots in ancient Greek medicine, including much of the terminology and philosophy that is commonly used among modern-day physicians and medical practitioners. Ancient Greek civilization and medicine date back about 5,000 years and were the result of war and conquest, when various tribes of adventurers overtook the many islands and territories that comprised ancient Greece. With conquest came new influences and insights that came from exposure to various cultures, including those of Egypt and Mesopotamia. One of these influences designated all illness to supernatural forces. Conversely, these same outside influences are credited with the innumerable discoveries and accomplishments of the Greek philosophers, astronomers, and engineers, whose curiosity and questioning led to the ascendancy of Western civilization.

The Aesculapians

The ancient Greeks had many gods, who they believed inflicted disease and injury as punishments for the weakness of the masses. In turn, the ancient Greeks also had gods who were capable of curing illnesses. Most prominent were Apollo, the god of both disease *and* healing, and Aesculapius, who later supplanted Apollo as the god of healing. Historians do not know with any certainty whether Aesculapius (sometimes called Asclepius) was an actual living person who was deified after death, or simply myth. Legend has it that he fathered many children, and his sons were the gods of surgeons, while his daughters were the goddesses of health and healing. His daughter, Panacea, supposedly had a cure for every ailment, while his other daughter, Hygeia, concerned herself with the health of the public. Whether real or imaginary, the cult of Aesculapius became quite prominent in the first millennium B.C.E. and built many sanctuaries dedicated to the worship of their deity. It was at these sanctuaries, including those at Cnidus, Kos, Athens, Pergamum, and Cyrene, that the Aesculapians (priest-physicians) treated patients suffering from diseases, such as breast cancer, tuberculosis, and arthritis. Among the treatments were various religious rituals consisting of baths in mineral-rich waters, fasting, herbals, and purgatives, all of which were "revealed" by the gods, including those of the underworld. It is believed that Aesculapius adopted the ancient caduceus symbol (see Figure 7.3), which still signifies the medical profession, as an homage to the forces of the underworld responsible for illness and death and his ability to vanquish them with his ministrations, symbolized by the long staff around which the snake is intertwined.

Knowledge of anatomy was limited because dissection was forbidden—it might offend the gods. Patients who recovered offered money to the priest-physicians of the sanctuary, who in turn recorded the person's name, illness, and treatment on clay tablets that were hung on the walls, presumably for the benefit of future patients. For those who did not recover, the Aesculapians blamed a lack of faith on the patient's part, and/or a lack of adherence to the prescribed course of treatment. Sometime during the midpoint of the first millennium B.C.E., a more secular approach to the practice of medicine became more influential but did not entirely supplant the religious practices of the Aesculapians, whose belief system was popular among certain parts of the population until about the fifth century C.E. Thus, the religious approach to healing was contemporaneous with classic Greek Hippocratic medicine.

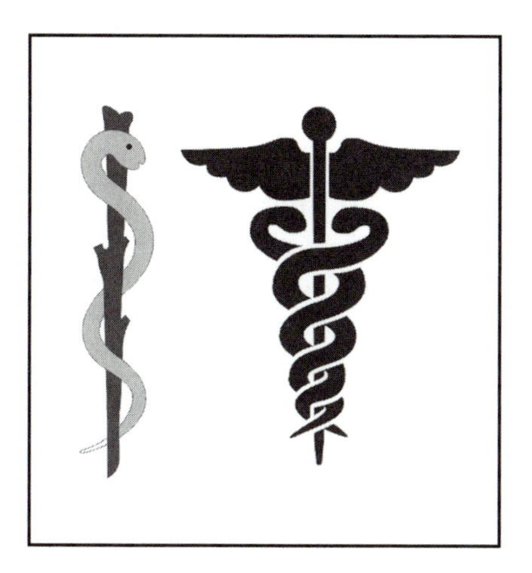

Figure 7.3 Aesculapius's Caduceus and the AMA Caduceus Symbols Greek legend purports that Aesculapius's sons were the gods of surgeons, while his daughters were the goddesses of health and healing. The Greeks believed that Aesculapius was able to vanquish the forces of the underworld (represented by the snake) with his ministrations (represented by the long staff). The American Medical Association (AMA) continues to use a variation of the ancient *caduceus* as a symbol of the medical profession.

Philosophy and the Four Humors

Before examining Hippocratic medicine, it is necessary to explain the foundation of ancient Greek secular or lay medical practices. They, too, were founded in war and conquest, as recounted by the famous Greek writer Homer in his epic poem, the *Illiad.* Supernatural or religious ascriptions notwithstanding, wounds received in battle required quick and effective treatment. Even then, the mortality rate of soldiers in battle was high, primarily from infection and loss of blood. Homer describes the process by which battlefield physicians removed spears and arrows, applied pressure to stop bleeding, and used compresses and herbal balms, as well as wine, to revive the injured. While the dying may have recited prayers, the physicians of Homer's experience were independent and paid, and relied on skill and surgery rather than on the blessing or curse of the gods. Thus, the population was exposed to the practical applications of medicine. Coupled with this was the rise in about the sixth century B.C.E. of medical schools that trained and licensed students, who performed medical services for a fee. These students then imparted to their patients the philosophy that recovery based on lay practices was possible. These medical schools were quite different from the modern-day schools that teach clinical pathology and surgery. Rather, the ancient medical schools were actually schools of philosophy that were concerned with addressing the question of humanity's place in the universe. The founders and teachers were philosophers who taught biology and the natural sciences through the prism of their own beliefs. Their rise in popularity by about the seventh century B.C.E. meant that the ancient Greeks began to believe and emphasize the principles of the natural world and to reject supernatural explanations for illness and health. Among the most famous founders, teachers, and pupils were the following:

Pythagoras, the founder of the Graeco-Italic medical school at Croton, introduced new concepts in the existent medical school at Croton. He founded a brotherhood or cult called "the Pythagoreans," dedicated to a life of speculation and contemplation that was shrouded in complete secrecy. While this cult believed in mysticism and religion, they also attributed scientific causes to the natural world, including medicine.

Alcmaeon, a contemporary of Pythagoras and a physician at the medical school at Croton, was one of the first to dissect human cadavers. He discovered that the optic nerves connect the eyes to the brain and that the Eustachian tubes connect the ear to the pharynx (mouth). He was also the first to declare that all the sense organs are connected by nerves to the brain and that the brain is the site of human intellect. His book, entitled *On Nature,* described how disease was caused by natural rather than supernatural influences, and suggested various methods of prevention and treatment. He believed that well-being and illness were dependent upon pairs of fundamental opposites—for example, hot and cold, wet and dry, sweet and sour. Any imbalance in these pairs was likely to cause illness, while correcting the disturbance was likely to result in a cure.

The Four Humors—Empedocles of Agrigentum (ca. 490–430 B.C.E.), a student of the Graeco-Italic medical school, was influenced by the philosophy of the Pythagoreans. Empedocles believed that four elements governed the universe, namely, earth, air, fire, and water, and that these four elements corresponded with four fundamental humors in the human body, namely, blood, phlegm, yellow bile, and black bile. These humors originated, respectively, in the heart, brain, liver, and spleen. From this belief, he taught that the heart was the center of the circulatory system and that blood flowed continuously to and from the heart. *Pneuma* or respiration was distributed throughout the body by the blood vessels as well as through the pores in the skin. Not all ancient Greek physician-philosophers subscribed to humoral theory. For instance, Greek *physiolo-*

gists believed that the arteries carried "vital spirits" that were distributed to parts of the body. They concentrated on the anatomy of the brain, nerves, veins, and arteries to explain illness. *Empiricists* were surgeons and pharmacologists. Challenges to humoral theory waned over the centuries and ceased altogether when Galen of Pergamum, the famous Roman physician who accepted humoral theory in toto, became the reigning medical authority for hundreds of years. The misguided belief in the four humors concept was significant in that it was the prevailing medical theory for nearly 2,000 years and served as the justification for a variety of "evacuation" treatments, such as bloodletting and **cupping.**

In addition to the close association between medicine and philosophy, ancient Greece was also known for its creation of the **gymnasiums.** The populace was encouraged to maintain their good health through a regimen of diet and athletic pursuits at these facilities. The benefit to medicine was that it enabled Greek physicians to learn more about the anatomy of the human body while treating injuries sustained during athletic exhibitions. The study of anatomy became a branch of medicine about the time of Hippocrates of Kos, the most famous of the ancient Greek physicians, who is also known as the father of medicine.

Hippocratic Medicine

The widely traveled (for his time) Hippocrates of Kos (ca. 460–377 B.C.E.) was the descendant of a family of physicians who founded a medical school in Kos (also known as Cos), where he taught until the time of his death. The practice of modern medicine is based on the principles that Hippocrates and his followers espoused, which are contained in a multivolume medical and ethical treatise known as the *Corpus Hippocraticum.* Upon students' graduation from accredited medical schools, the famous Hippocratic Oath, modified and approved by the American Medical Association, is recited by new physicians, who pledge to practice medicine based on the ethical standards that were taught nearly 2,500 years ago in ancient Greece. The principles that were taught at Kos by Hippocrates and others emphasized service to the patient, diet, and hygiene. The prescribing of drugs and the performing of surgery were done only after diet failed. Surgeries, including trephinations, venesections (bloodletting), empyemas (gallbladder removal), and so forth, were done conservatively, almost as a last resort. Hippocrates is credited with writing only a few of the nearly 70 volumes of the *Corpus Hippocraticum,* but all of the works stress the same fundamental approach to diagnosing and treating illness. The components of this approach are (1) belief in the natural causation of illness; (2) practical observation; (3) prognosis; (4) treatment of the patient as a whole, rather than merely the ailment or disease alone; (5) belief in the theory of the four humors of the human body; and (6) nontreatment of those considered incurable. Detailed descriptions of the clinical pathology of diseases, symptomatology, procedures (medical and behavioral), and treatment are given throughout the volumes, in addition to the "sayings" that are widely associated with the practice of Hippocratic medicine. About 406 of these rather cogent dictums are contained in one of the volumes, entitled *Aphorisms.* Among the most famous are "First, do no harm" and "Extreme ills need extreme cures." Notwithstanding its contributions to patient care, anatomy, and physiology, Hippocratic medicine did foster the misconception that the human body (as well as all matter) was formed of the four elements of earth, water, fire, and air (aether). On the other hand, it also fostered the rational belief in the natural causation of disease and eschewed the magical or mystical approach to diagnosis and treatment. Upon the death of Hippocrates, the medical school at Kos declined and no further medical canons were written. Consequently, the material contained in the *Corpus Hippocraticum* hardened into accepted medical doctrine that promotes honesty, vigilance, and dedication, and a respect for the patient's right to privacy.

Medicine in Alexandria (Egypt)

In the third century B.C.E. the practice of medicine, as well as Greek civilization itself, became centered in the newly erected Egyptian city of Alexandria. (Alexandria was named in honor of the Macedonian king, Alexander the Great, who conquered Egypt around 332 B.C.E.) The excellent library and museum that were housed in the city encouraged the pursuit of scholarship. Thus, a number of renowned Greek noblemen settled in Alexandria and influenced generations of followers, who preserved and propagated their theories on mathematics, astronomy, and medicine. In Alexandria the medical community was comprised of three major groups: the *anatomists,* the *physiologists* (also called therapists), and the *empiricists,* each with a distinct approach to theory and practice. The three most famous representatives of these groups were the following:

Herophilus of Chalcedon

This anatomist is credited with establishing human anatomy as a branch of medicine. Since the ban on dissections had been lifted for the first time in the history of Greek medicine, Herophilus (fl. 300 B.C.E.) was the first person on record to perform anatomical dissections in public. He correctly noted that blood coursed through the arteries from the heart with a pulsating motion, unlike the blood that returned to the heart through the veins. He accurately described the anatomy of the eye, brain, male and female genitalia, and other vessels of the body, including the duodenum and the prostate, both of which he named.

Erasistratus of Kos

This physiologist (or therapist) (ca. 300–260 B.C.E.), founder of another medical school at Kos, rejected humoral theory and incorrectly believed that air was pulled into the body by the blood as the blood moved upward in the veins. He also believed that the arteries carried "vital spirits" throughout the body. Nonetheless, he made significant contributions in the field of anatomy, including the pathological anatomy of the brain, heart, veins and arteries, and liver. However, his acceptance of the Aristotelian concept of all things moving in straight lines hindered further insight into an accurate understanding of the systems of the human body.

Heracleides of Tarentum (fl. 100 B.C.E.)

Heracleides was the most famous empiricist. Empiricism was not as dominant a medical philosophy as anatomy or physiology, but the empiricists did contribute to the fields of symptomatology, pharmacology, and surgery, including the use of the ligature as a hemostatic agent.

Medicine in the Roman Empire

The forerunners of the Romans were the Etruscans, whose medical practices were based on the supernatural. In fact, in the early days of the Roman Empire, it was Etruscan priests who served as "physicians" to their Roman conquerors, since the Romans themselves viewed the practice of medicine as coarse and inferior. The Etruscans built a number of sanctuaries where Aesculapian medical principles were followed. They made some significant contributions to the fields of public health and dentistry. The Etruscans believed in the healing powers of certain water sources and the advocacy of public hygiene measures. (It was the Romans who took these beliefs a step further, when they built the famous *Cloaca Maxima* in the sixth century B.C.E. The great drain, as it came to be known, diverted brackish standing water in the marshes of Etruria [Italy] into the Tiber River.) In the practice of dentistry, the Etruscans were skilled at making dental bridges using gold and also extracting teeth, as well as fashioning dental crowns and gold fillings.

At about the same time that influential Greek scholars and scientists traveled to Alexandria to study and practice, other Greek nationals, including freemen, slaves, and physicians, began to relocate to Rome. The expansion of the Roman Empire had been ongoing for centuries and the intermingling of cultures was a natural consequence. Throughout their history, the Romans were contemptuous of "all things Greek." Nevertheless, they were a practical people in all respects and borrowed and adapted a wide variety of ideas and technologies from other civilizations and populations, often improving on them to make them uniquely Roman. However, the practice of medicine was somewhat different. By about 148 B.C.E. the Roman conquest of all the Greek city-states had been accomplished, and the Roman armies, in particular, were often patients of Greek physicians. One of Rome's most famous generals, Marcus Cato (234–149 B.C.E.), specifically warned against the influence of Greek culture and medicine, fearing that the intent of the Greeks was to "murder all foreigners with their medicine." His fears, however, were unjustified and were based on his rigid beliefs in the past glories of the Roman Empire. His prescriptions for disease and injury consisted of religious incantations and cabbage (cooked and raw), which, according to Cato, was a near-universal panacea. Less stubborn Roman leaders saw an opportunity to improve the condition of Roman health and invited Greek physicians to practice in Rome, often bestowing upon them the honor of Roman citizenship. Throughout the existence of the Roman Empire, the medical profession was held in rather low esteem (that is, until a noble Roman needed medical care). There are a number of reasons for this. First, physicians were considered tradesmen rather than nobles. Second, the number of successful treatments and cures was low, so confidence was understandably low as well. And third, many of the physicians practicing in Rome were freed Greek slaves, and Romans always had a mistrust and skepticism of anything or anyone associated with ancient Greece. Thus, "Roman medicine" is really a misnomer, since it was primarily physicians from Greece or Alexandria who practiced and dominated the field of medicine. In time, the Romans came to appreciate the training and skill of physicians, although by and large the Romans preferred the fields of art, architecture, politics, and law to the practice of medicine. Most, but not all, Romans who chose to associate themselves with the medical field were essentially historians or recorders of data, rather than physicians who learned the craft at a medical school and subsequently practiced among the people. Like the physicians who had practiced centuries earlier in

Greece, physicians who practiced in Rome and who made the most significant contributions, discoveries, and experiments adhered to individual theories of illness and health. Among them were the following:

Asclepiades of Prusa (ca. 124–40 B.C.E.)

This Greek physician was the first to achieve great success and recognition as a healer in Rome. Rejecting Hippocrates' theory of the four humors, Asclepiades was an *atomist* who based his medical pathology on the belief that the human body was composed of atoms, or tiny unseen particles, that coursed continually throughout the body. According to him, illness was a result of a disturbance in the movement of these particles, and only with a restoration of the proper movement could a cure be accomplished. His was a holistic approach that emphasized diet, exercise, and fluids in the form of baths and by mouth. He performed the first tracheotomy and was the first physician to treat those with mental illness with respect and kindness, using music, herbs, and exposure to the outdoors. Heretofore, mental illness was treated either by trephination or by confinement in dark cellars or rooms. Asclepiades' main contribution to medicine lay in the hands of his followers and successors. He is credited with recognizing the natural periodicity of a number of diseases, along with differentiating between acute and chronic stages. His theories and successful treatments meant that future physicians treated illness with the expectation of actual cures. His theories were widely popular and practiced as recently as the eighteenth century C.E.

Themison of Laodicea (fl. ca. 50 B.C.E.)

A pupil of Asclepiades, Themison was a founder of a medical sect known as the *methodists,* who believed that illness was caused by *status strictus* (a restriction or narrowing of the internal pores of the body) or by *status laxus* (an extreme relaxing of the same pores). Treatment involved baths to induce sweating or the reverse—astringents and tonics to prevent it. This oversimplification of the concept of illness and treatment was popular in Rome, since it accommodated the **formalism** of Roman beliefs in religion and medicine. It was an easy method or approach to treating large numbers of confined patients—for example, Roman slaves. In time, many methodist physicians rejected this simplistic approach to diagnosis and treatment and employed a number of other processes to treat illness, including pharmacology and surgery. However, they continued to rely on the dominant symptomatology of *strictus* or *laxus* as the indicators.

Soranus of Ephesus (98–138 C.E.)

A member of the methodist medical sect, Soranus is commonly referred to as the "father of obstetrics and gynecology." He trained and practiced in Alexandria before moving to Rome, where he authored the book *On the Diseases of Women,* which remained popular for the next 1,500 years. The book describes in detail the female genital system and the physiological and pathological conditions that are exclusively female, including contraception and childbirth (natural and Caesarean section). Caesarean section was named after Roman laws that demanded that (a) upon death, the body of a pregnant woman could not be buried until her unborn child had been removed; and (b) in a living woman, the procedure could not be performed until her tenth month of pregnancy, which, in most cases, meant that the woman died, along with her child. In reality, physicians rarely performed the procedure. Soranus also discussed the care of infants and children, as well as the broader areas of acute and chronic illnesses and bone fractures in both children and adults. Other books written by Soranus discuss mental illness, including hysteria, which is referred to as a "disease of the uterus," and hypochondria, a "disease below the diaphragm." Copies of books attributed to Soranus still exist, but not under his name. Rather, they survive under the name of their translator, Caelius Aurelianus.

Pedanus Dioscorides

This Greek physician classified about 600 medicinal plants and thousands of drugs. Although his alphabetical classification and many of the accompanying drawings were inaccurate, nonetheless he is considered the father of pharmacology. (Pharmacology is from the Greek word meaning "the study of drugs.") He is also the author of *De Materia Medica,* a five-volume treatise that contained all the pharmacological information available during the time of the Roman Empire. He was the first to describe opium and what was later known as laudanum, an extract of alcohol and opium-plant blooms.

Aulus Cornelius Celsus (fl. first century C.E.)

A Roman nobleman, Celsus is considered the most recognized and accomplished of the Roman medical writers. His famous work, entitled *De Artibus,* contained massive amounts of information on agriculture, warfare, philosophy, and law, as well as medicine. The last was incorporated into the section of the book called *De Re Medica* and is the only part of *De Artibus* to survive. Lost for centuries, it was rediscovered in 1426. It has the distinction of being the first ancient medical text to be printed (1478) by the Gutenberg printing press. *De Re Medica* is an encyclopedic work, with a systematized organization of diseases and treatments. It is divided into three parts according to the suggested treatment—dietary, pharmaceutical, or surgical. Among them are the four signs of inflammation (heat, pain, redness, swelling); facial reconstructive surgery, where skin grafts from other parts of the body are used; heart disease; insanity; fractures (closed and open); the methodology of splint-making using bandages stiffened by wax and paste; and the use of ligatures to stem arterial bleeding. *De Re Medica* also describes a number of surgical instruments, including forceps, scalpels, hooks, probes, tongs, and a meningophylax that was used to hold the meninges (the membranes enclosing the brain and spinal cord) back after trephination. A sharp needle was used in the removal of *cataracts,* an ophthalmological condition that obscures the lens of the eye and that, untreated, can lead to blindness. Celsus described how surgeons removed the cloudy lens, restoring the patient's sight. He emphasized the importance of hygienic conditions at all times. Wounds were to be cleansed with substances that can be considered as "antiseptics," such as vinegar and oil of thyme. Although not a physician but rather a compiler of doctrine, both Celsus and his work have great importance, since he assembled what is judged to be an accurate historical record of Greek and Alexandrian medical and surgical practices.

Claudius Galen of Pergamum (ca. 129–216 C.E.)

Galen is often regarded as the greatest physician in antiquity, most likely because his works were the most detailed and based on both observation and experience. Galen's medical treatises fill 22 printed volumes or about one-half of the surviving medical data from antiquity. Born in Pergamum in Greece, he attended the schools of philosophy and medicine in Smyrna, Corinth, and Alexandria. Galen was both a humoralist and a *pneumatist,* meaning he believed in the three different forms of action in the body: (1) *pneuma psychicon* or animal spirit, which controlled sensation and movement; (2) *pneuma zoticon,* which was centered in the heart and controlled blood and temperature; and (3) *pneuma physicon,* the physiological action that was centered in the liver and controlled nutrition and metabolism. These were not the only medical misconceptions that he fostered. One was that the blood flowed from the liver, which according to Galen was the main organ of the entire cardiovascular system, through the veins to the heart where it was purified. Another misconception was that the nerves in the human body were empty tubes or ducts. He also mistakenly believed that the heart was a single pump with two chambers that allowed the blood to pass from one chamber to the other and then pumped the blood out to both arteries and veins. He incorrectly believed that the heart was the source of respiration. With all these misconceptions about the physiology of the human body, one might ask why his works are so important and he is personally so respected. The answer probably lies in the force of his personality, his position in the Roman Empire as a confidant and personal physician to two emperors, the abundance of detailed medical dictums meticulously transcribed by personal aides, and his own particular religious belief. He learned about anatomy from two sources: his experience as a surgeon to gladiators in the Roman amphitheater, and the dissection of animals, primarily pigs and monkeys. (Since human dissection was forbidden, he never dissected a human corpse.) Thus, he incorrectly assumed that what he observed in animals applied in the human body. He dismissed his critics as ignorant and resented any challenges to his medical authority and genius. His writings about the structure of bones, and kidney and bladder function, were accurate. His most important contribution was to disprove the belief held for over 400 years that arteries carry air. Just as important in solidifying his reputation was his own particular religious belief. Though he was born little more than a century after the origins of Christianity, he was not a Christian but did believe in one god and that the human body was an instrument of the soul or spirit. Thus, his belief in one god was important to the founders of the Church and later to Islamic scholars and accounted for their acceptance of his theories. In their belief systems, God had a plan, and Galen's interpretation of human anatomy and physiology followed this plan. As a result, his extensive writings were accepted as final word. This acceptance hindered progress in the advancement of anatomy and physiology for almost 1,500 years, well into the Renaissance. It was not until the sixteenth century C.E. that most of Galen's concepts were openly challenged and his mistakes corrected. This accounts for the phrase "the tyranny of Galen" when referring to his influence over the pre-Renaissance medical community. However, this should not diminish or overshadow his many contributions to medicine. Galen was a keen observer and an excellent diagnostician, as well as someone who recognized the value of experimentation. Upon Galen's death, Rome eventually retreated to its past, relying on magic and superstition, rather than on science, a factor that probably contributed to the Empire's collapse in about 476 C.E.

Public Hospitals

In addition to the concept of public health, demonstrated by the numerous aqueducts and water and sewer systems built by the Romans, the creation of public hospitals is considered to be Rome's greatest contribution to medicine. Roman hospitals were designed for two classes: soldiers who lived in permanent encampments throughout the Roman Empire and domestic slaves. (Artifacts from two of these ancient army hospitals, which were located in present-day Chester, England, and Inchtuthil, Scotland, still exist.) The Roman hospitals were operated much as present-day facilities, with wards, kitchens, staff dining halls, and apothecaries (pharmacies) for the preparation of medicines.

Summary

The significance of the discoveries, inventions, and experiments in medicine that took place from about 3500 B.C.E. to 500 C.E. should be viewed in relation to what has occurred in the past 200 years. The nineteenth and, particularly, the twentieth centuries saw astounding discoveries and advancements that have added both longevity and quality to human life. The ancients, on the other hand, did not have the benefit of superior technology. Microscopes and computers were unknown, as was the existence of germs, bacteria, and viruses. Science struggled to overcome superstition, and often failed when a dominant figure in the medical community died without a successor to carry on his endeavors (e.g., Hippocrates and Galen).

Nevertheless, antiquity was a time of curiosity and discovery. Medical treatises (describing diseases, treatments, anatomy, surgery, pharmacology, and diet) that were considered unimpeachable for centuries all were written during this period. Personal and public hygiene were recognized as vital to the maintenance of health. Holistic practices, such as acupuncture, have survived, essentially unchanged, for thousands of years and interest in them has been revived. The ethical treatment of patients and the respect for their privacy, which was deemed so important nearly 2,500 years ago, is every bit as vital today in a world filled with technological access and almost limitless possibilities. Care for the ill and injured in public facilities that was begun during the time is often described as being primitive. The prototypes of today's surgical instruments—scalpels, clamps, forceps—were invented during a period when inquisitiveness about human anatomy was constrained by laws based on superstition and fear. We know that after the fall of Rome, the pursuit of all knowledge, including medical, was stymied by the continuing struggles of religious and territorial wars. This is one reason that the period from about the sixth to the fourteenth or fifteenth centuries C.E. is called the "Dark Ages." Fortunately, not all of the documentation of the ancient physicians and medical recorders was lost during this time of conflict and neglect. Much of this material was rediscovered during the Renaissance, when scholarship was once again appreciated and valued.

8

Personal and Household

Background and History

If asked, many people would assume that most items sold today in stores were invented in the twentieth century. They would be wrong. The creation of articles such as home insulation, cosmetics, intimate personal care items, lighting, cooking utensils, rugs, eyeglasses, vitamins, gardening tools, hunting supplies, security devices, and decorative pictures, to name a few, can be traced back thousands and thousands of years. Some of these articles were invented by Neolithic humans endeavoring to survive in a hostile environment, while others were developed by more advanced agricultural civilizations that lived in relatively stable conditions with more leisure time. The imprint of human history can be seen on most things that are found in homes today—from chairs to toilets to windows to lamps, ceramic plates, and glassware, even to the earrings and tattoos that adorn the inhabitants of those homes.

It is easier to determine a time period when many of these inventions or discoveries took place than it is to establish the "science" that went into their creation. We can imagine the common sense that was involved in the making of a bowl that kept liquids in a confined space as opposed to a flat surface where spillage was inevitable. But what kind of thought and science preceded the development of such items as candles, glassware, locks and keys, contraceptives, and soap? We do not have the benefit of written texts that describe the processes that ancient humans used over 3,000 years ago to make the first candles from tallow and beeswax. There are, however, instructions inscribed on ancient clay tablets from Mesopotamia for the building of furnaces to make glass. Was it serendipity that produced the first item of glassware or was it trial and error based on the correct mixture of sand, lime, and metals? Did the Egyptians who perfected locks and bolts to secure their treasures employ the scientific principles of weights and balances or simple machines? What chemistry did the ancients use in the development of soap and contraceptives? We know that they collected the leaves from plants growing on riverbanks to scrub their bodies, but soap was not used for cleansing in the conventional sense for centuries. As for contraception, it was a common practice early in human history and most certainly in the patriarchal societies of Mesopotamia, Greece, and Rome, where limiting the number of births, particularly of females, was acceptable.

Despite the lack of documentation as to the why and how, we are nevertheless left with vast numbers of artifacts and information relative to how the ancients lived, which, surprisingly, is not so dissimilar from modern humans as to be unrecognizable. While advancements in industry and medicine, as well as new technologies, have greatly improved the quality and longevity of human life, the basic necessities to sustain that life have remained unchanged since the dawn of civilization—food, shelter, and companionship. The evolution of the everyday basic items that enabled early humans to enrich their lives with some degree of comfort and efficiency is the subject of this chapter.

Body Ornamentation

Tattoos

The derivation of the word *tattooing* is Polynesian and thus the term was not used as such in the ancient world. However, the custom of ingraining marks and/or designs on human skin dates back many thousands of years, although its region of origin is unknown. The oldest evidence of tattoos was found on the mummified body of a man discovered in the Italian Alps in 1991. Commonly referred to as "the Iceman," his corpse, dating to 3300 B.C.E., was marked with three sets of blue lines on his back, another set on his right ankle, and a cross on his left knee. Archaeologists believe these markings were made with needles and finely powdered charcoal. At its advent, tattooing in the ancient world had a twofold function. First, it provided a supernatural safeguard against evil, adversity, and disease. Second, it identified the station or rank of the person, or his affiliation with a certain group or tribe. In that sense, the tattoo was really a "brand." For example, the mummies of female Egyptian concubines believed to be 4,000 years old are adorned with dark blue facial tattoos. (On occasion, a *cicatrix,* which is the deliberate raising of scars on the skin without the use of dye or pigment, is referred to as a tattoo.)

The Greek writers Herodotus and Xenophon (ca. 430–355 B.C.E.) both describe tattoos among the various tribes of the Mediterranean and Balkans. For instance, the barbarian tribe of Thracians, who some historians believe learned this ancient art from the nomadic Scythians, equated tattoos with symbols of nobility, while those unadorned were considered to be of lower birth. A significant find in Pazyryk (an area on the borders of Russia, China, and Mongolia) bears this out. The well-preserved body of a man thought to be a chieftain of a migratory tribe was discovered in 1948. The body, which is about 2,400 years old, bears numerous and detailed tattoos of monsters and wildlife. In the Far East, tattooing was widespread. In China, it was used as a punishment for criminals during the more than four-hundred-year reign of the Han Dynasty (ca. 202 B.C.E.–ca. 220 C.E.). In Japan, it was a decorative and status adornment for men and boys alike. The Mayans of Mesoamerica also used tattoos as a way of establishing high rank among the community. As with all discoveries and inventions, the Romans appropriated the concept of tattooing and took it to another level. While the Romans branded all criminals and slaves, Roman soldiers were encouraged to elaborately tattoo themselves with designs of animals in an effort to project a menacing presence in battle. The soldiers used an herb, called *woad,* to produce the blue-colored symbols on their bodies. The custom of tattooing spread in all areas of the Roman Empire from about the mid-first century to the end of the fourth century C.E., when its popularity waned and was eventually banned by the Christian Emperor Constantine (ca. 280–337 C.E.). Although the first Christians tattooed crosses on their arms and faces as a sign of their faith, in time they were viewed as pagan symbols that bespoiled the wearer's image. After the fall of Rome in the fifth century C.E., the practice of tattooing, at least in the Western world, became less popular but did not disappear altogether, as evidenced by tattoos found on the bodies of English kings in the eleventh century C.E. (The Roman Church banned the practice completely in the eighth century C.E.) Commerce, particularly seafaring, with the Far East brought about a resurgence of the practice in the 1700s. For most of the twentieth century, tattoos were applied primarily to the male population engaged in the military or naval services. The last decade, however, has seen this ancient art applied more equitably between the sexes, although males, by far, apply larger and more numerous tattoos to their bodies.

Cosmetics

The oldest evidence of the use of cosmetics or makeup was found in the ancient Sumerian city of Ur. Archaeologists discovered a small gold, shell-shaped makeup case that is believed to be over 4,000 years old. We know from paintings and carvings that the ancient Egyptians (both sexes), Chinese, and Romans (primarily female) used a number of metals and minerals to accentuate their eyes, lips, and fingernails. The substances they selected were sometimes harmful, usually garish in effect, and always applied with the help of many servants. Thus, the use of cosmetics was limited to the wealthy classes. We can only assume that trial and error led to the discovery of chemical mixtures and compounds in appropriate proportions that achieved the desired result of enhancing the physical appearance of the wearer. Some of the most commonly used chemicals and substances in antiquity were the following:

Antimony—The mineral/ore antimony sulfide was the basis for *kohl,* a substance used to blacken the eyebrows and eyelashes as well as to outline the eye to make it appear larger. *Galena,* a lead ore, was also used for this purpose, but primarily in the Mesopotamian countries.

Copper—An oxidized copper ore, called *malachite,* was used as an eye shadow. It is a deep green to nearly black mineral that was and is a popular gemstone in jewelry.

Lazurite—A rare mineral that can be either blue, violet-blue, or greenish-blue in color, it was used as an eye shadow. The azure-blue gemstone, *lapis lazuli,* is actually lazurite.

Iron—An oxide of iron, called *ochre,* was popular as a skin lightener among the olive-skinned Egyptian nobility. Ochre is a powdery mixture of hydrated ferric (iron) oxides, sand, and clay, and can be brown, red, or yellow in color. Yellow ochre lightened the complexion, while an orange-colored ochre darkened the skin. Red ochre mixed with fat was applied as cheek and lip rouge.

Lead—Around 3000 B.C.E. both the men and women of the Indus civilizations used a cream made of white lead to lighten their complexions. The formula apparently spread both eastward and westward to China, Greece, Rome, and the British Isles. Despite its toxicity, which was known by physicians in the first millennium B.C.E., the preparation continued to be used—especially by Roman women.

Henna—Henna (botanical name *Lawsonia inermis*) is a flowering treelike shrub native to Asia and North Africa. When pulverized, the leaves of the henna bush produce a reddish-brown dye, commonly called henna. In antiquity, it was used to darken the fingernails. Later, it was manufactured as a hair colorant, a practice that continues to this day.

A number of other natural substances were utilized by both sexes, primarily in Egypt and in Rome, including bean-meal paste and lemon juice to bleach pigmentation spots (freckles) on the face and pumice to whiten teeth. The ancient Chinese emphasized the appearance and length of fingernails, using an herb to darken them to a deep red color and growing them to long lengths. Long fingernails were indicative of wealth, because it was impossible to work with nails of extreme length and thus servants were needed. The nails themselves were protected with silver shields. The use of cosmetics fell out of favor with the rise of the Christian Church. It is believed that the excesses of the Romans in their use led the early founders of the Church to condemn the application of makeup. Cosmetics ceased to be produced in the Western world for several centuries after the fall of Rome. Their popularity eventually returned in the late Middle Ages after the Christian Crusaders returned with bounty for their wives from the Islamic countries, including cosmetics and perfume. Except for a brief period during the Victorian Era when their use was deemed unseemly, cosmetics have remained popular and their manufacture is lucrative.

Perfume

The derivation of the word "perfume" is from the Latin *per,* meaning "through," and *fumum,* meaning "smoke," but the art of perfumery itself is much older, dating back at least 5,000 years to the ancient civilizations of Egypt, Mesopotamia, and China. The Bible even contains references to perfume materials and formulas. One can only assume that the original purpose for perfume was to cover up body and other unpleasant odors. In many ancient cultures, soap and water, though probably available, were not perceived as hygienic, but rather as medicinal or religious substances. The ancient Egyptians were particularly renowned for their manufacture of perfume using frankincense, myrrh, cinnamon, lemongrass, rose, lily, and cardamom in a base of ben or balanos oil. The choice of oil was important in that the oil needed to have a pleasing but bland aroma, so as not to interfere with the particular formula chosen. In Egypt perfume was applied as a solid, cone-shaped mass of scented fat that was placed atop the head or wig. The warmth of the evening, along with the wearer's own body heat, melted the fat, which dripped on the hair or wig and clothing, intensifying the aromatic qualities of the perfumed cone. The ancient Greeks also manufactured and used perfumes, but preferred oils that were stored in vases or jars and applied in a less unkempt manner. On the other hand, the city-state of Sparta outlawed the use of perfume among its citizens.

(This ban may have contributed to the derivation of the term *spartan* when describing a simple, healthy, no-nonsense existence or lifestyle.) Romans of both sexes used perfumes extensively to conceal body odors even with the availability of public bathhouses.

The Chinese are believed to have invented the "perfume burner" around 100 B.C.E. Using a device called a *gimbal,* the burner was placed in the bedchamber, presumably to mask or cover up odors on covers and pillows that were infrequently—if ever—laundered. (The gimbal is thought to be an invention of Philon of Byzantium [fl. ca. 100 B.C.E.].) It is constructed of two metal rings affixed on axes that are at right angles to each other. The purpose is to assure that the object being held by the gimbal (such as a compass or an inkwell) remains suspended in a horizontal position regardless of any kind of motion that may affect its support. This was particularly important for supporting navigational devices. (Today, the gimbal is used to direct the jets of space rockets.) As for the Chinese perfume burner, the gimbal allowed the burner to be moved in any direction without tipping either the flame or hot oil. Today, the popularity of perfume, both for personal use and for the home environment, appears to be inexhaustible, and though the manufacture of fragrances has been technologically advanced, the basic formulae, particularly in more expensive and exotic perfumes, are very similar to those that were popular in ancient Egypt.

Piercing

Body piercing is not a new and unique phenomenon peculiar to the younger generations of the 1980s, 1990s, and the twenty-first century. In fact, it has been in existence since the beginning of human civilization—in one form or another. In ancient times, body piercing was a sign of tribal affiliation and custom, social rank, virility, and wealth, as well as being an adornment of beauty and, in many cases, a ritual with more generative or amatory significance. There were apparently no constraints on which parts of the body were subjected to this procedure, including noses, lips, ears, and genitals. The cultural and social norms of each civilization determined what was acceptable and, apparently, attractive. The custom was prevalent in both sexes in ancient Mesopotamia, India, Greece, and Rome, as well as in Mesoamerica. Its popularity waned on the European continent after the fall of Rome and the rise of the Roman Church. It remained as popular as ever in the countries of the Indus civilizations and Islam, as well as among indigenous peoples in many areas of the globe, for example, Pacific Islanders and South Americans. The ancient art of body piercing, like the application of tattoos, is currently enjoying a renaissance, particularly among teenagers and young adults of all social and economic classes worldwide.

Contraception

In antiquity, the practice of limiting the number of pregnancies did not engender moral or social indignation. Rather, it was viewed as the sole responsibility of the female, who for the most part encountered little interference either from her husband, in particular, or the community, in general. (Condoms, though in existence since the time of the ancient Egyptians, were not used as contraceptive devices but as "protectors"—ostensibly from disease.) There were several reasons for this attitude. The first reason was economics. Most ancients lived a life of meager subsistence. Large families, while providing the bodies that performed the labors of daily living, also required the necessities of food and clothing ("another mouth to feed"). Second, medical care for a pregnant woman was essentially limited to midwives—not physicians—and it was not uncommon for women to die in childbirth, along with the infants. Third, all ancient civilizations were patriarchal in character, and contraception was viewed as a way of limiting the number of female births. (However, this seems a bit implausible, since the ancients were no more assured of a male birth than those of us living today.) Statistically, there have always been more male births than female, along with a greater mortality rate for male infants. To some extent, infanticide, particularly of females, was practiced in all ancient civilizations as a gruesome method of family planning. Fortunately, with the increase in commerce between regions, coupled with the rise and influence of Christianity, infanticide was less tolerated and eventually became a totally unacceptable form of population control on the European continent. India and China continued the practice well into the twentieth century. In fact, an ABC television network reporter, Mark Litke, presented a story concerning the way modern technology is apparently aiding the ancient practice of infanticide. On August 20, 2002, Litke reported that in China, ultrasound technology is making it possible to determine the sex of unborn babies. Many couples choose to abort when the infant is determined to be a girl. In other words, sex selection, abortion, and female infanticide are being practiced, particularly in rural provinces, where boys are preferred. The current ratio of boys to girls born in China is 120 boys for every 100 girls.

Contraceptive techniques were widely practiced in Mesopotamia, but the Egyptians, Greeks, and Romans, using various plants and minerals, actually created specific formulae designed to prevent conception. As examples, the Egyptians formulated a "stopper" or pessary made of honey and sodium bicarbonate, as well as a paste that was made from sour milk and camel or crocodile dung. The chemical content of both these preparations would have decreased sperm motility. Another Egyptian contraceptive contained the finely ground leaves of the *acacia* tree mixed with honey. The lactic acid in acacia is a natural spermicide. The base for contraceptives used by ancient Greek women was olive oil, mixed with other minerals, including frankincense and poisonous white lead. Roman women also relied on olive oil, which decreases the motility of sperm, mixed with honey, cedar and balsam resin, and white lead. The famous Roman physician, Soranus of Ephesus, who is called the "father of gynecology," created suppositories made from wool that was moistened with resin along with *alum* (aluminum potassium sulfate) or wine, both of which are natural astringents. The Mediterranean country of Cyrene in north Africa (present-day Libya) was famous for its export of silphium, a now extinct plant that, reportedly, grew there exclusively. However, other records indicate that it was also grown in ancient Egypt. Silphium, a natural contraceptive and **abortifacient,** became Cyrene's official symbol as well as its main source of income until the end of the first millennium B.C.E. Oral contraceptives were also available. Dioscorides and Galen, two of the most famous Roman physicians, prescribed a number of plants and herbs that could be administered orally, including pennyroyal, juniper, and wild carrot (known today as Queen Anne's lace). The unpleasant and often dangerous side effects made these potions less popular with the women of antiquity than ointments, creams, and suppositories. For instance, pennyroyal contains the compound *pulegone,* which is a natural abortifacient in both animals and humans.

We can only assume that trial and error, undoubtedly with some unfortunate results, eventually rendered some effective contraceptive formulae and methods. And, just as it is today in most underdeveloped countries, the women of these ancient civilizations assumed the responsibility of family planning, often with methods that effected permanent damage to the health of these women, as in the white lead used by Greek and Roman women, and the oral administration of mercury encouraged by the ancient Chinese. Nevertheless, many of these ancient contraceptives contained chemical agents that were safe. Their obvious drawbacks are the time-consuming and messy preparations.

Modern pharmacology has managed to synthesize these agents into more safe and convenient products that bear fewer side effects. After the fall of Rome and the rise of the Roman Church, which effectively banned contraception, family planning through the use of devices and preparations was practiced clandestinely.

Glass

The production of glass is one of the oldest inventions. Credit for its discovery is sometimes ascribed to either the potters of Mesopotamia, who experimented with sand and indigenous silicate materials when creating the colorfully designed glazes for their clay pots, or to Phoenician sailors who, when cooking a meal on a beach, placed their pots atop stones of *natron* or *saltpeter.* (Natron [$Na_2CO_3 \bullet 10H_2O$] is a complex salt [hydrous sodium carbonate] found in the dry lakebeds of Egypt. Saltpeter [KNO_3] is potassium nitrate.) The heat from the cooking fire reportedly melted the natron or saltpeter stones, as well as the sand below, forming a translucent liquid. Most likely, there is some truth in both accounts. For instance, detailed instructions for the construction of a furnace, as well as for glassmaking, were written on ancient Middle Eastern clay cuneiform tablets that were passed from generation to generation. As for the sailors, they probably realized that the translucent liquid was glass, based on their likely knowledge of *obsidian* (a natural, glassy volcanic rock) and *fulgurite* (a glassy rock formed when lightning strikes desert sand). The actual root of the word "glass' is unknown, but its formula has been around for about 5,300 years: blending sand, limestone, and natural salt (natron, saltpeter, or plant ash) under extremely high temperatures to produce an inorganic translucent or transparent liquid that, when cooled, hardens into a brittle substance impervious to the natural elements. Chemically, glass is considered a solid solution. Over the years, all glass windowpanes become thicker at the bottom and thinner at the top as the "solid solution" is slowly pulled downward by gravity. (This is particularly noticeable in ancient glass windows that still exist.)

Adding other chemicals creates colors of varying hues; for example, iron gives glass a greenish color, while cobalt turns it a deep blue. Ancient Chinese glass contained massive amounts of lead, making the colors exceptionally brilliant. However, glassmaking in China was unknown until about 300 C.E., when it was brought to the Far East by Western explorers. Originally, glass was not used for drinking vessels, but merely as ornamentations, such as beads and statues. The same types of molds were used for both pottery and glassmaking. The Egyptians even carved massive chunks of molded glass, as evidenced by the discovery of an elaborately carved glass headrest, dating back to circa 1350 B.C.E., found in the tomb of the boy-king Tutankhamen.

Beginning in about 1100 B.C.E., glassmaking experienced a 300-year decline, the reasons for which are unknown. Around 750 B.C.E. the Phoenicians, an ancient Mediterranean tribe living in what would be present-day Lebanon, successfully revived the art. They are renowned as the most skilled glassmakers of antiquity. Glassblowing—blowing molten glass with a pipe to form specific shapes—was invented around 100 B.C.E. by the people of the Syro-Palestine region in the Middle East. At first, the glass was blown into a mold. Later, glassblowers, also called gaffers, learned how to achieve spherical shapes without the use of molds. This ancient technique essentially engineered the hollow-shaped form—that is, the drinking glass. Glassblowing has remained basically unchanged for some 2,100 years, although the tools of the trade have been improved with the aid of technology. It was at about this time that the use of glass drinking vessels became popular. This may be a result of the influence of the Romans, who reportedly used more glass than any of the other ancient civilizations. The Romans were the first *glazers* of antiquity, placing bronze-framed glass windows in their buildings to let in the light and keep in the heat. Prior to this, windows were merely iron or wooden grills. In the Far East, translucent oiled paper was a popular material for a windowpane. The fall of Rome brought with it another decline in glassmaking, this one lasting until the eighth century C.E., when Islamic artisans began to manufacture beautifully colored and exotically shaped glass images. The ancient art of glassmaking has never been more popular—or lucrative—than it is today, with seemingly limitless personal, scientific, and technological uses.

Heating, Cooling, and Refrigeration

Although it may be disputed by people living in extreme tropical or desert regions, it is more important to stay warm than it is to stay cool. Cold temperatures, particularly accompanied by rainy or damp conditions, not to mention snow, can result in ill health and death. Extreme heat, on the other hand, may be uncomfortable, but most inhabitants of hot climates have learned how to manage with protective, loose-fitting clothing, shelter, and water. The cold was an ongoing problem in antiquity, particularly in those countries where the climate could be inhospitable—for example, the Far East and the European continent. Even today in tropical countries (for example, India), a cold snap can mean death from exposure for many people unable to find adequate shelter. Central heating is not a modern invention, having been invented in about the second century B.C.E. Natural refrigeration has been in existence since the time of the ancient Egyptians.

Central Heating

This is primarily a Roman invention, but both the ancient Chinese and the ancient Minoans of Crete also experimented with diffuse central heating systems. (The Minoans' central heating experiment was limited to the famous palace of Knossos.) The Romans invented an underfloor heating system that may have been based on an earlier design developed by Caius Sergius Orata, a fish and oyster farmer. He built a tile and furnace system that heated large water tanks in which the fish and oysters were kept. In order for the fish and shellfish to thrive, it was necessary to maintain a constant water temperature in the tanks. By about 150 B.C.E. the Romans had manufactured the *hypocaust* system for central heating. (*Hypocaustum* is the Latin word meaning "to light a fire beneath.") It worked in this manner: Roman buildings had mosaic tile floors that were buttressed by ceramic tile pillars approximately three feet in height. The "basement" below the floor was an air space around which warm air would flow from a furnace connected to this area. Additionally, some Roman buildings were constructed with flue systems within the walls. The warm air flowed upward through the hollow flue tiles, resulting in an even warmer indoor environment. The Romans also engineered a *piping* system consisting of a central furnace connected to metal pipes that were set in the floor and walls and through which hot air from the furnace flowed. Though a less expensive system, it was also less popular because the heat distribution was less uniform and more difficult to control. Roman bathhouses were famous for their elaborate designs as well as for their comfortable indoor and water temperatures, both of which were maintained either through the hypocaust or piping systems. Central heating in China, though invented roughly at the same time as the Roman methods, was more primitive. As with most ancient Chinese inventions, it was not well documented. The ancient Chinese central heating systems of this period (ca. second century B.C.E.) were simply raised sleeping areas under which an oven or furnace was lit. The popularity of central heating waned after the fall of the Roman Empire, when widespread poverty and conflict ushered in the period of the Dark Ages. While most engineers would agree that the hypocaust system is quite efficient, its usage is limited even in the modern world, where duct systems are routed primarily through walls and hot air is either vented or channeled to individual metal radiators. Even today, worldwide, most houses lack central heating systems. In fact, central heating was not extensively used in British homes until after World War II.

Cooling and Refrigeration

Cooling was a much less successful endeavor in the ancient world—both for comfort and the preservation of fresh food. However, during the Han Dynasty of the second century C.E., the Chinese engineered a "mechanical fan" that essentially moved the air in the Emperor's Palace. Since it was powered by servants, its true mechanical worth is questionable. The Chinese were successful in about the eighth century, when they invented a water-powered fan. For most, a breeze through an open window was the only relief from the heat. As for the refrigeration of perishable foods, the ancient Greeks and Romans used ice that was brought down from the mountains into the cities. Available primarily to the wealthy classes, ice was stored in pits or cellars that were insulated with wood and straw. This was fairly effective, since snow could be stored in these ice pits for several months. These were the precursors to the icehouses that have existed for centuries and can be found, in some parts of the world, even today. The ancient Indians and Egyptians used the principle of evaporative cooling to obtain ice, although on a very limited scale. Water placed on shallow trays set out on cool nights would evaporate rapidly—even if the temperature remained above freezing—causing ice to form. By about 500 B.C.E., the Egyptians were successful in making large blocks of ice, presumably for the pharaohs and not for the populace at large. Reportedly, ice cream was even available in Persia around 400 B.C.E., made from ices flavored with wine, and later with honey, fruit, and nuts. Effective refrigeration and air conditioning were not developed until 1834 and 1902 respectively. With the discovery of electricity, both of these systems have been highly engineered and perfected. Air conditioning is available in all developed countries, although perhaps not to the extent that is enjoyed in the United States. The availability of refrigeration is becoming more widespread even in underdeveloped nations, making perishable foodstuffs easier to maintain and safer to eat. Nevertheless, many regions of Africa, Asia, and Latin America still do not have adequate refrigeration systems; thus the shopping for and maintaining of perishable food items is a daily task.

Ink

The derivation of the word "ink" is from *enkauston,* the Greek word meaning "to paint in encaustic." *Encaustic* is an ancient paint formula made of beeswax and pigment that is set with heat after application. Ink was invented in Egypt and China approximately 4,500 years ago. It is not known which of these ancient civilizations was the first to create this substance, but the use of ink in both countries was limited due to the high cost of its production. The ancient Egyptians were a practical race of individuals who documented in great detail many of their projects and pronouncements on scrolls made from papyrus. The Chinese valued scholarship and wrote extensively on silk and paper, a Chinese invention, although they were much less inclined to share their discoveries with those outside their country. Thus, it is not surprising that these two civilizations would create a substance that would essentially preserve their knowledge. The earliest inks were made from lampblack (a fine black sooty material) mixed with gum (a plant exudate or resin), such as acacia, that were formed into sticks or cakes and then dried. Ink was created when mixing the dried stick with water. The Chinese used *inkstones* to mix and apply ink. Inkstones, as the name implies, were black or dark blue in color, and were used for mixing the ink. Reportedly, it took many years of practice before being able to use them adroitly in transcribing. The Chinese scholar or scribe could deftly alter his technique in ways that would vary the shades of ink applied to either silk or paper. Chinese ink was costly to manufacture, and had limited availability, but it was superior to that made in the Western world. It did not fade and, even today, ancient documents retain their legibility. With the increased accessibility of paper, ink became more widely produced. Ink makers experimented with various natural substances—plants, animal material, and minerals—to create more durable and colorful inks. Some of these same materials were also used in fabric dyes. The invention of the printing press necessitated the creation of a different formula and consistency of ink—more viscous.

Lighting

It must have been a dark world without fire, the first source of light (and heat) apart from the sun. Archaeologists believe that over 1.5 million years ago, *Homo erectus* "invented" fire, a discovery that unquestionably altered the course of human civilization. We can assume that the first "lights" were campfires, later to be augmented by torches that were dipped with tallow and affixed to the walls of caves. The ancients experimented with various designs and fuels to produce light in their homes and communities, most notably oil lamps and candles.

Oil Lamps

The earliest oil lamps, dating back to at least 70,000 B.C.E., were either hollowed-out rocks or seashells in which moss or some other permeable

Figure 8.1 Clay Pottery Lamp—circa 800 B.C.E.
Clay pottery lamps, such as the one depicted, used vegetable oil, most probably olive, as a source of fuel.

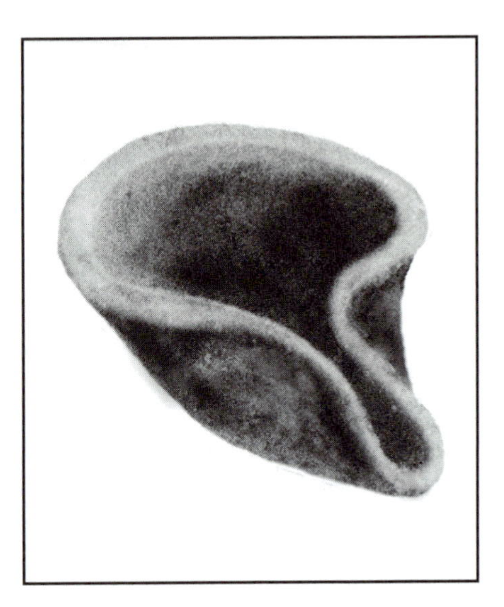

material was soaked with tallow and then ignited. Modifications to this basic design provided for a wick, fashioned from some sort of vegetable fiber that floated in the tallow itself. For thousands of years tallow and whale oil, where accessible, were the only fuels available for both light and heat. After the first deposits of coal were unearthed in Wales about 2000 B.C.E., the ancients used the black lumps in *braziers* to produce both light and heat.

A brazier is merely a metal plate or pan that holds the burning coals. Pottery lamps, such as the one shown in Figure 8.1, did not come into use until about the eighth century B.C.E., when vegetable oils, including olive oil, replaced tallow as a fuel. The specified design of the pottery lamp allowed for the wick to be stationary, presumably providing for a longer and more efficient burn. The word "lamp," meaning "a torch," comes from the Greek word *lampas.* Interestingly, the Greeks did not use lamps until at least the seventh century B.C.E., preferring to rely on braziers and torches. Pottery lamps were often designed with religious and other symbols, but the plain, undecorated Herodian lamp was the most popular style of lamp for common usage for over 300 years. (See Figure 8.2.) The Romans manufactured elaborate terra-cotta and metal

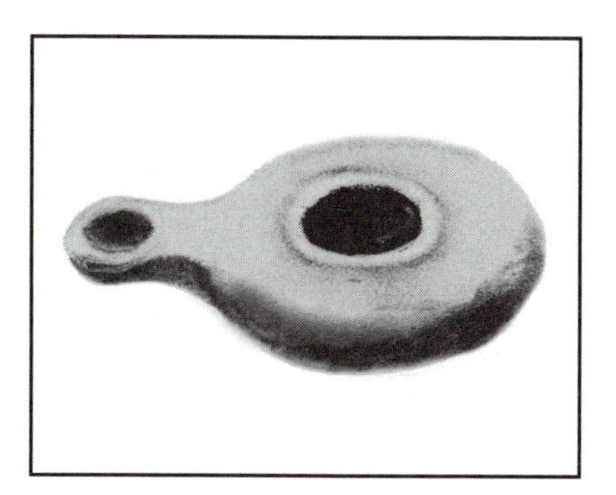

Figure 8.2 Herodian Oil Lamp—circa 100 C.E.
The plain, undecorated *Herodian lamp* was the most popular in common usage for several hundred years.

lamps, some of which were quite large, to accommodate the lighting needs of large public areas.

Candles

Candles were invented at least 5,000 years ago in Egypt and/or Crete, based on archaeological evidence unearthed in both of these regions. Tallow and/or beeswax were molded into long sticks that contained a vegetable fiber wick, usually flax. There are three chemical principles involved in the burning of a candle: *capillarity, vaporization,* and *combustion. Capillarity* is the attraction between molecules that results in the upward flow of the liquid tallow or wax on the wick, resulting in the *vaporization* of the liquid from the heat of the flame that is the *combustion* of the tallow or wax vapor. The essentials of candlemaking remained unchanged, and the profession itself secure, for many hundreds of years. By employing the principle of *saponification,* in 1825 the French chemists Michel-Eugene Chevreul (1786–1889) and Joseph-Louis Gay-Lussac (1778–1850) separated the stearic acid from the fatty substances in tallow, producing longer-lasting, less odiferous candles. Today, the function of candles is more decorative, aesthetic, religious, and often necessary in emergencies.

Locks

The Egyptians are credited with inventing the first lock-and-key system sometime around 2000 B.C.E. In 1887 archaeologists unearthed a

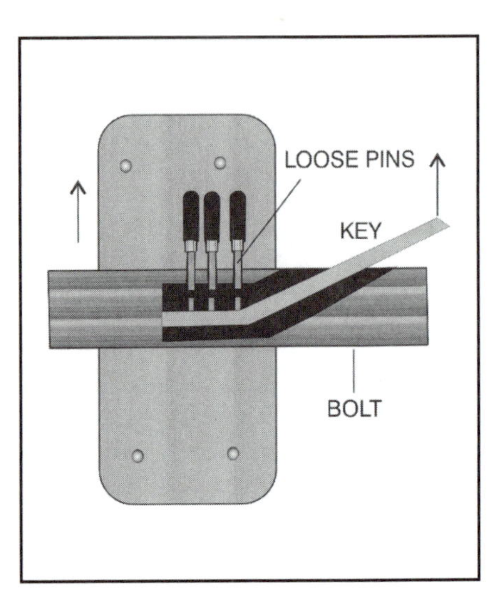

Figure 8.3 Egyptian Lock
Similar tumbler-style locks have been unearthed in China and parts of Africa and the Middle East, raising speculation that this was a simultaneous invention of other civilizations.

strange cylindrical device from the tomb of Kha, chief architect to the pharaoh, and his wife. Since it was located near her remains, they assumed the unknown object was most probably an item with a strictly female usage. Later archaeological discoveries in other regions contained similarly shaped objects that, in 1907, were finally identified as keys. (The only reason that the identification could be made was that a similar-style key/bolt system was discovered to be in active use in Ethiopia.) A wooden lock, fitted with tumblers, that required a pegged key to open it appeared about a thousand years later, around 1000 B.C.E., and its popularity spread rapidly. Some historians ascribe its invention to the Assyrians of Mesopotamia; others believe it was the Egyptians who built the tumbler. Tumbler-style locks have also been found in Chinese archaeological sites dating back to the first millennium B.C.E. Despite this controversy, the lock itself became known as the "Egyptian lock." (See Figure 8.3.) In this primitive-style lock, the latch or bar is secured by wooden vertical pins that fall from a box into their slots. The latch is unsecured by inserting a key with pegs or dowels, then pushing the pins upward, which allows the latch to slide away and the door to open. The Greeks used a more primitive, less secure lock consisting of a sliding bolt. The Romans, as with so many other inventions, manufactured highly secure, often beautifully decorative and ornamented iron tumbler locks with bronze keys. They also introduced *rotary-style* locks, which required a specific fitted key to release the bolt. The Romans, as well as the Chinese, developed *padlocks,* which were part of a chain system of security that remained popular until the Middle Ages. Chains and padlocks were the primary choice of merchants and traders, who used them to secure their shops and cargo. Civil authorities used them to secure prisoners. Although technological improvements have been made to all three styles, the basic engineering principles of these ancient locks continue to be valid in our modern-day world. The Yale lock is very similar to the Egyptian lock; the rotary lock with fitted key remains the most popular lock in homes and apartments; and the padlock safeguards cargo and private and commercial storage.

Mirrors

The stillness of an ancient pond acted as a reflecting pool and the first "mirror" into which our most ancient ancestors gazed. The very human traits of curiosity and vanity may have been the impetus for the manufacturing of mirrors that came much later in human history, about 6000 B.C.E. Archaeologists discovered a number of mirrors made from obsidian in the ruins of Çatal Hüyük in Turkey. Dating back at least 8,000 years, the mirrors were skillfully made and highly polished, considering that this volcanic glassy rock is extremely hard and difficult to work. Obsidian was also used by the Mesoamerican Aztecs in the "eyes" of their stone or clay idols, as well as in the mirrors used for religious rituals. By far, the most popular materials used by all ancient civilizations (Indus, Egyptian, Mesopotamian, Chinese, Greek, and Roman) in the manufacture of mirrors were bronze and copper. Dating back to around 3000 B.C.E., handheld mirrors of this period were often inlaid on the reverse side with gold, silver, and other precious metals, and were deemed a necessity that was carried by each woman, often on the belt of her garment. (Presumably, only women of the upper classes—not the peasants—carried these mirrors.) "Mirror" comes from the Latin word *mirari,* meaning "to wonder at." Aside from the personal, mirrors also supplied a practical solution to problems encountered by the miners of the Bronze Age, allowing them to shed light into darkened shafts.

The Greeks invented the first compact mirror, composed of two highly polished bronze or silver discs that were fastened together by a hinge. Except for the mirror, all other surfaces were engraved with various images and scenes. The Romans, as well as the Celts in Britannia, the territory conquered by the Romans, were famous for the intricate designs engraved into their bronze, copper, and silver mirrors. And the mirror is but another example of Roman ingenuity. At the height of the empire, mirrors were used for security, decoration, and for a variety of narcissistic practices that can be thought of as "typically Roman." The Romans were among the first to manufacture the silvered glass mirror. However, historians generally believe that the first true glass mirrors were made around 400 B.C.E. in Sidon, the ancient Phoenician city located in present-day Lebanon. The Phoenicians were renowned in antiquity as master glassmakers. (Although glass mirrors had been used by the ancient Egyptians a thousand years before, these were merely dark glass panels with little reflective quality.) Silvering is the technique whereby a reflective liquid metal—gold, silver, or copper leaf—is applied to one side of a sheet of glass. The first silvered mirrors were made from unpolished glass, and thus the images projected tended to be uneven and shaky.

The ancient Chinese are famous for their "magic mirrors," dating back to around 500 C.E., which are made of bronze and are engraved on the reverse side with various patterns and Chinese characters. When one of these mirrors is held up to sunlight, the design engraved on the reverse side is "magically" cast onto the wall. The mystery of how light could pass through the solid bronze of the mirror continued to puzzle scientists for centuries. Finally, the British physicist, Sir William Bragg (1862–1942), determined that the magic mirrors, though appearing to be smooth surfaces on the reflective sides, were actually curved (convex). During the manufacture of the mirror, the Chinese craftsmen would "scrape" into the reflective side the same design—in the same position—that was engraved on the reverse side. The reflective side was then polished, and an amalgam of mercury was applied that filled in the minute changes in the surface structure. The mystery could only be solved after the invention of the microscope. Prior to this, the naked eye could not possibly detect the extremely faint design that was actually engraved on the shiny side of the mirror as well as on the reverse. No magic, just an illusion.

Pottery

One of the most ancient inventions, pottery dates back nearly 13,000 years. There are three basic types of pottery: *earthenware* (crude, baked clay), *stoneware* (clay or composite that is "fired"), and *porcelain* (made from white clay [kaolin] and feldspar fired at extremely high temperatures). The oldest pottery finds were discovered in the Far East in Japan and China, dating back to around 10,500 B.C.E. and 8000 B.C.E. respectively. Archaeological excavations in Çatal Hüyük (present-day Turkey) have uncovered crude, unfired pottery figures from about 7000 B.C.E. These earthenware figures were made by simply pressing clay into a mold, then drying or baking it in the sunlight. There have been three significant inventions in the manufacture of pottery: *firing*, the *kiln*, and the *pottery wheel.*

Firing

Firing is the application of fire or heat to the clay (pottery or brick) and/or glaze to achieve a hardening of the product. Trial and error and serendipity played important parts in the history of pottery. The ancient Japanese fired pottery nearly 13,000 years ago, some 4,000 years or so before their counterparts in the Middle East. Only when Mesopotamian brickmakers learned to subject clay bricks to high heat did the durability and porosity of the brick improve markedly. Brickmakers also learned that the application of a glaze made from sand and other naturally occurring minerals, such as quartz mixed with sodium and potassium, resulted in a harder, stronger, and more waterproof brick. Up until this time, both potters and brickmakers encountered the same problems with their wares: breakage, cracking, leaks, and chipping. So in effect, the solution for one became the same for the other.

Earthenware that is dried or baked in the sun is porous and fragile. By some measure of ingenuity or luck, ancient potters realized that the higher the temperature, the harder brick or pottery became. So the very first firing process was merely subjecting the clay to the heat of a wood fire or hearth and then allowing it to cool slowly in order to prevent cracking. Firing temperatures that ranged from 450 degrees to 700 degrees Celsius (850 degrees to 1300 degrees Fahrenheit) produced fairly durable earthenware and bricks, particularly if the items were also glazed for waterproofing. Ancient brickmakers and potters continued to experiment and improve firing methods; thus, the invention of the kiln.

Kiln

A *kiln* can be as primitive as a pit dug into the ground or as sophisticated as a gas or electric oven specifically designed for hardening, burning, or drying materials, such as clay (pottery, bricks) or porcelain, at a controlled temperature. *Pit kilns* have been in existence since the beginning of settled farming, probably for 10,000 years or so. The crude pot-

Figure 8.4 Beehive Kiln
Kilns, in one form or another, have been in existence for at least 10,000 years. A modification of the vertical kiln, the *beehive kiln* provided more uniform circulation of the heat—the result of which was a better quality of pottery.

tery was simply set on top of and around the burning material (wood or charcoal). Once the fire died down, the objects were cleaned and allowed to cool. Pit kilns are still used by potters in some underdeveloped countries. The *vertical kiln,* where the clay pottery surrounded a centralized fire, was in existence around 4000 B.C.E. However, the heat did not circulate uniformly, and the quality of the fired pottery was uneven. Another type of ancient kiln was the *beehive* design. It looked just as the name implies—a beehive-shaped brick oven in which the pottery is stacked within the rounded chamber that is set on top of a firebox. On the center top is a flue/damper device that controls the amount of hot air and gases that escape from the kiln, and consequently the temperature of the fire. Air enters through the open bottom of the firebox. (See Figure 8.4.) The *horizontal kiln* that was developed sometime later distributed heat more efficiently. This design placed the fire atop the kiln on its roof, thus forcing the heat from the fire to be reflected downward onto the clay pots or bricks. It was necessary for the potter or the kiln operator to maintain a controlled temperature in the kiln of around 1,000 degrees Celsius or 1,850 degrees Fahrenheit, otherwise the articles would vitrify (turn into glass) when the temperatures reached 1,600 degrees Celsius or 2,900 degrees Fahrenheit. (The exception to this is porcelain, a combination of kaolin [white clay] and feldspar that undergoes a process of vitrification at temperatures below 1,600 degrees Celsius.) Stoneware is pottery that has been kiln-fired. Because of its hardness and durability, stoneware was popular for household or domestic use. Large amounts of wood and charcoal were required to fuel the kiln, along with the acquired knowledge and skill to control the fire at the desired temperatures. For example, some ancient kilns acted as *reducers*—that is, they removed oxygen from the interior of the kiln. Other kilns worked on the principle of *oxidation,* which added oxygen to the kiln over a smokeless fire. This specific type of kiln and the skill of the potter produced pottery that ranged in color from buff to red to black.

Pottery Wheel

For thousands of years, clay was worked by hand and pressed into molds or worked freehand into the desired shape. Then it was either baked in the sun or fired. Historians generally believe the wheel was invented in Mesopotamia around 3500 B.C.E.—not for transportation, but for two other purposes: first, as part of a windlass to raise water from wells, and second, as a *pottery wheel.* Two people were required for the pottery wheel's operation; one worked the clay while the other maintained the spinning of the stone. (See Figure 8.5.) Later pottery wheels were fashioned from a stone horizontal disk that contained an axle and a flywheel. The lump of clay was "thrown" onto a smaller disk that was positioned atop the same axle as the flywheel. The first pottery wheels used sticks to control the "spin." Later, foot treadles controlled the momentum of the wheel, thus eliminating the necessity for a two-person operation, as the potter could control the spin while working the clay. Though technology has improved upon the basic design, the physical principles of the pottery wheel, which was invented 4,000 or more years ago, are as valid as ever.

Porcelain

As the art of pottery making progressed, each country produced its own particular style of pottery, which was reflective of its skill, its culture, and the type of raw minerals that were indigenous to that region. For example, *porcelain* is a Chinese invention dating back to about the third century C.E. The process involves combining feldspar and kaolin

Figure 8.5 Potter's Wheel—circa 1900 B.C.E.
It is generally accepted that the *potter's wheel* was the forerunner of the "wheel," which has been so essential in the fields of engineering and transportation. The potter's wheel pictured above required two people to operate—one to "work" the clay and a second to "spin" the wheel.

and then firing them at high temperatures. Feldspar, which is potassium aluminosilicate ($KAlSi_3O_8$), is an abundant mineral found in many types of rocks (igneous, plutonic, and some metamorphic). Additionally, the ancient Chinese discovered significant deposits of "decayed feldspar" in one of their highly elevated provinces. They subsequently named this feldspathic material *kaolin,* after the region of Jiangxi province where it was first discovered. (The Mandarin Chinese word for kaolin is *gao ling,* which means "high place." Westerners referred to kaolin as "China clay.") Chinese porcelain makers also incorporated another feldspathic white clay into the manufacture of their wares. This was called *petuntze,* or China stone, and produced a translucent, somewhat softer porcelain compared to the porcelain consisting of kaolin paste, which was hard, smooth, and extremely glassy. Chinese porcelain was intricately painted, decorated, and glazed. Each dynastic period is represented by particular, often symbolic, designs that to this day maintain their value and popularity. Today, fine porcelain is referred to as "china," although it need not bear the intricate oriental patterns favored by its discoverers. It simply refers to a particular formula and technique.

Glazing

The discovery of various glazing techniques most likely occurred simultaneously and independently in Mesopotamia, Egypt, and the Far East. *Glazing,* which is defined as an opaque, colored, or transparent substance that is applied to the surface of an object, has a twofold purpose. First, it waterproofs the piece; second, it adds an often distinctive color. Ancient potters used a variety of glazes. Some examples are the following:

1. Mesopotamian brickmakers used a tin glaze on their bricks and brick panels.
2. A combination of copper and lead (used as a **flux**) produced a blue glaze, also popular in the Middle East.
3. Cobalt, manganese, and copper oxide were used in various proportions to produce the turquoise and green artifacts that are distinctively Egyptian.
4. Chinese potters used two different types of glazes. One contained feldspar. Another mixed quartz and/or sand with lead (used as a flux) and wood ash. Adding the wood ash produced a dull brown to gray-green color to the glaze.

In addition to the different types of clay material used by ancient potters, the individual designs, shapes, paints, enamels, and glazing were all indicative of certain cultures and civilizations. For instance, the highly stylized, beautifully colored Egyptian pottery was distinctive for its animal and daily-life motifs painted in white slip on red. (*Slip* is a thin mixture of clay, water, and a variety of colored minerals.) Grecian pottery is very symmetrical and often depicts scenes of nature, along with those of daily human activities. The most popular colors of ancient Greek pottery are red with black figures, or the reverse, black with red figures. The pottery of the ancient Minoans is decorated with various mythological representations, which were an important aspect of their lives. Because of pottery's durability, the number of ancient pottery artifacts that have been found probably exceeds that of any other type. It also provides the modern world with a true glimpse into the culture and lives of now-extinct civilizations.

Sanitation

For a period of about 5,000 years, from the time of the ancient Mesopotamian and Indus civilizations until the fall of the Roman Empire, the hygienic practices—both personal and civil—of humanity were remarkably advanced, given the limitations of technology and knowledge about disease pathology. We can only assume that a combination of resourcefulness and practicality, as well as aesthetic sensibilities and discrimination, led to the invention of toilets, showers, bathhouses, sewers, and aqueducts thousands of years ago. This section deals with the more mundane, but no less important, aspects of sanitation in the lives of ancient people until the mid-fifth century C.E., when religious and territorial conflicts and an increase in the population led to widespread poverty, disease, and a breakdown in the infrastructures of these sanitary systems.

Middens

Middens are prehistoric refuse piles that contain artifacts and lifestyle remnants of humanity from the late Mesolithic Period, or Middle Stone Age (8000–4000 B.C.E.). The word "midden" is Scandinavian in origin and means "dung heap." (Not surprisingly, the first middens were examined and studied in Denmark in 1848.) Much like today, the ancients accumulated waste in two types of middens: the kitchen midden and the village midden. As the names imply, *kitchen middens* contained animal and fish bones, shells, broken pottery, and other artifacts that individuals either consumed or used and then discarded. The *village middens* were ancient landfills that contained broken and discarded tools and implements, some animal and food remains, and other kinds of refuse that accumulate within a large, settled community. The study of middens has enabled archaeologists and historians to document the numbers of inhabitants within a community or village, their lifestyle, diet, length of habitation, and climate. Middens are in themselves a dichotomy. They are indicative of an awareness that garbage should be separate from living areas. On the other hand, it was not uncommon for early humans to move to another settlement once these open areas became befouled with the waste of daily living from themselves as well as the animals destined for domestication. Buried trash pits arrived much later in the course of civilized societies. However, the Dark Ages after the fall of Rome ushered in a period of ignorance and disregard for such measures. Simply throwing refuse out the window or door undoubtedly encouraged vermin to thrive and disease to spread.

Pest Control

Most likely, our ancient ancestors treated insects, rodents, and other scavengers not as pests but as an accepted part of life. They were unaware that these creatures carried disease or that they could be eliminated or even controlled in their environment. Some of them had natural enemies; for example, cats and some domesticated dogs were fervent rat and mice catchers. Improved construction of houses also helped to contain pests. A tile roof was preferable to the thatched or mud roofs that, in addition to leaking, were infested with insects and their wastes. And it is a fair assumption that most ancient humans probably were infected, at one time or other, with fleas, ticks, lice, and other parasites. Ancient herbalists may have discovered that certain natural plants will repel or kill these pests, but not in quantities sufficient to render them harmless. *Camphor* is an example of a natural pest (wool moths) repellent. It is obtained from the aromatic wood of the east Asian *camphora* tree. The insects and other animals that attacked crops were particularly bothersome, because they could limit, or sometimes completely destroy, the harvest of a hardworking farmer. Evidence exists that some of these ancient civilizations experimented with both chemical and biological processes in an attempt to eradicate the pests and minimize the damage. For example, over 4,000 years ago the Sumerians used sulfur compounds for insect control. The Romans, in the last centuries B.C.E., built granaries that supposedly were "ratproof" and experimented with various chemical oil sprays to control insects. White arsenic was applied to the roots of rice plants in China in an effort to ward off insects that fed on the plants. Since the oxide of arsenic is water-soluble, the poisonous arsenic was not deposited in the rice kernels. Thus, it was a safe and effective pesticide. (Because they are relatively inexpensive to produce, arsenic compounds continue to be used as pesticides/insecticides, particularly in undeveloped countries.) It was the Chinese, however, who first advocated a form of biological pest control in the third century C.E. They discovered that a particular species of large carnivorous ant, *Oecophylla smaragdina,* also called the citrus ant, would devour the black ants, beetles, and wormy caterpillars that preyed on the fruit of the mandarin orange tree, which grew in a number of the Chinese provinces. And best of all, the citrus ants would not eat the oranges themselves—only other insects. The Chinese actually built bamboo bridges between the mandarin orange trees so that the citrus ants would move quickly between them. They also experimented with other crops and have successfully utilized this method of agricultural/biological pest control for 1,700 years. The method is quite controversial, particularly in the West (where it was first introduced in the last decade of the nineteenth century) because of the potential for unanticipated consequences. In other words, introducing a natural predator into an environment where it can breed and populate unchecked may result in the natural predator itself becoming a pest.

Soap

Soap is the product of a chemical reaction (saponification) between the metallic salt found in a variety of fatty acids and an alkali. (In antiquity, fats were obtained from animal tallow and the alkali from wood ashes or soda.) When combined with water, which acts as an absorbent, soap disperses dirt from the surface into the water, where it is suspended and prevented from being redeposited onto the cleaned surface. In other words, it is a **surfactant.** A soap molecule is unique. The head of the molecule is *hydrophilic,* meaning that it attracts water; the tail of the molecule is *hydrophobic,* meaning that it repels water but is attracted to oil and grease. Prior to its invention, ancient humans scrubbed their bodies with leaves from plants that grew along riverbanks and lakes. These included *soapwort* (which produces a soaplike substance when the leaves are rubbed), soapbark (from the bark of the *saponaria tree,* the source of *saponin,* an ingredient in detergents), *horsetail,* and *yucca.* Soap has an interesting history going back to about 2800 B.C.E. in Mesopotamia. The Babylonians made soap by boiling animal fat and wood ashes in clay cylinders. Soap was not really used as a personal cleansing solution until about 200 C.E. Rather, it was considered an antiseptic for cleansing wounds, a hair preparation, or a product for laundering clothing. Other body cleansing substances, such as soda (sodium carbonate), pumice, sand, wood ashes, and olive oil, were preferred by the ancient civilizations of Mesopotamia, Egypt, Rome, Greece, and China. The Romans and Greeks used an instrument called a *strigil* that scraped off oil used to anoint their bodies. An unintended outcome was that along with the oil, dirt also was scraped off. Legend has it that Cleopatra never used soap—only a mixture of mare's milk, honey, oil, and fine white sand—to cleanse herself. The ancient Chinese did not manufacture soap, preferring to mix saponin, flour, minerals, and perfumes and then shape the mixture into balls for personal bathing and laundry. Saponin is an organic glycoside with detergent properties that produces a soaplike foamy substance. Soap derives its name from Roman legend. Supposedly, women laundering clothes in the Tiber River realized that a "soapy" residue that was found on its clay riverbanks removed dirt and stains more easily from the soiled garments. This residue was actually a mixture of melted animal fat and wood ashes. When it rained, the residue was washed down from Mount Sapo, the sight of animal sacrifices. (Though the terms are often used interchangeably, soap and detergent are different. For thousands of years, but not necessarily in modern times, soap was made from organic products. Detergents, introduced in 1916, are manufactured from synthetic substances.) Personal hygiene declined in the West after the fall of Rome in the fifth century C.E. Many historians believe the surge in disease, especially the plagues of the Dark Ages, was a result of this lack of cleanliness. Today, soaps and detergents are both plentiful and affordable, and the importance of personal and community cleanliness is unquestioned.

Weaving

For thousands of years, our most ancient ancestors wore animal hides as clothing. Furs and tanned leather provided warmth and protection in the colder regions, but lighter-weight garments were obviously a necessity in the desert regions of the Middle East and Egypt, and the warmer climate in parts of the Far East. Historians cannot date with any certainty when woven textiles were first developed, but most agree that the ancient art of basket weaving was the most likely forerunner of textile weaving. Archaeologists have radiocarbon-dated baskets found in Egypt to 10,000 B.C.E., and the techniques in basket weaving are similar to the weaving of cloth. Ancient baskets were woven from unspun plant fibers, such as flax, reeds, grasses, and cereal straws, which were abundant in Mesopotamia and Egypt; from bamboo in China; and from willow on the European continent. Evidence of woven cloth can only be dated to 6500 B.C.E., based on linen artifacts found in the Judean desert. However, it is believed, but thus far not proven, that textile weaving may actually have been invented several thousand years earlier, based on the quality of the cloth found in Judea. The texture and patterns of the various fabrics found in antiquity differed from country to country and were characterized not only by the cultural differences of the inhabitants but by the indigenous flora of the region.

Fibers, Textiles, and Cloth

With the exception of wool and silk, all textiles and cloth were woven from vegetable fibers and other wild plants. (*Textile* is defined as either the woven cloth [fabric], or as the particular fiber or yarn used in the weaving of a fabric. *Cloth* or *fabric* is the material formed by the process of weaving.) The coarse fibers used by weavers were extracted through a process called *retting*, which involves softening the fibers through the application of moisture, then beating the fibers until they are almost rotted. The most popular fibers used in antiquity were the following:

Flax—Generally believed to be the first plant to be grown and cultivated for use in the manufacture of cloth, flax (*linum*) grew abundantly in both Mesopotamia and the Nile Valley in Egypt. Linen is made from flax, and linseed oil is obtained from its berries.

Hemp—Reportedly, hemp was first grown in China and used primarily in the manufacture of ropes. Hemp is also the source of the narcotic hashish.

Jute—Grown primarily on the Indian subcontinent, jute was used for ropes and sacks, primarily burlap.

Cotton—Cotton was first grown in India 5,000 years ago, and then later in Egypt and China. Cotton is not retted. Rather, it is untangled from its naturally formed skein.

Silk—Silk originated in China with the famous silkworms that inhabited the Chinese mulberry trees. Although the Chinese manufactured silk for themselves at least 1000 B.C.E. (some historians believe it was earlier than that, around 2700 B.C.E.), it was not exported to the West until about 200 B.C.E. Even then, the Chinese did not reveal the origin of this expensive fabric, preferring to allow the West to speculate about how and where it "grew." The silk filaments do not need to be either retted or untangled. As formed, they are ready for the twisting process in preparation for weaving.

Wool—Wool generally comes from sheep, but goats and llamas are also a source. The wool is obtained by shearing, although a small portion is dropped during the natural shedding process. Wool fibers are also a tangled mass and must be separated before weaving.

The formation of fibers (threads) that are then used by weavers is accomplished by *spinning*. As the term implies, spinning involves drawing out and twisting the retted or untangled fibers into a thread. At first, spinning was done by hand. The invention of the spindle came much later at a date that is really unknown. The first spindles were merely short, straight sticks. They evolved, over time, into notched drop spindles with whorls (weights). The drop spindle, sometimes called the hand spindle, was used for thousands of years in the making of thread.

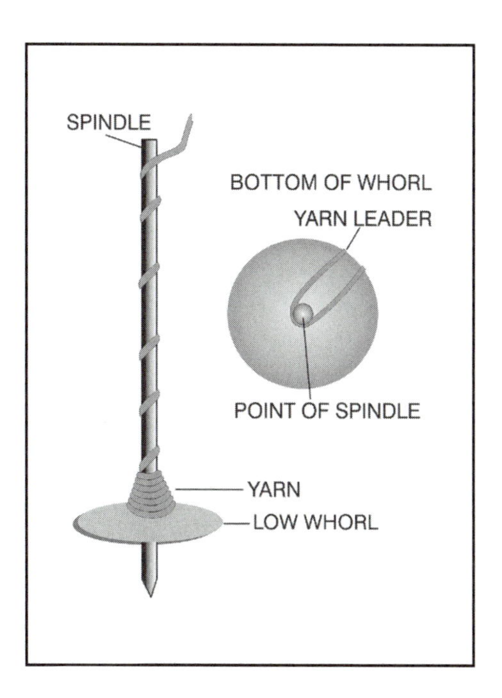

Figure 8.6 Low Whorl Drop Spindle
The *low whorl drop spindle*, with its symmetrical notches and whorl, evolved from the short, straight sticks used by ancient weavers thousands of years ago.

(See Figure 8.6.) The spinning wheel, most likely invented in India, was a much later invention.

Dyes

From antiquity until the nineteenth century, all dyes were a mixture of a variety of organic substances that produced subtle to intense colors on fabrics and other materials. (The first synthetically produced dye, mauve, was not discovered until 1856.) In addition to the application of dye, the fabric was prepared with a **mordant** that acted as a fixative to render the dye "fast." The ancients used aluminum sulfate for this purpose. Deposits of aluminum salts were plentiful in the Middle East and Italy and on the island of Sicily. Historians cannot date with any certainty when dyeing fabric was invented, but it is generally believed that it dates back to at least 3000 B.C.E., as evidenced by Egyptian paintings depicting green, red, and yellow clothing. Among the most interesting and popular plants and animals used by ancient dye makers and weavers were the following:

Madder—A plant whose fleshy root produces an orange-red juice called *alizarin.*

Indigo—A deep blue dye obtained from the *indigofera* plant.

Safflower—A plant, *Carthamus tinctorius,* whose orange flowers produce a yellow dye.

Cochineal insect (also called *kermes berries*)—A brilliant scarlet-red dye was produced after drying and pulverizing the bodies of the female of the insect species *Coccus cacti.*

Cuttlefish—This is a squidlike mollusk (genus *Sepia*) that secretes a dark, inky fluid from which the sepia pigment is manufactured. Sepia can range from grayish brown with yellow tones to a medium brown with olive tones.

Murex—A marine gastropod (snail) found in the Mediterranean Sea that exudes a yellow secretion, which becomes purple when exposed to sunlight. This purple dye was known as *Tyrian purple,* named after the Phoenician town of Tyre. It was commonly called "royal purple" because only the wealthiest individuals (i.e., royalty) could afford garments that were dyed with this exceptional color.

The craft of the dyer was arduous and unpleasant. The organics used in dyes often emitted strong and unpleasant odors, and the mixing process involved a form of fermentation that used alkalis and fruits—even urine—which further added to the pungency of the batch. Today, most dyes are produced synthetically from either petrochemicals or coal tar.

Loom

The prototype for the modern loom was invented about 4400 B.C.E. in the Middle East. These are the looms that contain the basics: (a) some type of *beam* or *frame* around which the warp threads are wound; (b) the *warp,* which is the vertical threads; (c) the *weft,* which is the horizontal threads; and (d) the *shuttle,* which is a beam of wood that carries the weft back and forth between the warp. After the weft was passed through the warp threads, the weaver used a comb, called a *sley,* to press the threads firmly against each other. (See Figure 8.7.)

Prior to the vertical loom's invention, ancient people pegged out a framework on the ground and worked the warp and weft alternately. At some point, ancient weavers realized the ground loom could be placed upright against a wall or a roof beam, making it easier to work the threads. Over time, improvements were made to the fundamental mechanics of the loom. For instance, by about 3000 B.C.E. *weights* made

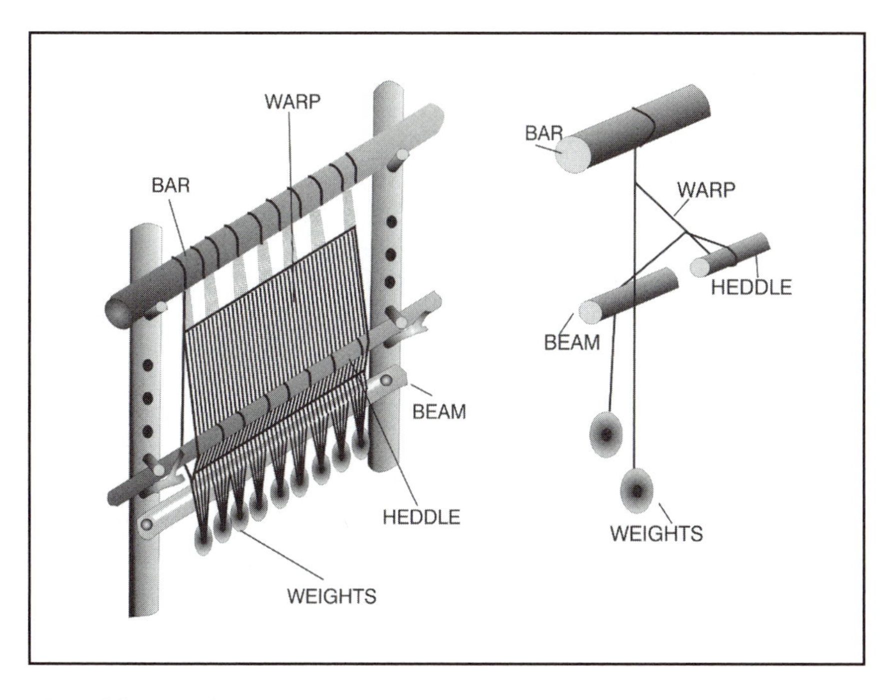

Figure 8.7 Vertical Loom
The *vertical loom* is the prototype of the modern loom. It was invented over 6,500 years ago and contains the basics—the beam or frame, the warp, the weft, and the shuttle—necessary for weaving the first ancient cloth, linen.

from clay or ceramicware were added to the warp, allowing the threads to remain taut and easier to control. Another important addition was the *heddle,* the horizontal, moveable wooden rod that separated and raised the warp threads and provided a pathway for the shuttle. In later designs, cords or wires replaced the wooden heddle, and several could be used simultaneously on the loom. *Treadles,* or foot pedals, were also incorporated into the loom, thus increasing the speed of the operation. (Based on silk damask artifacts that date back to about 2500 B.C.E., historians believed the Chinese may have used a loom with a foot treadle at this time to weave elaborately patterned material.)

The intricate patterns found in antiquity were dependent upon the skill of the weaver and the quality of the textile yarns available, as well as the number of additional features, such as the number of heddles, or a duplicate warp, on a particular loom. For example, a *draw loom,* reportedly used in China, advanced the intricacy of patterns woven into Chinese silk by enabling the warp threads to be raised in groups, which was necessitated by the often complicated patterns that are so characteristic of fine Chinese silk. Prior to the manufacture of the draw loom, young boys, aptly called *drawboys,* manually raised the warp threads. Although modern industrial looms with flying shuttles are power-driven, and possibly computerized in some instances, they all retain the fundamental principles of the first looms that were invented nearly 7,000 years ago. The late Middle Ages and Renaissance in Europe saw a resurgence in the art of weaving with the production of elaborate tapestries. Weaving fabric on a vertical loom remains popular among some primitive cultures, as well as with many individuals in the developed world who are interested in preserving this ancient craft.

9

THE PHYSICAL SCIENCES

Background and History

The encyclopedic definition of physical science divides the examination of the inorganic (nonliving) into the four areas of astronomy, chemistry, physics, and Earth science—a logical division, given the depth and breadth of our current knowledge. On the other hand, for most of our history, beginning some 40,000 years ago with the evolution of *Homo sapiens sapiens,* humans were oblivious to any deeper context concerning their environment other than the immediacy of survival in a hostile and dynamic world. However, their ignorance of the physical processes and forces that have always governed our universe did not mean they were unaware of their existence or that they did not avail themselves of the phenomena that unfolded around them. Quite the contrary, for two of the attributes that distinguish us from other species are our extended memory and a consciousness of ourselves and our environment. Ancient humans remembered the physical event that brought about some change and, over time, came to not only depend upon it but to utilize it and manipulate it for their own advantage. An example is the natural progression of night and day. The setting of the Sun signaled the time for rest, while sunrise heralded the activities of hunting and gathering food. Our most ancient and primitive ancestors were as involved with the physical sciences as is modern humanity. The major difference is that they lived at the genesis of discovery and experimentation, while we are somewhere in the middle.

Next to spoken language, the understanding of the physical sciences—the interrelationships between matter and energy (forces) and the resultant processes and effects—is probably the most important intellectual achievement of humanity. All technological discoveries, inventions, advancements, and improvements (in other words, the total accumulation of our knowledge of our universe) have only been possible because of the positive exploitation of the Earth's matter and energy. Accident and serendipity, however, cannot be discounted in the process. The understanding and use of fire undoubtedly was humanity's first important breakthrough in the physical sciences. But fire was not an invention or even a discovery in the conventional sense, as its existence and presence were apparent. Warmth, the cooking of foods, the art of metallurgy, and pharmacological and chemical processes are among the seemingly limitless discoveries that resulted from the use of fire as a controlled source of energy.

The tangible or practical application of knowledge relative to inorganic matter was a small part in humankind's understanding of the universe. The impetus for all invention and innovation is really dissatisfaction. Evolution produced a species that not only remembered an event or outcome but also questioned why it occurred and, more importantly, how it could be altered or utilized to maximum benefit. For example, the first farmers realized the predictable natural events of the vernal and autumnal equinoxes were important to the planting and harvesting of their crops. It would be many millennia before equinoxes would be named as such and their underlying physics understood. But their reliability was never discounted.

An outgrowth of the questioning of specific events and behaviors was the development of philosophies, religions, and, subsequently, the methods of scientific research. If we assume that settled farming, which took place between 10,000 and 12,000 years ago, was the beginning of human civilization, then this philosophical questioning of the known world evolved over thousands of years and culminated in the dissertations and pronouncements of mathematicians, astrologers/astronomers, and physician-scholars of ancient Greece. While the ancient civilizations of Mesopotamia, Egypt, China, and India availed themselves of the practical elements of the physical sciences in the form of cities, civil projects, and monuments (such as ziggurats and pyramids), they ascribed mystical and religious explanations to many events that defied their logic or knowledge. On the other hand, the ancient Greeks of the first millennium B.C.E. openly questioned the natural progression of astrological events, the basis for chemical processes and reactions, the concepts of motion and energy, and the very essence of matter, organic as well as inorganic. They sought answers from the natural world, believing that heretofore unexplained phenomena could be interpreted using observations, measurements, and principles based on predictable forces and processes. Thus, in a very real sense, the ancient philosophers of Greece were "physical scientists" who laid the foundation for succeeding generations of astronomers, chemists, physicists, and geologists. This chapter will focus on the discoveries, philosophical theories, advancements, and experiments of the ancients, who grappled with the complexities of the inorganic world around them.

Fire and Its "Discovery" as a Tool

The French structural anthropologist, Claude Lévi-Strauss, wrote, "the domestication of fire makes man 'transgress' into culture; it severs him from nature and impels him towards the solitude of history." This is an apt description of the history of our relationship with fire, since, in itself, it was neither an invention nor a discovery. Rather, humans learned to use it as a tool. Fire is defined as the rapid burning of combustible material with the evolvement of heat and, usually, a flame. Presumably, fire was present during the time of the "big bang," but the human relationship with it began in the form of lightning and volcanic activity. History cannot know with any certainty when humans first realized that the terror of fire could be controlled and utilized in positive ways. Archaeological evidence, in the form of charred animal bones found in ancient hearths from nearly 1.5 million years ago, suggests humans cooked animals for food. Thus, it is believed that these primitive humans somehow were able to maintain fire, probably by keeping it alight at all times, by this time in history. However, humans discovered how to "make" fire only about 9,000 years ago, when our Neolithic ancestors used flints or dried wood to create friction, sparks, and, ultimately, fire. The obvious benefits of fire were warmth and the cooking of food, which contributed to the longevity of life. Cooking not only improved the taste of food but also aided in the digestion of animal protein, and killed harmful bacteria and organisms that previously had caused numerous zoonotic illnesses and death. Of course, the last was an unknown benefit to early humans as they had no concept of the pathology of disease. Recent research suggests that the practice of cooking food may have begun about 1.8 million years ago and contributed to the evolvement of the species. Cooking plant food, such as tubers, resulted in smaller teeth and bigger brains and provided early humans with more calories and thus more energy for hunting. Fire as a tool was absolutely necessary in the progression of human civilization. Nearly every implement that spurred technology—from bricks to pottery to glass to weapons to medicine—was developed by the application of fire, flame, and heat.

The Three-Age System

The human experience in the physical sciences is often delineated in terms of the three-age system, also called the Metal Ages, which chronicles the technological evolution of our ancient ancestors. These are the *Stone Age*, the *Bronze Age*, and the *Iron Age*. Within these periods are subperiods and categories. Many archaeologists, anthropologists, and historians have genuine disagreements as to specific dates for these periods based on the interpretations of various findings and discoveries. Another reason for the variances is that the three-age system occurred on all continents and among disparate cultures and civilizations; thus the discovery of metals and their subsequent use in the manufacture of tools and weapons was not a simultaneous event. Also, there is evidence that some regional civilizations did not always advance in sequence from one age to another. For example, the widely varied cultures of *both* India and the European Finlanders skipped the Bronze Age entirely, going directly from the Stone to the Iron Age. There is, however, general agreement as to the technology developed during each of these ages. There were only seven metals with which the ancients had a working knowledge. They were copper, tin, gold, silver, iron, lead, and mercury. Some other metals, such as manganese, antimony, and nickel, were present in mined deposits or meteorites, but were not identified as metals by the ancients.

The Stone Age

The Stone Age, defined by the fact that, obviously, stone tools and weapons were the basic technology of the period, is itself divided into three periods, based on the developmental level of those implements—in other words, whether they were crude stones, polished flints, or developed axes. Thus, there is no sharp delineation between these periods.

The *Old Stone Age*, also called the Paleolithic Period, is characterized by the use of rudimentary stone tools. The dates are generally believed to be between 2.5 million years and approximately 10,000 years B.C.E. Within this period are the Lower and Upper Paleolithic Periods.

The *Middle Stone Age*, also called the Mesolithic Period, existed between the Paleolithic and Neolithic periods, from approximately 10,000 years ago to about 6000 B.C.E.

The *New Stone Age*, also called the Neolithic Period, is generally believed to have begun about the time of settled agriculture and animal domestication, around 9000 B.C.E., lasting until about 3000 B.C.E.

The Bronze Age (also the Copper Age)

Alluvial copper is generally believed to be the first metal ore that was fashioned by Neolithic humans into tools, weapons, and ornaments approximately 10,000 years ago. Nuggets of gold, too soft to use for tools, were also fashioned into jewelry and statues. Thus, the early part of the Bronze Age is referred to as the Copper or Chalcolithic (Copper/Stone) Age, beginning by at least 8000 B.C.E. (The term "Chalcolithic" is a combination of the terms "chalcocite," a copper ore, and "lithic," referring to the use of stone.) The beginning of the Copper Age was a transition period during which stone tools gradually gave way to those made of this newly discovered metal that was also used for personal adornments. Obsidian, an extremely hard, lustrous volcanic rock, was the popular choice for tools and weapons during this period. Though more difficult to shape than copper or bronze, it was more available to the general masses. Another metal ore, tin, was discovered around 3600 B.C.E. in the Middle East where it was added to melted copper to make the hard metal alloy, bronze. Bronze was a very important metal in ancient times because of its hardness and ability to maintain a cutting edge, thus making it suitable for weapons and tools. Bronze was widely produced in the Middle East, China, and the Aegean regions of the Mediterranean. However, it did not become popular in the British Isles until approximately 1900 B.C.E. Brass, a combination of copper and zinc that produces a goldlike metal, was first developed about 900 B.C.E.

The Iron Age

Historians generally accept that iron ore was first extracted sometime between 1400 and 200 B.C.E. and then smelted and forged to manufacture extremely hard and durable tools and weapons. Prior to the development of smelting, prehistoric humans used the iron that they found in meteorites to shape both tools and weapons. The technology spread rapidly in the Middle Eastern and European nations, but did not reach the Far East until about 600 B.C.E. The Iron Age production of weapons and other implements created opportunities and benefits that resulted in the building of permanent towns and cities. More efficient tools made agriculture and industry more profitable. And while sophisticated metal weaponry was not in itself responsible for aggressive behavior, it did make war more efficient for armies: fortified with highly developed and nearly indestructible weapons, they could advance against enemy settlements. Thus, along with tremendous technological advancements, including the development of steel, the Iron Age effectuated conquest and war.

Metallurgy

In a very real sense, the ancient art of metallurgy and the science of alchemy (chemistry) are intertwined inasmuch as the practical applications of one would not be possible without the experimentation of the other. Metallurgy is defined as the chemical technology that allows for the extraction of metals from their ores, the purification of those metals, and finally the creation of useful items. After the discovery of copper sometime during the Chalcolithic Period of 8000 B.C.E., ancient humans managed to produce copper tools through the technology of cold metalworking. Copper is an extremely **ductile,** naturally occurring metal that is relatively easy to work even without the application of heat. This was verified with the discovery in the Italian Alps in 1991 of the 5,300-year-old mummified remains of the "Iceman," who was carrying a cold-forged copper axe at the time of his death.

Smelting, Casting, and Forging

The development of the pyrotechnology of the kiln, either pit or vertical, between 8000 and 4000 B.C.E., serendipitously advanced the art of metallurgy. Historians credit the Sumerians with eventually realizing that the same kiln that produced a more durable brick or piece of pottery also created a change (i.e., a chemical reaction) in the metal-bearing rocks and ores subjected to the high temperatures of the kiln, thus allowing for the extraction of pure metals. This is the process known as *smelting,* in which the ore or deposit is heated beyond the melting point in the presence of an oxidizing agent, such as air. Kilns, and later forced-air draft furnaces, that produced the high temperatures necessary for metal extraction were built. Charcoal was generally used from about 5000 B.C.E. until sometime during the eighteenth century, when the English introduced coke in their blast furnaces. However, ancient craftsmen experimented with other fuels and ores, both of which may have been more locally available, often with less than satisfactory results. Smelting requires temperatures in excess of 982 degrees Celsius (1,800 degrees Fahrenheit). Following the smelting process, the extracted metal was once again heated to a molten state, at which time it was *cast* (poured) into various molds. Copper is believed to be the first metal to undergo the smelting process. However, its production was slow, uneven, and limited due to the fact that much of the population had neither access to the raw material nor a kiln in which to smelt it. *Forging* refers to the process of hammering and shaping the metal object over a furnace, called a forge. The term *wrought* applies to those objects that have been beaten or hammered with tools—for example, wrought iron.

Bronze, Gold, Silver, and Brass

Another example of the intertwining relationship between metallurgy and chemistry is the discovery of *bronze,* which is an **alloy** of copper and tin. History can never know with any degree of certainty whether bronze was discovered by a resourceful metalworker or by an ancient alchemist who believed the correct combination of metals would produce gold and/or a potion called the *elixir of life,* which would ensure immortality. What was produced was a harder substance consisting of 90 percent copper and 10 percent tin. In fact, this alloy was so durable that ancient miners excavating copper ore used picks and axes forged from bronze. During the latter part of the Bronze Age, metalworkers on the European continent discovered that adding approximately 7 to 10 percent lead, another of the newly discovered metals, to the bronze alloy produced a mixture that was easier to pour and also improved the integrity of the subsequent casting process. Sheet metal was also a Bronze Age invention. Many copper and bronze objects underwent a forging process in order to fashion them into useful tools, weapons, armor, and artifacts.

Other metals mined in far less quantities than copper, tin, and lead during this time in history were *gold* and *silver.* While copper and bronze were used for tools, weapons, and jewelry, the softer metals of gold and silver were primarily used for personal adornments, artifacts, and coinage. Gold can be found in an alluvial state in rivers and creeks as well as in buried veins of quartz. For example, the Egyptians mined for gold in the African kingdom of Nubia (present-day Sudan), while the Romans mined gold extensively in the province of Spain. Silver can be found as a free metal in nature, combined with other ores, or as a by-product of copper, lead, zinc, and gold. In fact, at various times throughout ancient history, silver was deemed more valuable than gold, primarily because it was less abundant. The ancient tribe, the Hittites, controlled silver mines in Asia Minor and what is present-day Syria.

Although historians disagree on the exact dates, the Chinese are credited with discovering *brass,* the metal alloy containing copper and zinc, as well as experimenting with other metals, such as silver, mercury, and tin. The Chinese, in fact, used the amalgam of tin and silver for dental repair. The Indus civilizations were also renowned for their metallurgical expertise. Their brasswork was and is among the most intricate ever seen. Most notable is the eight-meter (26-foot) wrought iron pillar that

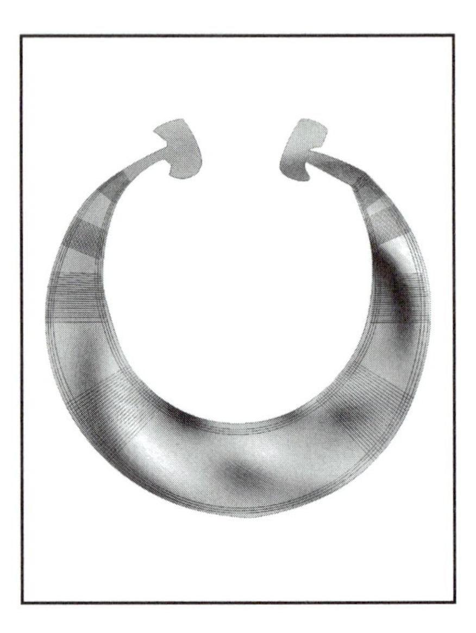

Figure 9.1 Bronze Age Collar
The first uses of metals, such as copper and bronze, were as adornments, such as the *necklace* or *collar* shown above, which were often buried with the deceased.

stands today near Delhi, India. Believed to date back to 1400 B.C.E., there is no evidence of rust or corrosion on its surface, supposedly due to the expertise of the ancient Indian metalworkers. The Romans used bronze for military weaponry and brass for coins, cups, plates, and decorative armor.

Another notable significance concerning the metallurgy of the Bronze Age was its role in establishing varying echelons of communities based on material wealth. There has always been a defined hierarchy within each community or society of humans, beginning with the most primitive hominids. However, this hierarchy was based on size, strength, or territory. Metallurgy provided certain members of the community with property that for the first time indicated wealth. Those who controlled the actual mining of the ores were themselves wealthy, as were those who could afford to purchase the artifacts and adornments produced by metalworking craftsmen. This has been verified by archaeological findings of this period all over the globe. The wealthy of the community were buried with bronze, gold, and silver objects, while pottery and stone artifacts surrounded the poorer members of the same community in their passage into the afterlife. (See Figure 9.1.) This particular burial practice continued unabated until the fall of Rome, when Western civilization fell into a deep decline. However, the stratification of hierarchical societies based on wealth and property has never abated and is well established universally.

Iron and Steel

Iron is the fourth-most abundant element on Earth, constituting about five percent of the Earth's crust. It is believed that the Earth's core consists primarily of molten iron. Bronze Age humans knew that iron existed but were unable to extract it because of limitations in existing smelting processes. They could not create fires hot enough to smelt the iron from its ores—that is, at temperatures of at least 1,482 degrees Celsius (2,700 degrees Fahrenheit) along with carbon. The exact history of the discovery of iron is unclear. Some archaeological evidence suggests that a crude form of iron may have been produced in the Middle East as early as 3000 B.C.E. and that a foundry in southern Africa was established around 2000 B.C.E. The discovery of iron was most likely serendipitous. For instance, iron-oxide meteorites may have been added to the copper or bronze smelting process purely by coincidence. The iron oxide acted as a **flux**—that is, a chemical that assists in the fusing of metals while preventing oxidation and absorbing impurities in the form of a slag that can be removed. Bits of iron were contained within this slag. However, it was not until sometime between 1400 and 1200 B.C.E. that the technology advanced to the point that a consistent quality of iron could be produced. The ancients experimented with a smelting process fueled by charcoal that could attain temperatures of only 1,200 degrees Celsius (2,192 degrees Fahrenheit) and that produced a pulpy material, later referred to as *bloom*. The bloom was repeatedly heated and hammered in an effort to remove the slag, a time-consuming process with less than satisfactory results. Over time, the proportions of fuel and ore were adjusted, and larger, more efficient leather bellows increased the air drafts into the furnaces, thus increasing the temperature. Eventually, carbon, usually in the form of charcoal, was added to the iron to make it hard and brittle (cast iron). Iron that contains little or no carbon is soft and malleable (wrought iron). The smelting process was difficult and dangerous. The quality of the resultant metal products varied depending on the raw materials, the efficiency of the furnace, and the skills of the metalworker. An expert metal craftsman could produce varied grades of iron that were designated for particular uses. For example, carburized iron or *steel* was manufactured by heating malleable iron for long periods, during which time the carbon would soften and spread within the mixture of molten iron. The result was case-hardened iron—steel coated on the exterior with a softer iron interior that was particularly desirable for armor plates and shields.

Probably the most important and durable iron product, both in terms of its physical properties and use, was *steel.* Its exact origin is often traced to ancient India, where Hindu physicians, supposedly, used carburized iron for surgical instruments. This may or may not be accurate, since all ancient civilizations participated in the metallurgical technologies of the Bronze and Iron Ages—although not necessarily at the same time nor at the same levels. In its pure state, iron is soft and therefore not useful as a tool or weapon. It must be **tempered.** In their quest for the hardest and most durable iron substance, ancient metalworkers experimented with the chemistry of the molten mixture. Steel is an alloy of iron dependent upon the nature and proportion of the alloying agent which, in most cases, was carbon. (Nickel, which was known to the Chinese, can also be an alloying agent in the production of stainless steel.) Metalworkers also discovered that after forging the red-hot iron and then plunging it into a vat of cold water, the result was steel. It is probably a safe assumption that ancient metalworkers did not understand the chemistry behind the production of low-, medium-, and high-carbon steels, but they did comprehend that regulating the ratio of the ore to the carbon (charcoal) produced varying grades of steel. These differing grades were then selected for the manufacture of nails; agricultural, medical, or domestic implements; machinery; or weaponry. Again, the Romans exploited the discovery of steel. The steel sword of the Roman soldier became the most important weapon in their military arsenal and is often credited with the success of Rome's conquests in Europe, Asia, and Africa, and its subsequent dominance of the civilized world for nearly a thousand years. The production of this ancient discovery continues to be among the most consistently profitable industries. For instance, steel is used in the manufacture of the smallest precision surgical instruments and the strongest, heaviest beams that support skyscrapers the world over.

Settled farming may have been the genesis of civilization, but the technological discoveries of the Metal Ages enabled those civilizations to flourish. Metallurgy created a class of artisans and with it a division of labor. In addition to the farmers and brickmakers and potters of the later Paleolithic and early Neolithic Periods, the demand for, along with the support of, other craftsmen was a necessity in ancient communities. Towns and cities grew with a stabile yet increasing population that relied on the abundant and plentiful crops and harvests resulting from improved agricultural implements. Blacksmiths and other metal craftsmen worked with domesticated animals used for farming, commerce, and the military. In fact, metallurgy is probably most responsible for advancing the effectiveness and the horrors of war. With the discovery of metal horseshoes, steel swords and lances, and metal helmets and armor, armies had the ability to attack and subsequently conquer enemies and survive while sustaining manageable numbers of casualties. With conquest came occupation and the dissemination of information and knowledge from one civilization to another and from one continent to another—and so it continues.

Alchemy and Chemistry

In order to comprehend the relationship between ancient alchemy and chemistry, it is necessary to understand each of these disciplines. *Alchemy* had two components but one goal—perfection. Alchemy was an *art* that attempted to transmute base metals (copper, lead, etc.) into precious metals (gold, silver), as well as a *science* that endeavored to cure illness and achieve immortality through the ultimate understanding of the transmutation process. In other words, ancient alchemists continually sought the *philosophers' stone,* the catalyst for transmutation that would also produce the medicine to cure disease and effect immortality—the *elixir vitae* or the *elixir of life.* Its origins are unclear, but it is generally accepted that its roots are founded in the craft of metallurgy and the subsequent philosophical examination of life, illness, and the physical world. Neither term (alchemy or chemistry) was used by ancient practitioners; they were not coined until sometime after the fall of Rome in the mid-fifth century C.E. when alchemy, based on Greek science, rapidly advanced within the Muslim world. In fact, it was not until the twelfth century C.E. that the knowledge of alchemy was introduced on the European continent by Arab invaders.

The word "alchemy" is derived from the Arabic *al,* meaning "an object," and *kimia* or *khem,* which was an ancient term for Egypt, where the genesis of alchemy was supposedly founded in the writings of Hermes Trismegistus, the Greek designation for Thoth, the Egyptian god of the sciences. The Hellenistic Greeks also influenced the early phases of alchemy. The Greek word *chyma* means "to melt or cast metal." Alchemy is often referred to as a medieval craft or pseudoscience, since it flourished in Europe during the Middle Ages and Renaissance with many alchemical practitioners of medicine. However, history is not discrete. Alchemy evolved over eons of time, often in fits and starts, and discovery occurred simultaneously and independently in many regions and countries. The ancient Egyptians, the Chinese, the Indians, the

Mesopotamians, and the Greeks all were knowledgeable about metals, medicines, and herbs. In addition, they all had their own philosophies, religions, and theories relative to life, death, and the unknown. The experimentation with metals often met with disaster. While a potion mixed with alcohol (a chemical fermentation by-product) could produce a sedating or effusive effect on a patient, a mixture containing excessive levels of arsenic or mercury could mean death. The ancients were not immune to cheats and charlatans among the authentic practitioners, who genuinely believed that immortality or, at the very least, improved health and well-being could be effected by the transmutation (chemical) process. Modern medicine, physiology, and pharmacology are really extensions of ancient alchemical practices, including pseudocures, empirical evidence, and other procedures based on actual scientific findings.

On the other hand, *chemistry* is an actual science that deals with the properties, composition, and structure of elements and compounds, the reactions they undergo, and the energy released or absorbed during these transformational processes. The most primitive humans first encountered the physical sciences in the form of fire. Their mastery of this physical/chemical process led to innumerable inventions, among them cooking, pottery, glass, dyes, metallurgy, and herbalism (primitive medicine and pharmacology), and the procedure known as mummification. These early inventions all entailed chemical processes in order to effect the tangible result or product. Our most ancient ancestors did not theorize on the physical principles or laws that governed the chemical reactions that occurred when clay was baked over a hot kiln or why sand melted when subjected to high temperatures. They did, however, rely on the consistency of the reactions. For instance, potters, glassblowers, dye makers, metalworkers and the early priest-physicians of antiquity were, in a real sense, chemists. First by trial and error, later with more sophisticated techniques, they refined and improved both processes and products by scrutinizing and experimenting with the proportions of the inorganic materials used in their crafts.

For instance, two of the most important classes of chemical compounds are **acids** and **bases,** both of which have been used by humans for thousands of years. (Acids and bases found in our own homes include fruit juices, aspirin, milk, baking soda, vinegar, and soap.) Most likely, the first acid produced by humans was acetic acid ($HC_2H_3O_2$). Vinegar is a dilute aqueous solution of acetic acid, an organic acid that forms when naturally occurring bacteria, called *acetobacter aceti,* convert alcohol to acetic acid. The use of bases (or alkalines) also dates back thousands of years. For example, bases are key ingredients of soap that date back to Babylonia around 2800 B.C.E. The Egyptians combined lime (calcium oxide) and soda ash (sodium carbonate) and evaporated the product to produce caustic soda (NaOH) or lye, which they then used as an ingredient for cleansers, dyes, and the preparation of papyrus. The Chinese also used lye to prepare paper. "Alkaline" is derived from the Arabic term *al qualy,* meaning "to roast." Evidently, the term referred to roasting or calcinating plant material and leaching the ash residue to prepare a basic carbonate solution.

Among the many actual inventions that are by-products of a chemical process, two ancient discoveries can be considered as stand-alone or independent and have basic formulae that have remained unchanged into modern times. They are salt and the fermentation process.

Salt

The importance of common *salt* (sodium chloride, NaCl) both as a chemical compound and a commodity can be traced back at least 10,000 years to the ancient Middle Eastern town of Jericho, situated on the banks of the Dead Sea. The original site was one of the first Neolithic settlements around 9000 B.C.E., becoming a wealthy town in about 8000 B.C.E. Its source of wealth is believed to have been trade in bitumen, a naturally occurring petroleum that oozed out onto the surface of the ground, and salt. Salt can be obtained through the solar evaporation of seawater in shallow ponds, or by the mining of salt deposits that are found worldwide. (These deposits are the dried remnants of ancient seas.) The mining of these deposits could be worked in two ways: (a) like any other mineral, in which case the salt could be contaminated with clay and other impurities; or (b) by pumping water into the deposit to form a brine. The brine was then either fire- or solar-heated, causing evaporation of the liquid with the resultant salt crystals. Another method of producing salt was invented by the Chinese sometime around 500 B.C.E. They discovered hundreds of brine wells, some of which were over 100 meters (330 feet) in depth, within the provinces of their country. The Chinese erected bamboo towers, similar in style to modern-day oil derricks. An iron bit was hung from a bamboo cable attached to a lever on a platform constructed atop the tower. The bit fell to the ground and was designed to "drill" into the earth at the location of the underground brine well. The source of power was human: two or more men jumping on and off the lever that moved the iron bit. Eventually, the bit dug deep enough into the ground to hit the brine, which was then pumped up to huge iron saltpans, where it was heated to evaporation. The source of the heat for these huge pans was natural gas, as brine wells were often located near natural gas deposits. Through trial and error, the Chinese learned to tap the natural gas deposits in the ground using bamboo pipes that led to a series of pipelines, which fed the gas under the iron pans containing the brine. Openings in the pipelines—similar to a burner on a gas stove—allowed the gas to escape, and it was then lighted. The result was fires hot enough to evaporate the brine's water, thus leaving salt deposits. Villages near these saltworks used the natural gas for cooking and lighting as early as 100 B.C.E. This technology was limited, however, to China. In Europe, Asia, the Middle East, and parts of Africa, salt was mined conventionally, mostly through slave labor.

Salt is a requirement for life, a fact that was not lost on the ancients. For example, in parts of Africa, now as in antiquity, salt is a commodity that is too expensive for all but the wealthiest of the community. (People who live on diets of milk and raw or roasted meats generally receive adequate supplies of naturally occurring salt in these food products. However, diets that consist of cereals, vegetables, and boiled meats are deficient in natural salts, and supplements must be provided in order to maintain health.) Nevertheless, community leaders and government officials exploited this need, and salt taxes were common during ancient times and constituted a major portion of revenue in the Middle East and in China. Salt was a lucrative trading commodity, even to the point that in ancient Greece slaves were bartered for portions of salt. The word "salary" is derived from the Latin term *salarium argentum,* meaning "an allowance of silver to purchase salt." Nomadic tribes carried salt with them as both an offering of goodwill and a health necessity. Caravans transported salt along with other precious metals, jewels, and textiles, providing a lucrative commodity for the seller. In addition to improving the taste, salt was the only preservative for food for centuries until the advent of refrigeration. The Egyptians used large quantities of salt in their mummification rituals. The earliest potters used it in the making of glazes.

The ancients also discovered three other saltlike chemicals—*natron, nitre,* and *borax,* all of which were extensively traded for use in ancient rituals and crafts. *Natron* ($Na_2CO_3 \bullet 10H_2O$), a mineral of hydrous sodium carbonate, is a complex salt found in dry lakebeds in Egypt. The Egyptians, who had a state monopoly on natron in the western desert region of Wadi Natrun, used it in the mummification process, as well as in glassmaking, pottery glazes, and textile cleaning. Today, its most common form is sodium carbonate, or baking soda. The use of *nitre* (KNO_3)

(potassium nitrate or saltpeter) was limited to that of an ingredient in the glazes produced by the Akkadian potters of Mesopotamia until the Chinese discovery of gunpowder between 900 and 1000 C.E. The third chemical, *borax* ($Na_2B_4O_7 \bullet 10H_2O$) (sodium borate), was an ingredient in the making of glass and ceramics. Its main worth, however, was as a metallurgical flux. It is also called *tincal.*

Fermentation

Fermentation is defined as an enzymatic chemical change induced by single-celled organisms (bacteria and mainly yeast) that causes the decomposition of starches and sugars, the result of which is the production of the gas carbon dioxide and the liquid ethanol (alcohol). For the ancients, the discovery of the fermentation process was probably serendipitous, most likely in the form of overripe fruit or honey that had come in contact with an airborne organism. While fermentation is a natural chemical change that has been occurring since the very existence of plants and bacteria on the earth, the deliberate manipulation of fermentation to produce alcohol is about 10,000 years old. The first beverages produced were beer and wine. Beer is made from grains with a high starch content (such as barley, wheat, or millet), which must be converted to sugar in order for the fermentation process to occur. Wine is manufactured from fruits (usually grapes) that have a high sugar content. Which of these beverages was the first to be manufactured is unknown. However, it is generally believed that beer was more popular. In general, grains were easier to grow and thus more plentiful than grapes, which required more moisture and a cooler climate than the hearty crops of barley and wheat. At first, the use of beer and wine was limited to Mesopotamian and Egyptian priests, physicians, and pharmacists, who prescribed them as intoxicants in ancient religious rituals and as calmatives for the ill. However, it is assumed that the euphoria that resulted from ingesting these new beverages was seductive to the masses, and their popularity beyond religious or medicinal purposes spread quickly among the general population. Public drunkenness was a problem even for the ancients, as evidenced by the regulation of drinking houses that was written into the *Law Code of Hammurabi* around 1770 B.C.E.

Approximately 5,000 years ago, the Egyptians employed a two-step process in the production of beer. They divided the selected grain into two batches, one of which was coarsely ground and dried and then heated in water; the other was made into malt. Malting means that the grain has been allowed to sprout. Most likely through experimentation, the ancients learned that sprouted grain (usually barley) was more effective in breaking down the starches in the grains to sugars, with the result being a more potent mixture. This is an example of the reliance on the chemical process without the actual physiologic knowledge of the ingredients. Malting increased the enzymes needed to break down the starches in the cooked grain. Wine was made nearly 7,500 years ago. Archaeologists unearthed a jar found in a Neolithic settlement in the Zagros Mountains in present-day Iran. The jar, believed to date back to 5400 B.C.E., contained a liquid residue, later analyzed as wine containing a resin from the terebinth tree. The purpose of the resin is believed to have been to inhibit bacteria, that is, to prevent the wine from turning into vinegar—another example of the utilization of a chemical process. The resin that came from the terebinth tree was also used as a liniment, commonly called turpentine.

The alcohol content for beer and wine produced in antiquity was relatively low by modern standards—probably no more than three to five percent—or about the same as most modern beers. This changed in about the eighth century C.E. when the distillation process was discovered and the *still* invented by Arab alchemists. At first, **distillation** was not used in the production of alcoholic beverages. Rather, it was employed to extract essential oils in the manufacture of perfumes. When the Muslims invaded the southern European continent and north Egypt during the early Middle Ages, bringing with them previously unknown technology, the West quickly learned to use the still to produce beer and wine with a higher alcohol content—around 16 percent. Fermentation is often misunderstood in the negative sense, inasmuch as it is associated merely with the production of alcoholic beverages (e.g., "demon rum"). However, the process, along with the refinement of distillation, is responsible for a great many products with far more beneficial and productive effects. The ethanol or ethyl alcohol that is a by-product of the fermentation of plant matter is an ingredient in solvents, dyes, pharmaceuticals, elastomers, detergents, cleaning products, coatings, cosmetics, antiseptics, antifreeze, octane boosters in gasoline, and explosives—to name just a few.

Philosophy of the Physical Sciences

Despite their ignorance of the chemical and physical basis of organic and inorganic *matter,* the craftsmen and artisans of antiquity continually discovered, invented, and improved upon concepts, devices, and techniques in metallurgy, medicine, agriculture, engineering, astronomy, and mathematics, as well as in the life sciences of biology, botany, and zoology. In fact, the ancients used the word *material* to mean matter, which was any substance that could be measured with three dimensions (height, width, and depth) and had characteristics such as weight, impenetrability, color, odor, texture, and so forth. Today, we define such matter or material as *chemical substances.* Gas was more or less ignored as a state of matter, essentially because the ancients could neither see nor measure it. Concepts of the "unknown" that seemed rational at the time emerged as mysticism and religious systems that, along with the accumulation of wealth, spurred on the development of the sciences. Philosophy is the interpretation of these concepts—that is, analyzing the underlying reality of matter and systems. The one thing that all civilizations, ancient and modern, have in common is their pursuit of philosophical ideas to explain the unexplainable. In antiquity, the mystical interpretations of the physical world often resulted in misconceptions that by today's standards seem silly and unreasonable. Yet their technological discoveries and inventions are quite remarkable. Modern science has the equipment, technology, and information to verify many discoveries and inventions in the physical sciences to an almost absolute certainty. Despite this, the trial and error of modern science with its built-in error-correcting system has not diminished human interest in the religious or philosophical pursuit of the "hows" and "whys" of nature.

The Mesopotamian, Egyptian, Chinese, and Indus civilizations pursued philosophical explanations for matter and the unknown while they engaged in the development of practical inventions and technological advancements. They attempted to turn each discovery into something with tangible benefits—even if it was a sharper knife or deadlier weapon. On the other hand, the ancient Greeks produced the most influential philosophy of the physical sciences. Though not always correct in their assumptions—in fact, most of them were misconceptions—the Greek theorists and philosophers who lived after 700 B.C.E. nevertheless established an ordered system in the form of "schools" that encouraged the examination of the natural world and the belief that explanations lay therein. More than any other civilization, the Greeks laid the foundation for the modern sciences.

The Presocratics

The philosopher-scientists who lived prior to the birth and philosophical influences of Socrates (ca. 470 B.C.E.–399 B.C.E.) are known as the *presocratics.* Aristotle characterized them as the first "Investigators of Nature," because they emphasized the rational unity of matter and the universe and rejected mythological explanations. For the most part, copies of their original works have been lost. Their philosophies are preserved through the writings of other Greek theorists, including Aristotle and Theophrastus.

The Ionian School

The first group of presocratics were the founders and pupils of the Ionian School, which was located on a group of about 40 islands in the Ionian Sea, west of Greece and east of Italy. (The term *school* refers to a group of persons, in this case, philosophers and theorists, whose body of thought reflects a unifying influence or common belief system.) The Ionians invaded Greece from the north and settled in the islands west of the mainland, as well as in southern Italy (Elna). In 530 B.C.E. the major Ionian cities were invaded and subsequently destroyed by the Persians, who brought with them new ideas that influenced the thinking of the islands' Greek inhabitants. After Athens defeated the Persians, it became a prosperous city-state that allowed its citizens to develop new skills and inventions and to concentrate on learning. Many historians consider Ionia to be the birthplace of Greek science and philosophy. The Ionian philosophers explained the world in physical terms—the science of *physics.* Although most of their explanations were incorrect, they were the first to separate the Greek gods from material things in nature. As is the nature of philosophy, one good idea often begets another. Disagreements and conflicts were a natural aftermath of the interchange of ideas, and several prominent Ionians formed new schools in other cities where their own prevailing theories were taught. Later philosophies expounded by *Eleatics, Neoplatonists, atomists,* and *Socratics* were built upon the foundational theories of the Ionians.

The most famous Ionian philosophers were *Thales of Miletus,* the father of Greek philosophy, who believed in the "mobile essence," that is, that all things were made of water; *Anaximander of Miletus,* who called primal matter *apeiron,* meaning "the indefinite," an idea that was further developed by the atomists; *Anaximenes of Miletus,* who believed that air was the defining principle that thickened and thinned into earth, clouds, wind, fire, and water; and *Heraclitus of Ephesus,* who believed in aetherial fire, from which all things originate and return in a never-ending process. This process was a constant state of flux structured by *logos,* the Greek word meaning "reason," from which the term "logic" is derived. Heraclitus is best remembered for his statement, "A man cannot step into the same river twice," as well as for his assertion that reason was primary to all systems and philosophies. *Pythagoras of Samos,* the famous mathematician, was also a follower of the Ionian philosophy.

The Eleatics

The Eleatic School was located in Elea, also known as Velia, a Greek colony in southern Italy. Its philosophy was characterized by a belief in radical *monism* or the doctrine of one. Eleatics believed that everything is a motionless entity and nothing exists that either contrasts or contradicts this state. They believed in movement (energy) and objects (matter) as separate units and recognized that, in nature, basic units of matter exist. However, they did not attempt to categorize or examine them. *Xenophanes of Colophon,* the father of *pantheism* (a belief that God is the eternal entity directing and governing the universe), was the founder of the Eleatic School. *Parmenides of Elea* stressed *idealism* and that numbers were paramount. He believed form or function was more important than matter (i.e., physical or chemical attributes). Idealism was the basis for Greek philosophy for centuries. *Zeno of Elea,* the famous atomist, was Parmenides' pupil.

Anaxagoras of Clazomenae and *Empedocles of Agrigentum* were two famous philosophers who do not fit into either the Eleatic or atomistic schools of Greek philosophy. Rather, they were influenced by the Ionians, Thales, and Pythagoras, respectively. Anaxagoras believed in the concept of *nous* (the Greek word for mind) or the doctrine of divine reason. He believed the world was formed by the mind in two stages: first, by a continuously revolving and mixing process, and second, by the development of living things. He challenged the philosophies on the divisible nature of matter, and criticized the Eleatic beliefs of Zeno and Parmenides. On the other hand, Empedocles was conflicted regarding Eleatic principles, believing in some tenets while rejecting others. He is credited with the development of the *four-element theory of the universe,* which he based on the beliefs of other Ionian philosophers. He asserted that all reality was forever, and all things were composed of a proportional combination of indestructible primordial matter: fire, air, earth, and water.

The Atomists

The word "atom" comes from the Greek word *atomos,* meaning indivisible. The atomists believed that life, including all animals and humans, developed from some type of primitive slime or swamp, and that each person contained at least a portion of every kind of atom. Some of these atoms were continually emitted, while others were continually assimilated into the body. Their concepts of chemistry and physics were mechanistic—everything was predetermined or preordained. In other words, everything is as it was meant to be. They believed in the existence of the ultimate particle, which was incapable of being divided. They demonstrated this belief in a most practical manner. Taking a portion of simple ground soil, they continually divided it into halves until only one particle remained that could no longer be divided. Previous philosophers had maintained that soil—or anything else for that matter—could be divided into infinity.

The most famous Greek atomists of antiquity are *Leucippus, Democritus of Abdera,* and *Zeno of Elea.* It is believed that Democritus conceived the atomistic theory that everything was composed of atoms—the ultimate, homogeneous, indivisible, and eternal bits of matter. But the credit should go to Leucippus, his teacher, who first proposed the atomic theory in the fifth century B.C.E. Democritus merely refined Leucippus's concept. This is significant because these men conceptualized atoms and elements millennia before their existence was clearly demonstrated. Zeno was primarily concerned with motion, space, and time, all of which are aspects of modern physics. However, many of his beliefs bear little relationship to modern science. He developed a number of paradoxes to illustrate his philosophy, the most famous of which is *Zeno's paradox,* which deals with the misconception of space/time. It is based on the concept that if you begin to travel from point A to point B, you must first go *half* the distance toward the end (to the middle). Once at the middle between the two points, you must next go *half* of the remaining distance toward point B. Once you reach this middle point, you must again travel another *half* of the remaining distance, and so on, but you will never really get to point B because it is impossible—at least according to Zeno's reasoning. Zeno's paradox told the story of a race between Achilles (a mythological Greek hero) and a tortoise. In modern terms, it is more commonly referred to as the race between the tortoise and the hare. No matter, for in both versions of the story, which give the tortoise a head start, as per Zeno's paradox, the faster runner must always reach a point from which the slower runner began. Neither Achilles nor the hare can ever overtake the slower tortoise; hence, the tortoise will always be ahead, and neither will ever arrive at point B. This may have been the beginning of the concept of infinity.

Socrates

Socrates, Plato, and Aristotle exerted the greatest influence on Western intellectualization. Socratic philosophy states that humans should be taught truth, justice, and virtue as guides for living. Socrates is considered to have been a great moral teacher but not much of a scientist. His famous aphorism, "Knowledge is virtue," meant practical knowledge. He developed a technique for teaching, called *dialectical reasoning,* which is still in use today. It is based on a continual in-depth form of questioning and the subsequent examining of contradictions. This questioning technique is considered to be Socrates' major scientific contribution to the exploration of natural phenomena.

Plato

A student of Socrates, Plato established the first university in Europe on land belonging to Academus, a hero of Attica—thus, the name *academy.* Plato's academy taught the sciences—biology, astronomy, mathematics—as well as philosophy and politics. His major contribution was a technique of writing and analyzing that involved a dialogue or conversation between two people as they attempted to solve problems. This dialectical process became part of scientific inquiry. Plato also believed that mathematics was a form that could define the universe. The *Platonic Idea* was more concerned with the form of a concept rather than the actual matter itself, and a great deal of what was taught at Plato's academy was based on misconceptions of reality. While copious amounts of time went into thought, talk, and argument, Platonics failed to analyze their own observations and make use of empirical experimentation. Aristotle, one of Plato's students, eventually changed this classical Greek attitude to a more modern scientific approach.

Aristotle

Aristotle's writings cover a whole range of knowledge, most prominently in medicine, mathematics, and biology. He believed that each individual existed in a fixed natural type (species), and that individuals have built-in growth and development patterns preordained by nature (genetics). In physics, he developed the concept of *causality* or *four causes,* as follows: (1) the *material cause* of which things are made—earth, air, water, and fire (chemistry); (2) the *efficient cause,* that is, the source of motion and energy that causes things to move (in other words, nothing moved unless it was pushed by some outside force—energy, inertia, momentum, force); (3) the *formal cause,* which relates to types or species (biology); and (4) the *final cause,* which is the goal of full development of an individual, invention, structure, or belief (growth, technology, engineering, religion). He considered his causal concepts the key to organizing knowledge and attempted to organize everything in nature by categories or classes.

While Aristotle espoused many erroneous beliefs in astronomy, biology, and physics, he made many contributions in the fields of philosophy, psychology, ethics, and logic. A major contribution was the concept of *deductive reasoning;* that is, a person should base his or her conclusions on observations—not just on beliefs. Many historians believe that Aristotle's deductive reasoning relied heavily on intuition, an approach that was easily understood and therefore easily accepted. Aristotle was also a teleological philosopher. In other words, he believed in an ultimate purpose or design as a way to explain natural phenomena, the same belief system that is the foundation of the three major religions: Judaism, Christianity, and Islam. He also believed in the five elements of the universe: fire, air, water, earth, and aether. The existence of this fifth element (aether), an unknown and unseen medium in the air, was accepted by scientists well into the nineteenth century and the early part of the twentieth century. Aristotle is considered to be an intellectual giant. Charles Darwin, the famous proponent of evolutionary theory, said of the scholars of his time that they "were mere schoolboys compared to old Aristotle." Aristotle's philosophy, with its subsequent influence on scientific and physical inquiry, made its way throughout the world because of his association with Alexander the Great. Commissioned as a tutor by Alexander's father, Philip of Macedon, Aristotle apparently sufficiently impressed the young boy, who consequently shared his lessons with those he conquered and later ruled.

Summary

There were a number of less popular philosophies espoused by other Greek scholars, among them *Epicureanism, Skepticism,* and *Stoicism.* Their importance lies in the ancient Greek concept of looking to the natural world for answers, rather than the supernatural. While not always able to ask the right questions, and certainly not able to arrive at the correct answers, the ancient Greeks nevertheless established the framework for the inductive reasoning of the scientific method that was later developed by Francis Bacon in the sixteenth century. (*Deductive reasoning* is defined as a process of reasoning in which a conclusion is based on a stated premise—that is, from the *general to the specific. Inductive reasoning* derives general principles from particular facts, events, or circumstances—that is, from the *specific to the general.*) Other civilizations developed their own belief systems to deal with the unexplainable elements of the natural world. In China, the *Naturalists* developed the dualism of the yin and yang, as well as *Taoism.* In India, the *Vedic* and *Tantric* philosophies professed to understand and interpret the physical world. These philosophies all began with questions on the existence, behavior, and destiny of humans and the cosmos. To the ancients, immortality was an achievable goal, and the physical elements in the form of precious metals (gold and silver) were the conduits. The catalyst for many of the discoveries and inventions in the physical sciences of chemistry, physics, biology, and medicine was the continuous quest for and challenge of accepted knowledge.

10

TIMEKEEPING

Background and History

If asked to provide a definition of time, how many of us would stumble around, unable to give a coherent description of this most abstruse concept? In other words, we know what it is, but could we explain it? Yet, everything in our lives is defined by time. It's time to leave, time to eat, time to attack, time for peace, time enough, it's time. Scientists divide the age of the Earth into units of geological time—**eons, eras, epochs.** Individuals divide their lives into years, months, days, hours, minutes, and seconds—past, present, and future. The concept of time has been debated for millennia. Monuments to time have been erected, such as Stonehenge. The major religions of the world—Judaism, Christianity, Islam, Buddhism—rely on the fundamental aspects of absolute time in their particular ideologies. Astronomers, both ancient and modern, have wrestled with time—from the conception of the zodiac to the theory of the big bang and an expanding universe. In fact, time is the *essence* of all sciences and disciplines. In actuality, time can be classified as being either *temporal* or *physical.* Temporal time is personal, psychological, and immeasurable. Physical time can be classified as either *absolute* or *relative. Absolute time* is measurable and always proceeds forward, that is, → from the past → to the present → and into the future. *Relative time,* or time related to space, was described by Albert Einstein in his special theory of relativity, in which time is really a fourth dimension in addition to the three dimensions of space (height, width, and depth). It can lengthen, shorten, and go forward or backward. In other words, it is bidirectional (↔).

All living organisms have a sense of time; not necessarily a conscious feeling of time—as is the human experience—but as part of their life cycles. They undergo time-induced changes as they are born, live, and die. All organisms experience physical changes related to the seasons, hours of sunlight, temperatures, and so on, all of which occur within a specified time frame. Insect metamorphosis is time-programmed, as is the gestation period and the ultimate age limit for mammals. It is reasonable to assume that once hominids evolved into conscious beings, they developed a keener sense of time: time past, time present, and time future, as well as the feeling of the passage of time.

Our primitive ancestors needed only to react to daily and seasonal changes, without considering time as we think of it. A natural consequence of the development of civilization was the tracking of time as it related to changing events. This required some sort of unitary standard and a means of measuring the passage of time. The most likely as well as the easiest methods of estimating time were found in nature. The passing of the Sun across the sky during daylight, the phases of the Moon, the motion of the planets and the stars, and, of course, the changing of the seasons, occurred in discrete units of time. (The above natural motions are caused as the Earth revolves around the Sun and rotates on its axis.) It was not until some 6,000 years ago that the ancient civilizations of Mesopotamia and Egypt devised methods to account for the passage of extended time periods, such as the periodicities of the Moon and seasons; hence the invention of calendars. Clocks were invented to divide the passage of time into shorter periods. Why? The most likely answer is related to either religious or civil activities. Before going into detail on the timekeeping methods and devices invented in the ancient world, it is important to accurately explain the scientific concepts of time.

Absolute Time

If people believe in the reality of motion, they must also believe in absolute time. Aristotle believed, as did Galileo some 1,800 years later, that if an object was already moving in a straight line, and if that object was proceeding at a constant speed, then it would have to cover equal distances in equal units of time. Sir Isaac Newton formalized this concept as one of his laws of motion. He also believed that time is absolute in the sense that it is possible to measure the interval of time that passes between two different events. Moreover, this time interval would be the same for all people who repeated the measurements for the same events, assuming that they used the same or similar accurate measuring device (a clock). Of course, the two measurements must also use the same units for the passage of time, or have some means of converting one unit used to another unit. We do know that it takes approximately $365^{1}/_{4}$ days for the Earth to circle the Sun, and one day is the period required for the Earth to make a complete rotation on its axis. These, and the phases of the Moon, and longer periods of planetary cycles, were natural motions used as time "standards" by ancient people. But are there any natural units we can use to measure the passing of time during the length of one day? The decisions to divide one day into 24 units of time (hours), and the hour into 60 units (minutes), and the minute into 60 units (seconds), unlike the revolution and rotation of the Earth, are somewhat arbitrary, based on the ancient Babylonian numbering system. In approximately 1800 B.C.E. the Babylonians developed the sexagesimal (base 60) system, most likely because 60 can be divided by 2, 3, 4, 5, 6, 10, 12, 15, and 30, thus eliminating the need for fractions. In addition, 6 times 60 equals 360, the number of units (degrees) assigned to a circle. This may have also been related to what the Babylonians believed were the number of days in a year—360. We still use this system for timekeeping.

Within the concept of absolute time, there are further subdivisions, as follows:

Chronological Time

Chronological time is linear absolute time. In other words, it organizes data as to the sequence of events over periods that have a beginning, continue or progress in a future direction, and may, at some point, reach an end. This is often referred to as the "*arrow of time,*" which can be presented as the arrows of time that describe the past → to the present → and into the future →. The three arrows of time are (1) the *thermodynamic arrow,* which describes the concept of entropy in the universe, in which order decreases and disorder increases, until equilibrium is reached; (2) the *psychological arrow,* demonstrated by the fact that humans can remember and keep a record of the past, but not the future; and (3) the *cosmological arrow,* which traces the origin of the universe and its expansion rather than its contraction.

Astrological Time

Astrological time is based on the chronological arrow for celestial phenomena (events), most of which are worked out backwards using mathematical computations based on astronomical observations and records. An example: The temperature immediately after the big bang is estimated to be a thousand million degrees Celsius. By measuring the radiation of specific wavelengths (microwaves) from space to determine the residual heat in space from the big bang (2.73 degrees above absolute zero cosmic microwave radiation), astrophysicists have determined that the universe's origin was about 13 billion to 15 billion years ago. Therefore, entropy in the universe continues.

Geological Time

Geological time covers many events and periods of the entire Earth's history, from its origin about 10 billion years after the big bang, to the past 4.6 billion years of its history, to the present. The examination of **fossils** and the sediment or rock strata where fossils have been found has enabled geologists to develop a chronology of the Earth.

Archaeological Time

This is the chronology of ancient cultures determined by observing, recording, and analyzing the differing layers of soil that contain human artifacts. Archaeological chronology is not based on records maintained by humans, but rather on the relics of life, the age of which can be established by carbon-14 dating techniques.

Historical Time

This is the chronology of life events determined by their particular sequence, documented by humans. Historical time usually begins with evidence of written historical records and may be divided into somewhat arbitrary periods based on human interpretation of events. The Mesopotamian culture, which includes several Middle Eastern civilizations, was one of the first cultures to record events, somewhat sequentially.

Relative Time

Albert Einstein once wrote, "The past, present, and future are only illusions, even if stubborn ones." This view of time supports his special theory of relativity, which rejects any absolute time or space. Rather, as Paul Davies in the September 2002 issue of *Scientific American* writes, Einstein believed that "simultaneity is relative." When on Earth doing earthly things, Newton's laws of motion still apply, as the senses of humans make it rational for us to consider absolute time and absolute space. For instance, we know that it takes *x* minutes to go from here to there on Earth, and that we have covered an exact space from the start to the end of our movement in a specific time period. But in the larger context of the universe, this absolute nature of time and space no longer applies. Einstein's special theory of relativity refuted the concept of absolute time in the sense that each observer traveling in space must have his or her own measure of time as recorded by his or her own unique clocks and position in space. Observers moving relative to each other in space-time will assign different times and locations to the same event. Even though all these measurements are different, they are all correct, and any one observer can tell exactly, in time and space, when another's observations occurred—if their relative velocities are known. In other words, each observer has a *unique frame of reference* in both time and space. This may seem somewhat complicated because it does not involve ordinary everyday experiences. An example of typical "earthly" relativity is the passing of a train in a station. A passenger on the train, from his or her unique frame of reference in space, or point of view, sees the station moving past the train. The person standing on the platform, with a different frame of reference, sees the train moving past the station. Scientists have long accepted that time is not completely separated from an independent space (or distance). Rather, it is now a new entity called space-time and leads to a fourth dimension in cosmological measurements—*height, depth, width,* and now *time.*

Calendars

A calendar is defined as a system that divides time over extended periods (days, weeks, months, years) and arranges such a division into a prescribed and established order. The word "calendar" is derived from the Latin *calendarium,* meaning "register" or "account book." *Calendarium* is a derivation of *calends,* the first day of the Roman month on which market days, feasts, and other special celebrations were declared. (A Roman priest, called a *pontifex,* who was designated to monitor celestial movements, declared the first day of the month based on the sighting of the new moon.) The human inclination is to compartmentalize events into measurable and manageable entities. A number of archaeologists and historians have collected fossils of mammoth tusks and eagle bones that contain sets of organized, calendar-type markings. Some of these fossils are over 30,000 years old, and these scholars believe these etchings are indicative of prehistoric humans' awareness of lunar and solar cycles and their need to physically illustrate their natural progression. Settled farming hastened the growth of civilization about 9,000 to 10,000 years ago, and the management and survival of civilization—then as now—was dependent on economics. The bounty of the crops determined prosperity. Thus, it was imperative that crops were planted and harvested at optimum times. Nature provided the framework for this in the seasons (spring, summer, fall, and winter), a fact that was not lost on primitive humans. Religious and agricultural rituals were based on the phases of the Moon. It followed, then, that civil ceremonies also were dependent on the various complex and independent celestial cycles. The early Mesopotamian, Egyptian, Chinese, Roman, Mesoamerican, and Islamic civilizations each developed calendar systems based on astrological observations and mathematical calculations—the equinoxes, the phases of the Moon, the solstices, and the positions of the stars, such as Sirius. In fact, the Stonehenge monument in Great Britain is believed to be an elaborate calendrical edifice. Despite the discoveries in celestial mechanics that have been made by astronomers throughout the ages, approximately 40 different calendar systems continue to be followed by various groups and countries. An in-depth examination of all or even most of these calendars is beyond the scope of this book. Rather, the following have been chosen because of their insight, impact, and controversy, as well as the continuing interest shown by modern scholars.

Egyptian Calendar

Egypt, like all ancient civilizations, was agrarian in nature. Thus, agriculture, along with the flooding of the Nile River and the celestial observations of Egyptian astrologers/astronomers and mathematicians, were the catalysts for the development of the Egyptian calendar in approximately 4236 B.C.E. It was the first and, by many standards, the most practical calendar to be devised. The flooding of the Nile River was a tangible event. In addition to the fertile farmland that was left in its wake, the regularity of the flooding provided a marker for which the Egyptians gauged the relationship between it and their observed movements in the skies. One of these celestial events was the apparent movement of the "Dog Star," also known as Sirius or, to the Egyptians, Sothis, in the constellation *Canis Major.* Sirius, the brightest star in the sky with the exception of our Sun, appeared, disappeared, and reappeared with regularity. The Egyptians determined that after being visible for long periods of time, Sirius completely disappeared, only to reappear a few days before the flooding of the Nile. This event, which occurred in mid-July, is now referred to as the *heliacal rising.* The Egyptian year contained three seasons. The months that began these seasons were named after certain aspects of the river, such as *flood, emergence,* and *low water.* The other months were named after lunar rituals. Interestingly, though the Sun was an important part of Egyptian life and was worshiped as the Sun-god *Re,* ancient Egyptians did not attribute any importance to the seasonal solstices, probably because of Egypt's geographic location near the Tropic of Cancer. In other words, Egypt is extremely close to the Tropic of Cancer, 23°27' north of the equator. In fact, the Tropic of Cancer passes through the southernmost part of Egypt. A solstice occurs twice a year when the Sun has no apparent northward or southward movement: in summer at the *longest* period of daylight hours, June 21, and in the winter at the *shortest* period of daylight hours, December 21. Solstitial events are more pronounced in areas farther away from the equator, such as in northern Europe, and less pronounced closer to the equator. Therefore, to the average observer in Egypt, the difference between summer and winter daylight hours would appear to be relatively insignificant.

However, their first calendar was fraught with problems. It was a lunar calendar, based on the Moon's $29^{1}/_{2}$-day cycle. This meant that in the 12-month Egyptian year, there were only 354 days. (The *sidereal year*—the time it takes for Earth to make a complete revolution around the Sun—is actually 365.242199 or approximately $365^{1}/_{4}$ days long.) Consequently, over a year's period, there was a discrepancy of approximately 11 days' time. Thus, the Egyptian calendar was unable to accurately predict the annual flooding of the Nile because it did not occur in the same month each year. In approximately 3500 B.C.E. the Egyptians devised a new calendar based on their observations of Sothis/Sirius. If Sothis/Sirius reappeared heliacally in the twelfth month, they added an extra month, named Thoth after the Egyptian god of science. This is known as an **intercalary** month. Egyptian months were standardized at 30 days, and the weeks at 10 days (decans), a purely arbitrary decision, not one based on any astrological principle. They were then able to more or less accurately predict the annual floods. This did not, however, entirely solve their problems, as Sirius's movements actually coincide with the *sidereal year* ($365^{1}/_{4}$) days. The "lost" five days were added as holidays designated by their god, Thoth. Even with the periodic addition of an extra month and the insertion of five extra days per year, the Egyptians were astute enough to recognize a real and future dilemma— that their calendar would be "off" by one day every four years, and that in just over 700 years their designated summer and winter months would be reversed. Nevertheless, the Egyptians, who regularly made calendar adjustments, held steadfast to the notion of a year with 365 days. The Egyptians actually employed three different calendars for over 2,000 years in order to accommodate the discrepancies inherent in their rigid 365-day calendar. They were the *stellar,* the *solar,* and the *quasi-lunar.* The *stellar calendar,* based on 36 designated stars known as the Indestructibles, was used for agricultural purposes. The *solar calendar,* based on

360 days plus the 5 extra intercalary days designated by Thoth, was used for civil purposes. And the *quasi-lunar calendar* was used exclusively for religious celebrations.

The Egyptians continued to resist any calendar modifications that had already been developed by other civilizations. In fact, Alexander the Great of Macedon, the conqueror of Egypt, attempted unsuccessfully to impose the Macedonian calendar on his new subjects. A compromise of sorts was reached during the reign of Ptolemy III of Egypt (247–222 B.C.E.), when an official decree added one day every four years to the Egyptian calendar. This practice, commonly called *leap year*, is one that we follow today. However, at the time of its inception in 238 B.C.E., leap year remained an unpopular rule among the Egyptians; hence it failed until about 45 B.C.E., when its imposition became law by virtue of their Roman conquerors, led by Julius Caesar.

Stonehenge

Located on the Salisbury Plain about 137 kilometers (85 miles) southwest of London, England, the Stonehenge monument is one of the most visited and mysterious Neolithic edifices built by the ancients. Controversy surrounds its age. Some archaeologists believe it was built between 2100 and 1500 B.C.E.; others contend it was constructed between 2800 and 1800 B.C.E. More recent radiocarbon dating, however, indicates that its age is at least one thousand years older than first believed, having been erected intermittently in three phases between 3100 and 1500 B.C.E. Another controversy involves its builders. A number of theories suggest that, using his supernatural powers, the mythical magician Merlin, counselor to King Arthur, moved the stones from Wales to the Salisbury Plain. Other theories propose that Stonehenge was built by the Romans, the Danes, or even the Druids. The last group has been totally discredited, since the Druids flourished about a thousand years after the completion of Stonehenge. Even after dating techniques had improved significantly in the latter part of the twentieth century, and it had been proven that Stonehenge was, in fact, built by Neolithic tribes living in Britain, a number of scholars vehemently denied these early inhabitants' engineering capabilities, preferring to believe that other people from other lands were responsible. Today, all reputable scholars ascribe its building to the Bronze Age people on the Salisbury Plain, who used it for religious rituals involving the Moon and the Sun. Stonehenge also was a site for a regional calendar.

Physically, Stonehenge remains one of the most impressive megalithic structures, even though a number of the original stones have fallen, eroded, or been broken or removed. The first phase entailed the construction of the henge, an approximately 106.5-meter-diameter (350-foot) circular ditch and embankment. Inside of the henge, 56 pits were dug, also in a circle, which originally held timber posts and the cremated remains of ancient people. These are called "Aubrey Pits" in honor of Sir John Aubrey, the English antiquarian who was the first modern scholar to identify them. A break in the henge, called "the Avenue," extends to the River Avon, 2 kilometers (1.2 miles) away from the site. During this phase, a number of massive stones were brought into the monument. The now-fallen "Slaughter Stone" was set inside the circle at the Avenue. The "Heel Stone" was placed 27 meters ($88^{1}/_{2}$ feet) outside the circle at the Avenue. It is 6 meters ($19^{2}/_{3}$ feet) high and weighs 35 tons (70,000 pounds). Four "Station Stones" were set rectangularly within the henge.

The second phase, around 2800 B.C.E., is considered the most mysterious because the exact method of construction is uncertain. It is believed a number of wooden timber posts were erected but later removed and possibly replaced by a double-horseshoe configuration of bluestones, also removed at a later time. Most of the stones at the monument site were placed during the third phase of its construction, beginning in approximately 2100 B.C.E. Five sarsen stone (sandstone) trilithons, consisting of a lintel stone atop two pillars, were erected in the shape of a horseshoe in the center of the area surrounded by the circular henge. These five trilithons range in weight from 30 tons (60,000 pounds) to 50 tons (100,000 pounds). The tallest trilithon is 7 meters (23 feet) in height. Thirty other sarsen stones, weighing an average of 25 tons (50,000 pounds) and standing 4 meters (13 feet) high and 2 meters ($6^{1}/_{2}$ feet) wide, were erected in a circular pattern around the horseshoe trilithons. A series of lintels (stones placed atop the rocks), weighing about 7 tons (14,000 pounds) apiece, connected the 30 stones. Thus, the horseshoe was enclosed by a perfect circle. Some historians attribute this perfect arrangement to the ancients' use of a geometric device known as the megalithic yard. In the latter part of phase three, a horseshoe of bluestones was placed inside the inner sarsen horseshoe, and an additional 60 bluestones were erected between the sarsen circle and the sarsen horseshoe. Stonehenge is considered an engineering wonder. The major part of its construction took place before the wheel was a viable tool. The sarsen stones are believed to have come from the area of Marlborough Downs, approximately 30 kilometers ($18^{1}/_{2}$ miles) south of the Stonehenge site, and the bluestones are believed to have been transported from the Preseli Mountains in Wales, about 385 kilometers (239 miles) away. Historians believe that the stones may have been floated on rafts on the local rivers and moved over land to the site. Others believe the process of glaciation during the Pliocene Epoch played a part in the movement of the stones—that is, the natural movement of huge ice sheets moved these stones over hundreds of miles.

Dr. William Stuckley (1687–1765), an earlier proponent of the Druids-as-builders theory of Stonehenge, was the first modern scholar to propose that the site had astronomical significance. Stuckley's measurements indicated that the midsummer sunrise is aligned at the center or axis of the monument—at the location of the Heel Stone—an event that is undisputed. Many modern-day astronomers and historians agree that the monument also served as an edifice to track the cycles of the Sun and Moon, inasmuch as the construction allows for the seasonal viewing of the Sun and Moon along the horizon as they rise and set. In other words, the summer and winter solstices can be observed, as well as the vernal and autumnal equinoxes. (See Figure 10.1.) Stonehenge was thus an accurate calendar and/or a megalithic timepiece that tracked, to the day, the solar year as well as lunar periodicities. In the absence of written records, historians, archaeologists, and astronomers continue to speculate on Stonehenge's astronomical relevance. Questions concerning its designation as a celestial observatory or a place for religious rituals are arguable. At the very least, the answers probably lie somewhere in the middle. For the ancients, there was an inescapable correlation between their dependence on the harvest and the natural manifestations of the elements, the movements of celestial bodies, and the quest to understand and explain the unknown.

Mayan Calendar

Of all the ancient Mesoamerican civilizations, the Maya were the only people who were truly literate during the pre-Columbian era, from about 100 B.C.E. to 900 C.E. This is the period when their culture and civilization flourished in the areas now known as Guatemala, Honduras, Belize, El Salvador, the Yucatán, and the eastern states of Mexico. The demise of the Mayan civilization has been attributed to drought, internal conflicts, and the grisly ritual excesses of human sacrifice of a civilization that worshipped fierce deities and lavished enormous riches on its rulers. Pockets of Mayan culture actually survived into the fifteenth century, particularly in areas of the Yucatán, site of the famous ancient city of Chichén Itzá, which was partially rebuilt by the Toltecs around 1000 C.E. The remnants of Mayan culture lay in ruin and obscurity until the nineteenth century, when explorers from Europe and the North

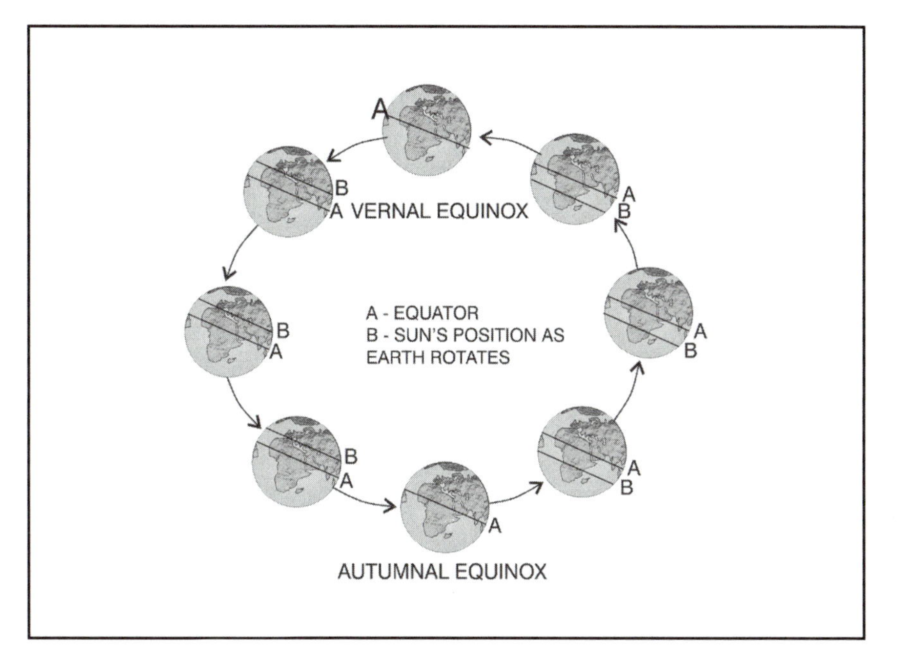

Figure 10.1 Vernal and Autumnal Equinoxes
Equinoxes occur twice each year when the sun crosses the celestial equator and the lengths of both night and day are approximately equal. The *vernal equinox* takes place in March and the *autumnal equinox* in September.

American continent realized the vast wealth of knowledge and treasure that was waiting to be uncovered. (The Spanish invaders of the fifteenth century looted their treasures and destroyed tremendous amounts of Mayan artifacts and codices, considering them paganistic and therefore either blasphemous or historically insignificant.) Along with temples, stone carvings, jewelry, and obsidian tools, these early explorers also discovered evidence of a highly institutionalized and regulated society, an example of which is the Mayan calendar. (To date, linguistic and archaeological scholars have only deciphered 85 percent of the approximately 5,000 Mayan texts in the form of glyphs and carved symbols.) The most famous Mayan book recovered is known as the *Dresden Codex,* so named because it is housed in a library in that German city. Late in the nineteenth century a German librarian deciphered the Mayan code, thus making it possible for other scholars to interpret the glyphs found on Mayan temples, buildings, stelae, and other artifacts.

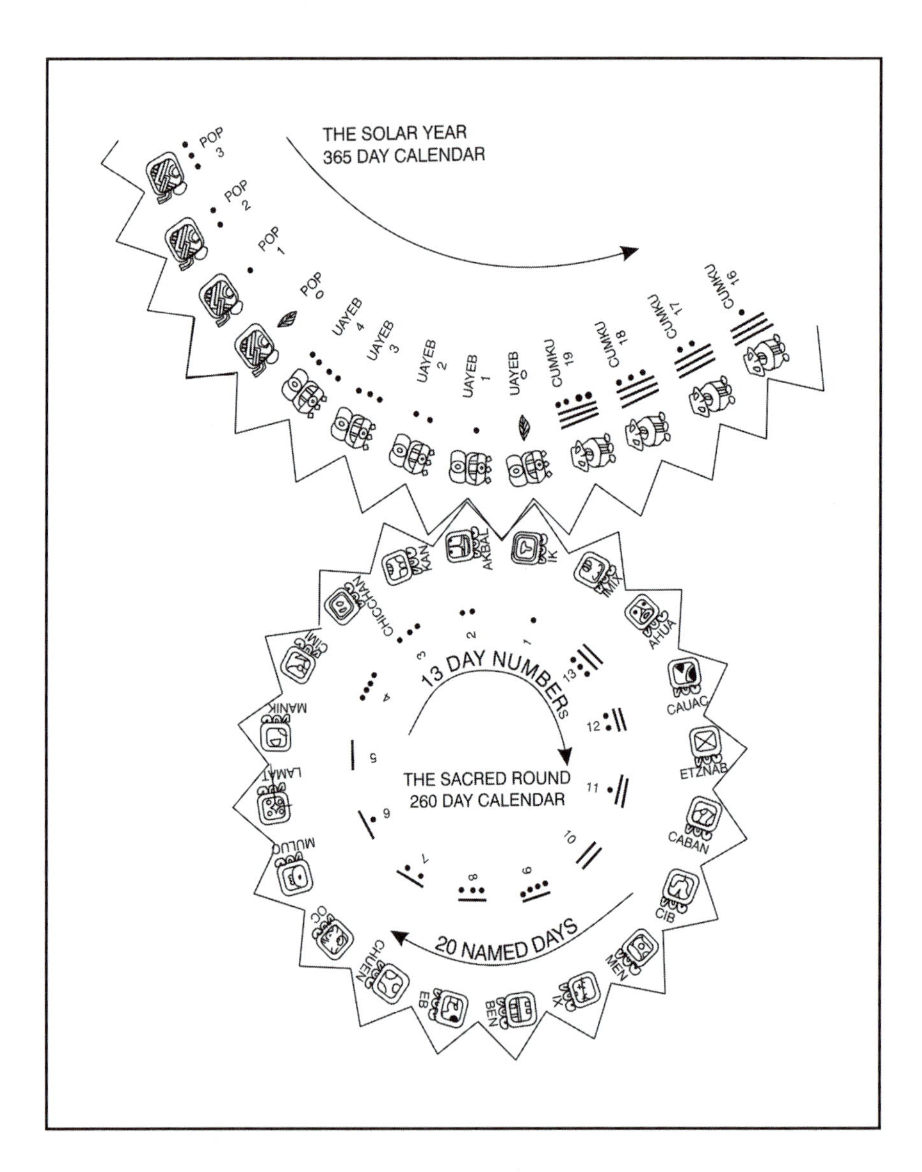

Figure 10.2 Mayan Calendar
The most complex calendrical system in ancient Mesoamerica, the *Mayan calendar* is actually four calendars: (1) the 260-day *Sacred Round*; (2) the *Solar Year*; (3) the *Calendar Round*; and (4) the *Long Count*.

The Maya were obsessively concerned with time and timekeeping, mathematics, and astrology/astronomy, all of which were interconnected. The Mayan calendar, however, was actually based on an earlier system that had been developed by the Zapotecs of Oaxaca, who along with the Olmecs, Aztecs, and Maya comprised the four advanced Mesoamerican civilizations. The calendrical system formulated by the Maya proved to be the most complex in Mesoamerica. It used many different calendars simultaneously. (See Figure 10.2.) To understand the Mayan calendar, it is necessary to have a basic understanding of Mayan mathematics, which used a base 20 place-value system that included zero. (Some scholars credit the Maya with inventing zero as a number, while others believe it was the people of the Indus civilizations on the subcontinent of India.) The number 1 was represented by a dot (•); units of 5 were depicted as a dash (—); and a shell, similar to the orb of an eye, was the symbol for zero. The four most important Mayan calendars were (1) the 260-day *Sacred Round;* (2) the *Solar Year;* (3) the *Calendar Round;* and (4) the *Long Count.*

The *Sacred Round* was a lunar calendar, called the *tzolkin.* It consisted of 13 20-day cycles or periods *(uinals),* each of which had names (e.g., *caban,* meaning "earthquake"; *muluc,* meaning "water"). The Sacred Round is believed to be related to the gestational period of humans, the length of a pregnancy. The *Solar Year,* also called a "Vague Year," consisted of 18 20-day months (360 days, called a *tun*), plus five ominous or bad luck days, for a total of 365 days (the *haab* or civil year). Even though they were aware that the solar year is approximately 365¼ days long, the Maya rejected the notion of adding a leap-year-like solution. Just like the Egyptian calendar, their seasons eventually reversed over a period of some 1,460 years. In addition, they chose to mesh the Sacred Round and the Vague Year calendars into one that was called the *Calendar Round.* The cycles of the Calendar Round turned in coglike fashion and returned to the point of origin every 52 years or 18,980 days. Its purpose was as a fortuneteller or augur of ominous events foretold by the Mayan priests.

The fourth calendar and timekeeping system was called the *Long Count.* It is the most controversial of the Mayan calendrical systems because it purportedly identifies the date on which the world will end. Using six units of time that range from one day *(kin)* to 400 years, the Long Count lists days in continuous linear succession, beginning from a base date that the Maya designated as the beginning of their civilization. In other words, the Maya began counting from the beginning of their existence until its end. The foundation of the Long Count was a *tun* (360 days), multiplied by 20 *tuns* (7,200 days), also called a *katun*. Twenty *katuns* made a *baktun* (144,000 days), with 13 *katuns* resulting in the "Great Cycle" (1,872,000 days or 5,130 years), at which time the Maya believed the world would come to an end. The controversy involves two aspects of the Long Count. The first is the actual date of origin. Scholars agree that it is written on the Mayan calendar as 13.0.0.0.0. Experts disagree about what that number actually corresponds to on the Julian and Gregorian calendars, an issue which is far too technical for this book. Nevertheless, there is genuine disagreement as to whether the count began on August 11 or August 13, 3114 B.C.E., or August 13, 3113 B.C.E. To believers of Mayan lore, this is significant because it would mean that the "end" would come either on the winter solstice of December 21, 2012—or December 24, 2012—or December 24, 2011. The other aspect, of course, is the accuracy of the calendar as a predictor. Many historians agree that the historical accuracy of the Long Count has been proven, at least in its ability to predict lunar eclipses and other astronomical phenomena, such as the cycles of Venus and Mars. Other researchers believe that the calendar was actually based on the behavior of sunspots that have always affected the magnetic fields of Earth and the other planets. They believe that dramatic sunspot activity could, in fact, result in cataclysmic events on Earth, ostensibly including its destruction. They use the drought-ridden centuries of 600–800 C.E. as an example of a localized disaster caused by sunspots. Scientists acknowledge that sunspots can affect global weather. Thus, it is reasonable to assume that the devastating drought in Mesoamerica during this period probably contributed to the demise of an already compromised civilization. In any event, historians and scholars will no doubt follow closely the predictions of the Long Count for the next decade of time.

Julian and Gregorian Calendars

As its name implies, the *Julian calendar* was named after the famous Roman general and statesman, Gaius Julius Caesar. By nature and temperament, Julius Caesar was organized, insightful, and logical as well as autocratic, impatient, and dictatorial. After his conquests on the European and African continents, he ruled an empire populated by people of varying cultures, customs, and canons, not the least of which was their calendars. The Roman republican calendar, like all others of the time, was rife with inconsistencies that related to its lunar foundation, as well as the politics of the day. For centuries, the Romans depended on lunar cycles, with the first day of the new moon being the first day of a new month, to determine their years. However, the Roman year consisted of only 10 months or 304 days. The 10 months of the Roman year began in *Martius*, or March. (See Table 10.1.) Lunar cycles are approximately 28 days, but most ancient calendars used 30-day months. During his reign, the Roman king Numa Pompilius (ca. 715–673 B.C.E.) reportedly added the months of February and January, in that order, between the months of December and March in an effort to appease Roman farmers, who needed a more accurate calendar. Farmers routinely ignored the Roman calendar and used the positions of the stars to determine the time of their plantings and harvests. The Roman year was now 354 or 355 days long, a figure that was approximately 11 days shorter than the sidereal year, a more accurate calculator. An intercalary month of either 22 or 23 days in length was periodically added in an attempt to correct the shortfall. The order of February and January was reversed in 450 B.C.E. During the time of his rule, Caesar recognized that the decision by the pontifices, the priests responsible for calendar adjustments, to insert an intercalary month was usually arbitrary and influenced by local or provincial politics. As a result, the civil calendar and the seasons did not correspond to each other; for example, *Floralia*, the Roman celebration of the advent of spring, fell in midsummer. Around 48 B.C.E., Caesar decided to end the caprices and abuses of the Roman calendrical system.

Popular fiction recounts that Caesar's motivation for the reform of the Roman calendar had less to do with altruism and more to do with romance. At the time, Caesar was in the midst of his affair with the Egyptian queen Cleopatra (69–30 B.C.E.). Thus, it is reasonable to assume that given the time he spent in Egypt, he became familiar with that country's calendar system along with its problems, some of which were similar to that of Rome's. He was also introduced to various Alexandrian scholars during this time in history. One such scholar was Sosigenes (fl. first century B.C.E.), a Greek astronomer, who recommended to Caesar that the Romans discontinue the use of a lunar calendar and replace it with a solar-year calendar $365^1/_4$ days in length, with the addition of one extra day in the month of February every four years. This leap-year solution was attempted earlier in Egypt and had failed. This time, however, Roman edict was far more persuasive. (See Figure 10. 3.) The new calendar had 12 months that were either 30 or 31 days in length, except the month of February, which was 28 days long—except every fourth

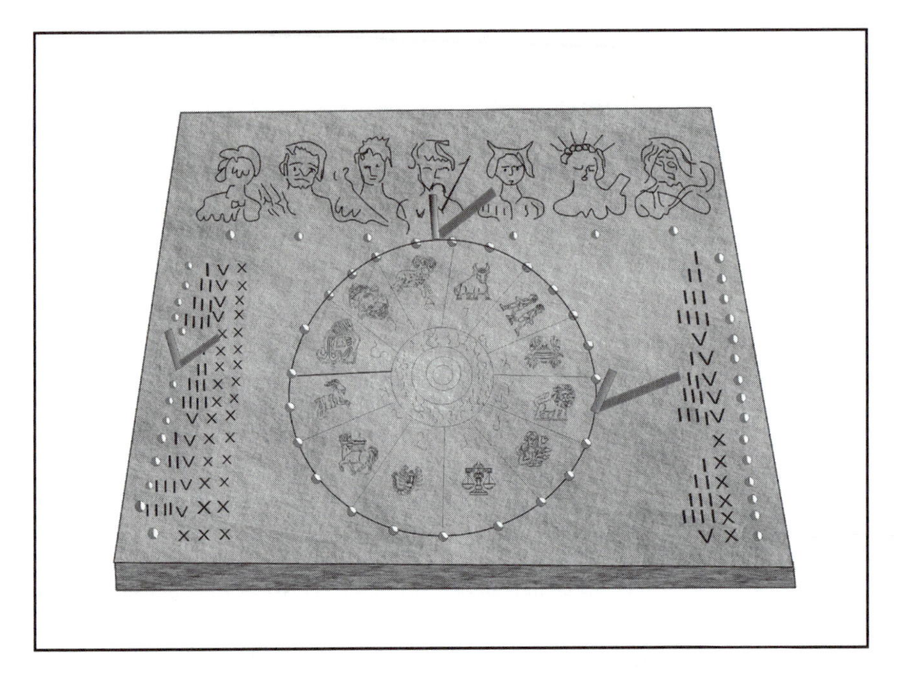

Figure 10.3 Julian Calendar
Imposed on the citizens of the Roman Empire in 45 B.C.E. by the Roman Emperor, Gaius Julius Caesar, the *Julian calendar* reconciled the discrepancies between the old civil calendar and the agricultural year and eliminated the abuses of the *pontifices,* the Roman priests responsible for calendar adjustments.

year, when the intercalary day made it 29 days. The length of the months has remained constant since 45 B.C.E.

Caesar imposed his new calendar on his citizens in 45 B.C.E. In order to reconcile the relationship between the civil calendar and the agricultural year, the previous year (i.e., 46 B.C.E.) lasted 445 days. Despite the logic and order of the new calendar system, the Roman pontifices continued to exert their will and added one extra day to the month of February more often than was decreed. Consequently, by the year 8 B.C.E. the Roman calendar again was in a state of confusion. Then-Emperor Augustus Caesar forbade the inclusion of February 29 for a number of years to allow for the necessary calendrical cycle to adjust itself. The Julian calendar, also called the Sothic or Old Style calendar, remained in effect for over 1,500 years until the adoption of the Gregorian calendar in 1582 C.E.

The *Gregorian calendar* corrected Sosigenes' error in calculating the length of the solar year, which is actually 11 minutes shorter than he had originally determined. This miscalculation was not significant at the time of the inception of the Julian calendar and for hundreds of years thereafter. (The inaccuracy amounted to a regression of one day per century.) However, it did become significant in the sixteenth century, when the calendar was off by 14 days. If left uncorrected, the same problem—reversal of seasons—would eventually ensue. The Roman Catholic Pope Gregory XII (1502–1585 C.E.) instituted a commission on calendar reforms to address the conflict. Meeting in Siena, Italy, they arrived at a solution: Restore the vernal equinox that fell on March 11 to March 21. The calendar was advanced 10 days subsequent to October 4, 1582—thus the following date became October 15, 1582. Further refinements were made concerning leap years. Today, the Gregorian or *New Style* calendar is accurate to within one day every 20,000 years. While it took until the early part of the twentieth century to gain near universal acceptance, the Gregorian calendar now is the primary calendar in virtually all parts of the globe. However, Islamic countries often retain their own calendar system.

The Months of the Year

In conventional terms, most people consider a month as a calendrical measurement—12 months make up one year. In scientific terms, there are actually seven measurements of time, each of which are considered months. However, *all* are based on the revolution of the moon around the Earth, and *all* astronomical months vary in length due to the **perturbations** in the moon's orbit. The seven measurements are the following:

1. *Synodic Month* – Length: 29.53 days. This is the complete cycle of the Moon's phases as observed from Earth.
2. *Sidereal Month* – Length: 27.32 days. This is the average period of time in which the Moon returns to a fixed direction in space.
3. *Tropical Month* – Length: 27.32 days. This is the period of time between the Moon's passage through the same celestial longitude. It is 7 seconds shorter than a sidereal month.
4. *Anomalistic Month* – Length: 27.55 days. This is the average interval between the Moon's closest approach to Earth.
5. *Draconic Month* – Length: 27.21 days. This is the average interval between the Moon's successive northwest passages across the eclip-

tic, that is, the apparent pathway of the Sun. It is sometimes called the *nodical month.*

6. *Solar Month* – Length: 30.44 days. This is one-twelfth of the solar or tropical year, that is, the length of time it takes Earth to orbit the Sun.
7. *Calendar Month* – Length: Varies from 28 to 31 days. This is an arbitrary measurement based on **lunation** (the time elapsed between two successive new moons). However, since lunation cannot accurately correspond to the solar year, astronomers and ancient timekeepers varied the lengths of each month.

The names of the months are based on the ancient Roman republican calendar, a complexity bordering on the confusing. For instance, the Romans did not number their days in sequence. Rather, they established specific points within the month and identified these with names. For example, *calendae* was the first day of the month; *nonae* was the ninth day before *idus; idus* was the thirteenth day of the month in January, February, April, June, August, September, November, and December. In the months of March, May, July, and October, *idus* fell on the fifteenth. Thus, the phrase "beware the ides of March" refers to March 15, the date of Caesar's assassination. The names of the months themselves were a combination of Roman gods and goddesses and the number sequence of the months in their calendar year. The ancient Roman calendar originally had only 10 months, with March being the first month. Thus, the months of September through December are designated by their original placement in the old republican calendar: They are called the seventh to tenth months of the year, rather than (in line with their correct placement) the ninth to twelfth months. Also, the months of July and August were originally called Quintilis and Sextilis (meaning the fifth and sixth months). July was named after Julius Caesar, who established the Julian calendar; August was named after his nephew Augustus Caesar, who in 8 B.C.E. made further calendar reforms to correct the maneuverings of the College of Pontifices.

The Days of the Week

There is no astronomical basis for the current seven-day week. It is purely an arbitrary measurement of time, most likely a compromise between the Roman eight-day week and the Babylonian and Sumerian seven-day week. There may be Biblical significance with the Old Testament's account of the creation of the world in six days, with the seventh as a day of rest. The actual names of the days of the week derive from the

Table 10.1
Months of the Year

Name (Julian Calendar)	Name (Gregorian Calendar)	Provenance and History
Januarius	January	Named afer *Janus,* the Roman god of beginnings and entrances who stood guard at the doorways of Roman houses.
Februarius	February	Named after *Februa,* the Roman festival of cleansing or purification held each year on the 15th of this month.
Martius	March	Named after *Mars,* the Roman god of war who was associated with agriculture and the rebirth of plantings.
Aprilis	April	Believed to be named after the Latin word *aperire,* meaning "to open" like the buds of flowers. Also may be named after *Aphrodite,* the Greek name for the Roman goddess, *Venus.*
Maius	May	Named after *Maia,* one of the Pleiades (daughters of *Atlas.*)
Junius	June	Named after *Juno,* the Roman goddess of women and queen of heaven.
Julius	July	Named after *Julius Caesar* who established the Julian calendar. (It was formerly named *Quintilis,* meaning the 5th month of the year.)
Augustus	August	Named after *Augustus Caesar* who made further adjustments to the Julian calendar in 8 B.C.E. to correct inaccuracies made by the Roman *pontifices.*
September	September	Means the 7th month of the year. (Latin prefix for seven is *septem.*)
October	October	Means the 8th month of the year. (Latin prefix for eight is *octo.*)
November	November	Means the 9th month of the year. (Latin prefix for nine is *novem.*)
December	December	Means the 10th month of the year. (Latin prefix for ten is *decem.*)

astrological beliefs of the Babylonians and the Romans, as well as from the Teutonic mythology of the Anglo-Saxon Europeans and the Nordic mythology of the Scandinavians. First, the Mesopotamians named blocks of time after the Sun, the Moon, and five planetary gods. The ancient Roman republican calendar originally comprised eight-day weeks, the names of which reflected their own gods and goddesses, who were essentially the same as those of the Mesopotamians. This eight-day week continued even after the reforms of the Julian calendar. However, in 321 C.E. the Roman emperor Constantine the Great (ca. 280–337 C.E.) officially decreed the seven-day week.

Just as the names of the week reflect the astrological influences of the ancient Mesopotamians, so too does the order and the length of each day (24 hours), at least in part. At the time of their existence, Sumerian and Babylonian astrologers identified the Sun, the Moon, Earth, and five other planets. However, they mistakenly believed Earth to be at the center of the universe with the Sun revolving around Earth in its own orbit. They were correct, however, in the order of the planets' distance from Earth. The Moon, then the planets Mercury and Venus were first in order, then the Sun (a mistake), followed by Mars, Jupiter, and Saturn. Babylonian religious beliefs dictated that the Babylonian gods, after whom these seven celestial bodies were named (excluding Earth), administered the 24 hours of each day. The complex arrangement of responsibility was based on Babylonian astrology. In descending order (from the farthest to the closest), the Babylonian god Ninurta (the Roman or Italian god Saturnus) controlled the first hour. Each of the other gods controlled hours two through seven, with Ninurta (Saturnus) then repeating the cycle at the beginning of the eighth hour, the fifteenth, twenty-second, and so on. Marduk (Jupiter) and Nergal (Mars) controlled the twenty-third and twenty-fourth, respectively. Then the first hour of the next day was given over entirely to Shamash (the Sun). The Babylonian concept for the number, length, and order of these hours was confirmed by an archaeological finding in the city of Pompeii in the form of a wall writing—a geometric septilateral drawing, called a *heptagon,* with the names of the gods written on each of the diagonal points. Obviously, the ancient order of the hours does not correlate with the current order of our days. It is reasonable to assume that over the centuries, with the adjustments and changes to the Roman calendar system, which is the foundation of the current Gregorian calendar, it made more sense to identify larger blocks of time (weeks) with specific names rather than smaller (hours). Even so, there is some illogic in the order of the days of the week, specifically with Sunday as the first day. In 321 C.E.

the Roman emperor Constantine, a Christian convert, proclaimed Sunday as the first day of the week and that it should be a day of rest, reflection, and worship. The Old Testament account of creation recounts that God created the world in six days and rested on the seventh. And of course, the beginning of each workweek is traditionally Monday, with the week's end on Saturday and Sunday. The following is the evolution of the individual names of the days of the week:

Sunday—Originally named after the Mesopotamian Sun god Shamash, who was concerned with justice. The Latin word for Sun is *sol.*

Monday—Originally named after Sin, the Sumerian Moon god. The Latin word for Moon is *luna.* Thus, Moon's day or Monday.

Tuesday—Planet Mars was named Nergal, after the Babylonian war god. Mars is the Roman god of war. Tuesday is a derivation of Tiu or Tiw, the Old English name for Tyr, the Norse god of war and the sky.

Wednesday—Planet Mercury was originally named Anu, after the Babylonian god of the sky. The Roman messenger of the gods was Mercury. The Norsemen's greatest god was Odin, who usurped the duties of Tyr. Woden is the Old English term for Odin. Hence, the current interpretation of Wednesday.

Thursday—Planet Jupiter was originally named after Marduk, the chief god of the Babylonians, who created order out of chaos. His Roman counterpart was Jove. In Norse mythology, after Odin, Thor was the mightiest god. Thus, the derivation of Thursday.

Friday—Planet Venus was originally named after Ishtar, the Babylonian goddess of fertility. The Roman goddess of spring and love was Venus. In Norse mythology, Frigg, the wife of the all-powerful god, Odin, represented beauty and love. Hence, the Old English Frigg's-day, which was abbreviated to Friday.

Saturday—Planet Saturn was originally named after Ninurta, the Sumerian god of war. Saturnus is the ancient Italian or Roman god of agriculture. Saturday is named in his honor.

Timekeeping Devices

Most people would correctly categorize clocks and watches as timekeeping devices. A calendar is also a timekeeping device for larger units of time, the origin of which was the agricultural heritage of the ancient Middle Eastern civilizations. The ancients needed an accurate tool in order to track the seasons and determine the most propitious times for planting and harvesting. Much less important in antiquity was a device to measure smaller portions of time between sunset and sundown. All humans, ancient and modern, possess an internal **biorhythmic** clock that governs our need for rest, sustenance, and activity. Therefore, our most primitive ancestors, the hominids, instinctively knew when to hunt and rest. Neolithic humans and their ancestors also knew that sleep came when the Sun set and activity began with the dawn. However, as civilizations and cities grew, the economic interests of artisans and craftsmen, as well as the religious interests of the priests and the community, combined. The necessity for more accurate short-term timekeeping devices was the inevitable result. The early inventions of the ancient civilizations of Egypt, Mesopotamia, and Greece were crude and merely measured the amount of daylight versus nighttime hours. The stars, the Sun, and water were nature's basic instruments. However, Earth's rotation on its axis as it revolves around the Sun, a phenomenon unknown to the ancients, affected the number of hours of sunlight on a daily basis. Thus, accuracy in minutes, a necessity of modern life, was not a standard in antiquity.

In order for any timekeeping device to be considered a "clock," two elements must be present. First, it must exhibit a repetitive and constant movement of some sort that gauges equal portions of time (for example, the movement of a shadow cast by the Sun or the flowing of sand in an hourglass). Second, it must contain a method or a means to track and display the increments of time (for example, the changing levels in a water clock). Following are the significant inventions in timekeeping:

Sundials (Obelisks and Gnomons)

Sundials are considered to be the oldest form of timekeeping, although not in the form that we think of in modern-day design. The Egyptian obelisk and the Greek gnomon were, in actuality, sundials inasmuch as they relied on the sun and its apparent movement to cast the shadows by which the ancients determined the time of day. As early as 3500 B.C.E., the Egyptians built four-sided, slender, tapered columns that cast a shadow from sunrise to sunset. At noon, the shadow is at its shortest, and at sundown, at its longest. Thus, the day was partitioned into two parts. The Egyptians added markers at the base of the column that acted as further subdivisions of time. Greek astronomers of the Hellenistic period (ca. 550 B.C.E.) used the gnomon, a pole placed in the ground in an open field or area, in a similar fashion. (*Gnomon* is the Greek word meaning "one who knows.") Aside from using the gnomon to tell the time of day, the Greeks used it to determine when the summer and winter solstices would occur, which in turn enabled them to calculate the time of the vernal and autumnal equinoxes. The Greeks and Romans are credited with developing more sophisticated sundials, particularly those that were bowl-shaped or rounded using gnomons (vertical or horizontal) as shadow-casters. The first Roman sundials were made of metal on which the dial face was carved with markings indicating the hours. An arrow-shaped marker or pointer (gnomon) was set into the center of the dial. When the Sun hits the marker, it casts a shadow, indicating the time. The Romans' first experience with the sundial was less than satisfactory, however. During one of their many military campaigns in Sicily in about 700 B.C.E., the Romans appropriated a round sundial that they installed in the Roman forum. For at least a century, they believed it was an accurate representation of the time in Rome. It was not. A sundial must be calibrated to the latitude of its location. In this case, the town in Sicily from which it was taken was about 530 kilometers (329 miles) southeast of Rome. The sundial was eventually recalibrated after Roman timekeepers interpreted the Greek writings on the subject and made the necessary adjustments. The Romans, who always had the reputation of borrowing ideas from other civilizations and vastly improving on them, further developed the sundial in varying complexities. They invented the pillar dial and the ring, both of which were portable, easy to construct, and depended on the altitude of the sun rather than its direction. In fact, in his writings, Vitruvius, the famous Roman engineer, described in detail the 13 different types of sundials that were being used across Italy, Greece, and Asia Minor by 30 B.C.E.

Shadow Clock

Invented by the Egyptians in approximately 1500 B.C.E., the shadow clock measured the passage of the hours of the day. The Egyptian day was divided into two parts, each of which contained 10 equal hours, plus 2 twilight hours in the morning and evening. A series of five variably spaced marks were carved onto the long stem of the shadow clock. An elevated crossbar atop this long stem cast a shadow over the marks from sunrise to sunset. Each morning the shadow clock was oriented east and west. At noon, the clock was reoriented north to south to track the afternoon hours. (See Figure 10.4.)

Merkhet

Ancient Egyptian astronomers also tracked the apparent movement of certain stars in an effort to differentiate the various nighttime hours. The Egyptians invented the *merkhet,* considered to be the oldest astro-

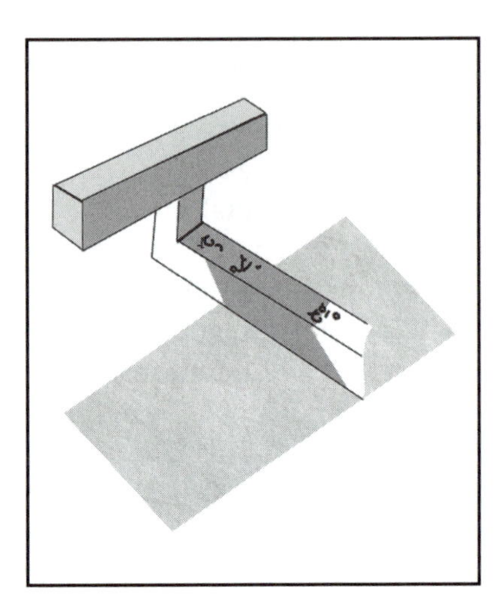

Figure 10.4 The Shadow Clock
An Egyptian invention dating back to 1500 B.C.E., the *shadow clock* measured the passage of the hours of the day that was divided into two parts. Each part contained ten equal hours, plus 4 twilight hours—2 in the morning and 2 in the evening.

nomical tool. As developed in about 600 B.C.E., the handheld *merkhet* was made of papyrus or stick, roughly the height of a 12-inch ruler, with a cutout notch used for sighting. It was used simultaneously with a *plumb line,* a thin piece of string from which a lead weight is suspended to determine verticality. The string is attached to a narrow strip of wood, called a *plumb rule.* Using these two instruments simultaneously, the Egyptians measured the position of the stars in terms of compass directions. In other words, by identifying the northern position of the Pole Star, also called Polaris, they could establish a north-south line. Egyptian astronomers were then able to mark off the nighttime hours by the "movement" of other stars that crossed the meridian (the imaginary north-south line). The plumb line, plumb rule, and *merkhet* were used by Egyptian engineers in the building of the pyramids at Giza, which are perfectly aligned directionally with the Pole Star.

Water Clock

An Egyptian invention from around 1500 B.C.E., the water clock was used mainly to mark off the hours during the night, the period during which religious rituals were typically performed. The first water clocks were stone or pottery bowls with sloping sides and a small hole in the

Figure 10.5 The Ancient Water Clock
Also called a *clepsydra* by the Greeks, the *ancient water clock* was invented in Egypt about 3,500 years ago. Water clocks were carefully engineered to regulate the flow of water, which changes relative to the water level in the bowl. Water clocks continue to be used in some Third World countries.

bottom, from which the water slowly dripped out. Hour marks on the inside of the bowl became visible as the water level ebbed. (See Figure 10.5.) Water clocks were carefully engineered. Water flows more rapidly from a bowl that is filled to the brim than it does from a bowl that is almost empty, because the rate of flow is determined by the pressure created by the depth of the water. Thus, the hour lines had to be carefully adjusted and placed to accommodate this differentiation in the flow rate. The Egyptians also developed another type of water clock that measured hours as water flowed *into* the bowl. One of the devices used by Egyptian timekeepers to regulate the water flow was the float. While the Egyptians divided the day and nighttime hours into 12 equal parts, they were aware of the seasonal changes in the number of daylight hours. Thus, their 12-hour days were shorter in winter and longer in summer, just as nights were longer in winter and shorter during the summer months. Therefore, Egyptian water clocks had specific markings for summer and winter: summer nights were 12 fingers high, winter nights were 14 fingers high. The Greeks began using water clocks, which they called *clepsydras,* sometime during the late fourth century B.C.E., but they also used them to measure daylight as well as nighttime hours. Over the years, various improvements were made to the *clepsydra.* Ctesibus of Alexandria (fl. ca. 270 B.C.E.), the famous Greek engineer, added a float. Water dripping at a constant rate raised a float that held a pointer (gnomon), which marked the passage of time. Greek timekeepers fashioned various designs, including a famous one designed by the Greek astronomer Andronikos of Kyrrhos in the first century B.C.E. Called the "Tower of the Winds," this was an octagonal tower constructed of white marble that stood on a base of three steps. Each of its eight sides, approximately 3.2 meters ($10^{1}/_{2}$ feet) high, bore a carved relief of mythological figures that represented certain types of winds. The interior of the tower was the water clock, and sundials were placed on the external walls. Other Greek astronomers and mathematicians, such as Thales, Eudoxus, Archimedes, and Ptolemy, designed and invented numerous astronomical tools—the stationary globe, the armillary sphere, and the astrolabe—to track the movements of celestial bodies that acted as markers of time. Hipparchus, often called the greatest astronomer of antiquity, is credited with the invention of stereographic projection, commonly called steriography, which depicted solid bodies on a plane surface. This technique was integrated into the design of *clepsydras,* rendering them more accurate timekeeping devices. The Romans constructed more intricate models of the basic water clock, although their timekeeping ability was not much improved over the Greek models. The Chinese also used water clocks that integrated mechanized escapements, but not until well into the second millennium C.E. The *clepsydra* or water clock continued to be popular until the development of mechanical clocks. Certain Third World countries continue to use water clocks, particularly those in northern Africa.

Summary

Since the beginning of civilization, the passage of time has shown us how far we have advanced, all the while demonstrating how much has remained constant. Today, we plan our days not by blocks of hours but by minutes. The atomic clocks in our homes are accurate to within a fraction of a second—more accurately, to within one-millionth of a second per year. (In 1967 the frequency of the oscillations of the cesium atom became the universally recognized international unit of time. Thus, the second is the physical quantity of time that can be most accurately measured by scientists.) The ancients were just as concerned with sustaining their economies as we are today. Religion was important in their lives. They worried about the future. Nature was unpredictable in many instances, but reliable in many others. They were concerned with time, but not necessarily to the precise minute or second. Modern life depends on accurate time. Ancient life did not. Nevertheless, the calendars in our homes and offices that regulate our domestic and business—even our leisure—activities were founded in Egyptian and Babylonian astronomy and Roman innovation. Twentieth-century physicists, like Albert Einstein, Stephen Hawking, and Steven Weinberg, have all challenged the conventional concept of time and have devised complex and intricate theories about the origins of and future direction for the universe—using astronomical and mathematical principles that date back thousands of years. Only time will tell what future discoveries will be made using the accumulated knowledge of the past.

11

Tools and Weaponry

Background and History

In the early 1970s, Stanley Kubrick's movie *2001: A Space Odyssey* depicted a group of primates posturing menacingly at a metaphorical black obelisk in an effort to exert superiority and control. The transformation from primates to humans ostensibly occurred when one ape picked up and then flung a large bone, sending it hurtling into space. Next scene: a twenty-first-century spaceship. The primate's aggressive gesture implied that this act somehow set the evolutionary process in motion. No doubt, many viewers believed this scene—minus the black obelisk—to be a relatively accurate account of how it all began. It is not. Biology and genetics are often accomplices on the path to the future. Evolution, however, is much more subtle, and our metamorphosis from primates to hominids to humans progressed over millions of years. The more intriguing aspect of the origin of humanity is whether the first implement was a tool or a weapon. Most likely, it was a combination of the two. Our most primitive ancestors killed other animals for food, that is, for survival. The exact moment when they began to kill each other out of a sense of fear because of differences or the need for dominance or territory is unknown. Nonetheless, once early humans grasped the essential concept that objects can have multiple purposes, the history of technology began to be written.

Hundreds of thousands of years before the crudest of agricultural tools were developed, archaic humans fashioned stones with flaked edges in what is present-day Ethiopia. The chopper, hand axe, cleaver, and knives of stone that were used for skinning and cutting up animals appeared over a million-and-a-half years ago. Neanderthals who lived perhaps 70,000-plus years ago used stone choppers and pounders. In our relatively short evolutionary history, succeeding species of hominids used the materials of nature (rocks, wood, bones, and shells) as tools. Then, at some point, probably in the Middle Pleistocene period (800,000 B.C.E.), tools began to be fashioned with a definite design or purpose by the toolmaker.

Dating the Origin of Tools and Weapons

The hunter/gatherer lifestyle of primitive humans persisted until roughly 10,000 to 12,000 years ago, when settled farming and the domestication of animals drastically altered the course of civilization. Paleolithic or Old Stone Age tools characterized the nomadic behavior of primitive humans, who hunted wild animals, fished in streams and rivers, and collected the seasonal grasses and berries needed to sustain their small groups. The demands of agriculture (the planting and harvesting of grains) created new uses for old tools, as well as the need for different tools for heretofore unknown tasks. Agriculture spearheaded civilization as we know it today. And civilization is responsible for the technological advancements that have taken us from the cave to the Moon and, someday, beyond. With each archaeological discovery, historians revise the dates of the origins of early tools and implements. Tools have been found in Africa that scientists believe are at least 2.6 million years old and suggest that tools older than this are likely to be found. Thus, we cannot know with absolute certainty when, for example, the axe was first crafted and by whom. It is probably safe to assume that when our ancestors became settled farmers, they used many of the same tools first employed by hunters. And as conditions dictated, human ingenuity forged the tools necessary to pursue this latest endeavor. For example, humans modified the axe by hafting—attaching a handle to the sharpened stone, giving the user more control and making the axe more efficient in the clearing of trees from farmland. Knives that were used to kill game were also used to cut stalks of grain.

As reported in the September 28, 2001 issue of *Science* magazine, some of the newest archaeological and anthropological research concerns the migration of primitive humans out of Africa, where our species first evolved, onto the Asian continent. Evidence suggests that early humans trekked longer distances and into colder, more inhospitable regions of present-day China, earlier in the timeline of humanity than was once thought. The evidence for this theory was the discovery in the northern provinces of China of stone tools (simple flakes, cores, and scrapers) that date back approximately 1.36 million years. Primitive tools and artifacts discovered close to the Arctic Circle in European Russia indicate that primitive humans established a camp in that region about 40,000 years ago, which is also earlier than researchers previously believed. Research continues on the Arctic findings to determine whether the tools were fashioned by Neanderthals or by modern humans. The significance of both of these discoveries is that the earliest humans were somehow able to traverse long distances into cold climates armed with relatively few of the crudest tools, and more significantly, to survive long enough to establish communities.

Ancient tools are generally divided into two periods: (1) the *Paleolithic Period* or the *Old Stone Age;* and (2) the *Neolithic Period,* which encompasses all of the *Metal Ages* (i.e., the *Copper, Bronze,* and *Iron Ages*). The Paleolithic Period is characterized by the invention and use of rudimentary, handmade stone tools that date from about 2.6 million to approximately 10,000 B.C.E. The Neolithic Period, beginning about 10,000 B.C.E., ushered in civilization, and with it the beginnings of agriculture, animal domestication, and technologically advanced ground and polished implements and weapons made from a variety of materials, most importantly, metals. Aside from the natural division of cultural periods, tools can be placed into several functional categories: (1) *survival tools,* those for hunting, fishing, chopping, scraping, killing and dressing prey, cutting wood and brush for shelter and warmth, etc.; (2) *colonizing* or *working tools*—agricultural implements, animal equipment, the wheel, etc.; and (3) *weaponry*—assault and attack weapons, as well as defensive implements that protected individuals and communities from marauders.

When humans discovered the Earth's metals, and subsequently learned how to smelt and forge hot metals into tools and weapons, the whole character of human society was transformed. Until the Metal Ages, humans were limited to the crudest of implements that could be formed from nature's resources. Metals and the smelting process enabled the ancients to fashion tools for specific purposes, such as the astrolabe to measure the altitude of the stars and planets, the copper saw that Egyptian physicians used to cut through the skull, and most especially, weaponry.

One fact has remained consistent from the time of prehumans up until the very present: almost anything can be used as a weapon. If the objective is to kill an animal or another human being, a precisely thrown rock can achieve the same result as a high-caliber bullet fired from a twenty-first-century polished handgun. The challenge has always been related to how swift, how accurate, and how extensive should be the damage inflicted by a particular weapon. Paleolithic humans fashioned knives, daggers, spears, and arrowheads from flints and stones. For thousands of years, these weapons were effective in killing game, from the smallest rabbit to the largest mammoth. These same weapons could and did kill people of rival clans, although we will never know when the first human took up arms against another.

The art of metallurgy brought another dimension to warfare. Weapons could be forged that were larger, stronger, sharper, more durable, and had the capability of causing greater destruction and massive casualties. It has been said that the Roman Empire's dominance of the then-civilized world was due in large measure to their military's use of the steel sword. The Romans did not discover how to make steel from iron, but they were the first to see its potential as a weapon and took full advantage of its possibilities. Even before the advent of the Roman Empire, ancient civilizations fashioned metal armaments (knives, swords) and armor (helmets, shields) that provided combatants with distinct advantages and a layer of protection that were integral in the success of their military campaigns.

Survival Tools

Choppers and Hand Axes

The *chopper* is believed to be the earliest tool invented by primitive humans, approximately 2.6 million years ago. It was merely a flat, weathered, fist-sized rock that had been chipped away on one side to create a jagged or serrated edge. The chopper, used exclusively by primitive humans for about 2 million years, cut through the skin and muscle of slaughtered animals. The *hand axe,* invented approximately 500,000 to 600,000 years ago, was an improvement or refinement of the chopper. Both sides of the hand axe were chipped and considerably sharper. Neolithic humans made further refinements to the axe, an important farming tool.

Spears and Harpoons

In addition to rocks and stones, wood, particularly in the northern climates, was particularly plentiful and useful. It provided the material for the fires that were often deliberately set to stampede animals toward the edges of high cliffs, where they fell, died, and were butchered. The unused portions of their carcasses were left there to rot and fossilize in many cases. A long stick picked up from the ground could be used as either a tool or weapon. A primitive *spear* was merely a long stick with a deliberately fire-shaped, sharpened end. It had many uses: as a fishing tool, a digging stick, and, of course, as a thrusting weapon to bring down an animal. At some point in their development, spears were affixed with flaked stones, enabling them to be sharper and more piercing instruments. Bronze Age refinements made the spear a deadly weapon of war. For the most part, early humans used as much of the carcass of a slaughtered animal as possible. Bones (large and small), horns, tusks, antlers, fur, and hides were materials with potential, including as tools. For example, *harpoons,* spearlike weapons with barbed ends used for spearing large fish or whales, were made from the antlers of deer that were much larger in size than the modern species. (See Figure 11.1.) Small bones or horns were carved into hooks and needles to sew leather garments.

Flints and Flakes

One of the most important rocks chosen by Paleolithic humans for tools and tool-making purposes was the *flint,* a hard, fine-grained quartz

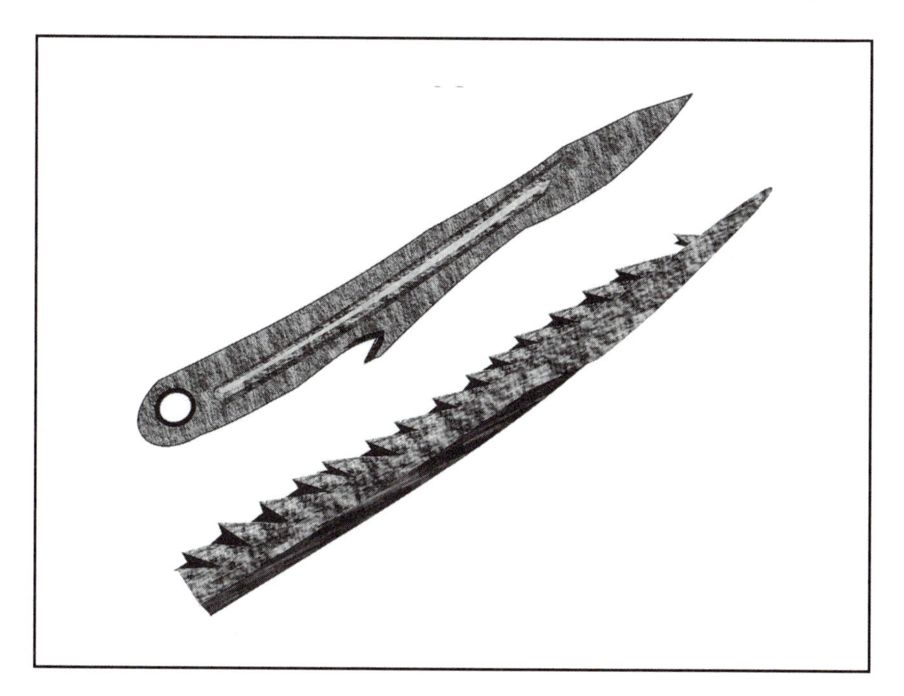

Figure 11.1 Antler Horn Harpoons, circa 6000 B.C.E.
Sometime during the Bronze Age, humans carved more sophisticated tools for specific purposes. The *harpoon,* a spearlike tool—or weapon—used in the capturing and killing of large fish, was made from the antlers (horns) of deer that were much larger than the present-day species.

Figure 11.2 Bronze Age Arrowheads and Flint Dagger
These *arrowheads* and *dagger,* made from flint, are about 7,000 years old. Found in Denmark, they are examples of flake tools.

that, when flaked, produced an extremely sharp cutting edge. Large-sized pieces of flint could be easily broken with a sharp blow, reducing them to smaller, more easily workable pieces. Using small pebbles, the smaller pieces could then be conchoidally fractured—a piece of the flint could be "flaked off," leaving a shell-like depression. Primitive toolmakers learned, over time, how to control the size and the shape of the *flakes.* An example of a *core tool,* the hand axe, a refinement of the chopper, was produced by flaking both ends of a large piece of flint. The scraper, a flaked tool that cleaned or abraded the flesh from animal hides, is an example of a *flake tool.* The remarkably shaped dagger and arrowheads found in Denmark, which date back to about 5000 B.C.E., are also examples of flake tools. (See Figure 11.2.)

Fishing Rods, Hooks, and Lines

Quite plausibly, the earliest and most primitive fishing equipment was merely a pair of hands that, when fast or lucky enough, grasped a slippery-skinned fish and held onto it long enough to bring it to dry land. About 40,000 years ago, primitive humans used hooks made from bone or thorns to catch fish. About 15,000 years later, early humans invented the *gorge.* Sharpened at both ends, the gorge was a short, spearlike rod, baited and tied in the middle with a short line. Since it could only snag one fish at a time, it was basically ineffective as a tool for catching sufficient numbers of fish to feed even a small community. In addition, fishing lines made from animal gut were simply not long enough to reach larger, bottom-feeding fish. Fishnets made from reeds, vines, and branches were used about 10,000 years ago, providing a larger catch, but necessitating that it be eaten almost immediately as fish spoils rapidly, especially in warmer climates. When the drying, salting, and smoking of fish was first discovered as a preservative about 5,000 years ago, the haul net made from a variety of natural fibers made fishing a popular and profitable vocation. Six thousand years ago, the ancient Chinese used bamboo rods and silk lines, while 2,000 years later the Greeks still used branches. As with so many other inventions, the Romans were the ones to perfect the original idea. They invented the modern version of the fly rod in the second century C.E. and the jointed rod in about the fourth century C.E.

Bow and Arrow

The *bow and arrow* is a prehistoric invention from the African continent, although there is a dispute among scholars as to whether it dates back to 20,000 B.C.E. or as early as 30,000 B.C.E. Regardless, it is an invention that allowed the hunter to kill prey at greater distances than with a thrown spear—and with less danger to the hunter. A bow is a rather simple device consisting of a *stave* (a narrow strip of wood that is bent) and held in tension by a string tied at each end. Strings were made from vines or animal gut, hide, or tendons. The arrow consists of a long, straight wooden shaft with a pointed head at one end and feathers or a tail affixed to the other. Presumably, the first arrows were carved sticks, later improved upon with the shaping of stone, flint, and metal arrowheads. A notch carved onto the shaft serves as the point at which the string and arrow fit together. The feathered end serves as a balance and a stabilizing and slow-spiraling directional device that results in a more accurate trajectory. The archer draws back the string (under tension), releasing and propelling the arrow at a tremendous rate of speed and distance. Extremely effective hunting tools, bows and arrows quickly became weapons, often the primary weapon of ancient warring factions that refashioned and improved upon the basic concept with deadly consequences.

Colonizing (Working) Tools

Axe and Adz

During the Neolithic Period or New Stone Age, ground and polished tools were in evidence and settled farming practices began to take hold. The stone axe was one of the greatest inventions. The handle with a weighted stone on the end made use of the physical principles of angular momentum and inertia, which made it a very effective tool and weapon. The polished axe replaced the smaller stone axes that were predominant during the Paleolithic Period. This improved axe provided an efficient tool for clearing land, leading to a "slash and burn" form of agriculture. The *adz* is a tool that is used to shape wood. In its earliest form it was a handheld stone that was chipped to form a blade. Early farmers used the axe and the adz to strip the bark in a circular pattern around the circumference of trees. This killed the trees, making it easier to cut them down. They then burned the materials from the trees in order to prepare a field for planting. Later, the adz was modified to resemble and function like a short hoe. After the burn, the field was prepared for planting by using axes, sticks, and then hoes to make depressions in the soil where seeds could be planted. This ritual was performed every few years as the community of farmers moved on to new areas. Ashes from the burn provided some additional nutrients to the soil, which was soon depleted, reducing its fertility. Since, at this time, fertilizers were not used, it was necessary to find and clear another field.

Hoe

The digging stick, the predecessor to the *hoe,* was a sharpened tree branch often hardened by fire. At times the stick was weighted with a stone. The more advanced hoe has a blade that is attached at a right angle to its long handle. Before the Metal Ages, the earliest hoes used stone or wooden blades and resembled an adz. Eventually, plows replaced hoes as the main agricultural tool to loosen soil for planting.

Plow (or plough)

The *plow* is the most important agricultural tool to be crafted by humans. Until its "invention," digging sticks and hoes were the only tools available to break up soil, dig furrows, and turn over weeds. The first plows were in all likelihood digging sticks with some type of attached handle that was either pushed or pulled by the farmer. Simple plows, called *ards,* date back to about 3000 B.C.E. in what is present-day

Iraq. A mural in a tomb that is more than 3,500 years old depicts Egyptian farmers plowing fields. It was a difficult task to maneuver the earliest plows (without wheels or animal power). The plowman needed considerable strength to dig the soil to make the appropriate depth of trench or furrow. This often meant having to plow a second or even third time over the same patch of field. At some time in its evolution, it became an advanced technological tool. A share (blade) was attached. The most significant innovation was using the strong backs of oxen and horses, under the direction of the farmer, to do the actual pulling of the implement. Sometime before 100 B.C.E., the plow was improved in Europe to include a coulter, which cut the soil, and a moldboard, which turned it over. When the Romans added wheels in the first century C.E., it became easier to control the depth of the furrow. These improvements led to the expansion of farming, particularly in the northwestern part of the European continent, where lighter, wheelless plows, in existence for centuries even during Roman times, could not adequately pulverize the heavier soils.

Before the Metal Ages, the wheels for plows were fashioned out of hardwood. With the advent of the Bronze Age, toolmakers fashioned the plowshares and later the plows themselves from metal alloys that were more durable and efficient. For instance, the agricultural output in China improved tremendously when metal farming tools first became available to the peasant farmers about 2,500 years ago. However, these early plows continued to be worked by hand; oxen-drawn plows in China were not used until about 100 B.C.E. The Chinese later developed a three-bladed plow and by the end of the thirteenth century C.E., they had made significant technological advancements in agriculture. This was also true for some but not all parts of the Asian and African continents, where the iron plow and other implements made it easier to clear and cultivate forestlands, and where, eventually, nonindigenous (not native) crops, imported from other regions, could be successfully grown.

As with all aspects of agriculture, its origin and the use of implements did not occur simultaneously on all continents and among all civilizations. For instance, in the Americas, Mayan farmers who successfully grew a number of crops did so for centuries without the aid of plows. Despite the innovations and improvements that were made to the plow over time, the basic techniques that ancient farmers used thousands of years ago continued to be practiced by farmers worldwide until the invention of the internal combustion engine, which made plowing a mechanical rather than an animal and human endeavor.

Figure 11.3 Roman Sickle and Vallus
Examples of colonizing tools, the Romans used the *sickle,* the most ancient of metal farming implements, to harvest cereals and grain, and the *vallus* as a reaping machine, powered by both man and animal.

Sickle and Scythe

The *sickle,* or reaping knife, is one of the most ancient of metal farming implements used to harvest cereals, grain, and rice. It has a sharpened curved blade that is set on a short wooden handle. (See Figure 11.3.) Because of its short handle, the farmer was forced to bend over or squat while cutting the plants, making it an arduous and difficult chore. The sickle was and still is widely used the world over, primarily because it is a low-cost and efficient tool. The *scythe* is an important metal agricultural tool whose exact origin is unknown. A hand tool for cutting grain, the scythe consists of a curved blade fitted at an angle to a long, curving handle. It was not widely used until the eighth century C.E., when hay grown on the European continent was harvested and stored to feed livestock through the winter months. The longer-handled scythe, which enabled the farmer to stand while cutting stalks of grain, is believed to have evolved from the sickle.

Vallus (Reaping Machine)

The Romans are credited with inventing the first reaping machine or *vallus,* probably about the first century C.E. (See Figure 11.3.) Prior to its invention, all farmers used the sickle or the scythe to harvest their plantings. The vallus resembled a cart, except that an animal, usually a donkey or mule, pushed rather than pulled the machine. The animal (who itself was pushed or guided by a laborer) pushed the vallus, which was crafted with metal teeth that cut the grain from the stalks. The grain then fell into a boxlike hopper where another laborer contained it. Although this was an efficient method for reaping the harvest, it was not widely used. The Roman aristocracy did not encourage technology for the slave laborers who worked the fields and farms, as they feared the advent of labor-saving devices would adversely affect the Roman way of life, which was basically a slave economy.

Flail

After the farmer harvested the crop, mainly grains or cereals, he needed to thresh (separate) the grains from the straw. This was accomplished by beating the grain with jointed *flails.* These were long-handled manual devices that contained a hinged shorter stick at the bottom end of a longer stick. The shorter stick would swing freely when the farmer or laborer swung it on a pile of grain in order to beat the grain loose from the stalks. As with other tools used in early agriculture, its exact date of origin is unknown, although once invented its use was widespread and consistent for centuries. After threshing, the grain was then tossed into the air and caught in open-ended shallow baskets while the wind scattered the chaff and blew it away. This process was called winnowing.

Quern

Various forms of the *quern,* a device used to grind grain, date back some 8,000 years. Most likely, the mortar/pestle was the forerunner of the quern. This ancient tool for pounding the grain in a bowl-like stone gave way to the saddle quern that was used in Egypt about 4,000 years ago. This was simply a flat stone upon which the farmer, laborer, slave, and even the housewife, while kneeling, pounded the grain with a rounded stone. (See Figure 11.4.) This was excruciating work that was quite debilitating, and especially hard on the knees, toes, and back of the miller. (In Egypt, for instance, the life expectancy of a miller was no

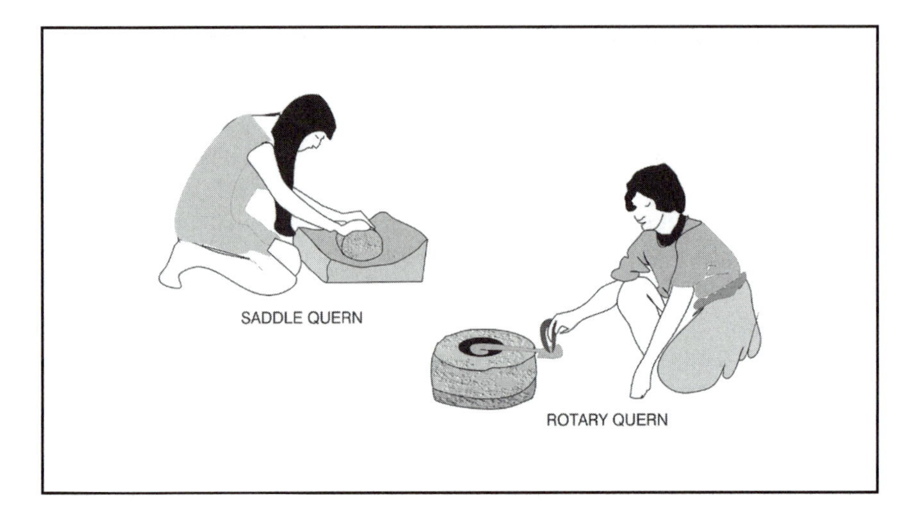

Figure 11.4 Querns
The *quern,* a device to grind grain, was invented about 8,000 years ago.

more than 30 years.) The Greeks improved on the saddle quern with the pushing mill, in which both the bottom and top stones were flat. The grain fell onto the bottom stone from a hollowed-out portion of the top stone through a slit, also on the top stone. The Romans used a rotary quern adapted from devices that came from the Near East. From the earliest days of farming, milling was a manual operation performed by slaves, laborers, and women. It was not until about the fourth century C.E. that watermills provided the power to replace the intense and arduous labor of humans as millers of grain and corn.

Weaponry

Daggers, Swords, and Spears

During the Copper and Bronze Ages, flaked, one-piece flint and stone daggers were replaced by more precisely designed bronze daggers. A *dagger* is defined as a short, pointed weapon with double-sided sharp edges, while a knife is a sharp cutting instrument with, usually, a single, finely tapered edge. (It is possible to have a double-edged knife.) In the earliest days of antiquity, the design and purpose of knives and daggers were most likely indistinguishable. Eventually, knives became more specialized and were fashioned for specific tasks—for example, longer, wider, heavier, more rounded blades to cut through animal hide, gut fish, or slash through vines, and so forth. On the other hand, the dagger remained essentially unchanged in its design as a deadly close-range weapon. Over time, the dagger's blade was modified to include a lengthwise ridge, rendering it more inflexible. A *hilt* (handle) was also added for a better grip.

While it might be said that a *sword* is merely a dagger on a large scale, the very length of its blade created great challenges for the sword's maker. Before the discovery of iron and the carburizing process, copper and bronze were the only metals available to the coppersmith for weapon making. Brass, an alloy of copper and zinc, was popular, particularly in China—but not necessarily for fashioning weapons. Copper, an extremely soft metal, was unsuitable for the sword's long blade. Bronze, an alloy of copper and tin, was somewhat stronger, and by tempering it on an anvil, a reasonably hardened blade resulted for both daggers and swords. It was, however, the discovery of steel that revolutionized weaponry. Of itself, iron is soft and malleable. The addition of carbon, as well as the tempering process, gives iron its strength. Steel is often said to be an invention of Hindu physicians, who used carburized iron (i.e., steel) for surgical instruments. The iron sword had been in existence since about 1200 B.C.E. and was designed to penetrate the copper or leather body-armor worn by foot soldiers. While the Romans discovered neither iron nor steel nor the sword, they did exploit the many ideas and inventions of other ancient civilizations, including the manufacture and use of steel. They perfected the process and made the steel swords of Roman soldiers the most important weapon in their arsenal. The military successes of the Roman armies on three continents were virtually unchallenged for nearly a thousand years.

As a weapon, the spear, which has been in existence for about 40,000 years, can be hurled at a distance or thrust at close range in order to inflict its deadly intent. Once humans moved beyond sharpening the ends of long branches, the spear evolved into a precise weapon. Prehistoric humans invented a *spear-thrower,* designed to give the spear greater thrust and acceleration. Sometimes referred to as a throwing-stick, it was constructed of a rod-shaped piece of wood (bamboo, bone, or antler horns were also used) carved with a groove on the upper surface and a hook or thong at the bottom that held the spear in place until released. The spear lay flat alongside the spear-thrower with the end resting on the hook or thong. When the spear was discharged, it flew great distances and was able to fell large animals, especially mammoths. The Mayas of Mesoamerica called the spear-thrower the *atlatl.* The ancient Greeks and Romans engineered an efficient and deadly spear-thrower, called the *becket,* that caused the spear to spin as it flew through the air. The modern-day *harpoon gun* is a variant of the prehistoric spear-thrower. The first spearheads were made, just as arrowheads were, from flints and stones. The first metal spearheads were made from copper that was fashioned with *tangs*—sharp pronglike projections designed to fasten the spearhead to the shaft. On the other hand, spearheads made from bronze, a harder metal alloy, were hollowed out and fitted tightly on the end of the shaft.

Along with the evolution of weapons from mere stones to those that were finely crafted, the conduct and demeanor of warfare also became more uniform and systematic with the rise of civilizations. Armies organized, and spears that were once thrown at the enemy somewhat randomly were now group weapons carried by foot soldiers, who held them in check until they were ordered to proceed with a thrusting assault. This formation, called a *phalanx,* is a military maneuver that is over 5,000 years old. The basic spear evolved into a number of weapons that were intended to inflict great harm while distancing the bearer from close contact with the opposing soldiers. These new spearlike weapons were the following:

1. The *pike,* a 2- to 3-meter ($6^{1}/_{2}$ to 10-foot) spear with a spike on the end.
2. The *sarissa,* a pike approximately 6–6.5 meters (20–21 feet) long, was invented by the Macedonians in the fourth century B.C.E.
3. Roman legionnaires used a *pilum,* a heavy javelinlike spear that was over 2 meters ($6^{1}/_{2}$ feet) in length.
4. The *lance* and the axe-bladed *halberd,* medieval weapons carried by knights on horseback, were innovations of the ancient spear.

The invention of gunpowder and the subsequent development of artillery weapons made the use of handheld and pole-armed weapons less desirable, though never quite obsolete. Even today, modern armies are equipped with variations of these most ancient instruments of war—for example, the *bayonet,* which is placed at the end of a rifle.

Body Armor

Long before medieval knights were outfitted head to toe in "shining armor," ancient warriors wrapped heavy leather around their torsos, arms, and upper legs as protection from the piercing wounds of spears, arrows, and daggers. The Sumerians of Mesopotamia are believed to have been the first to develop helmets and body armor (breastplates) made from copper and bronze in about 3000 B.C.E. However, body armor was an independent invention of many other ancient civilizations that used a number of natural materials for protective purposes. For instance, some historians believe the Scythians, a fierce, nomadic race who lived in the region of the Balkans, may have had bronze body armor as early as 5000 B.C.E., but there is little evidence to substantiate this claim. However, by about 2500 B.C.E. soldiers of the Middle Eastern civilizations were given rigid bronze breastplates, shields, and *greaves* (metal plates that covered the front lower leg) before going into battle. Not all civilizations recognized the advantages of metal body armor. As late as 1000 B.C.E., ancient Chinese warriors continued to wear armor made from tanned rhinoceros hide, as many as five to seven layers thick, while the body armor of fifth-century B.C.E. Greek soldiers consisted of the *cuirass.* This heavyweight fabric garment made from multiple layers of linen covered the back and breast/rib areas. The infantrymen of the Indus civilizations favored a similar type of fabric body armor, only quilted, which remained popular into the nineteenth century C.E.

The third type of body armor was called *mail;* it was a series of interwoven rings made from metal, usually iron or steel. The metal armor of the Sumerians 5,000 years ago was a design that resembled fish scales: small pieces of hammered copper or bronze were overlapped and sewn onto a backing of heavy leather. This metal breastplate of the Sumerians provided a bit more protection but could still be pierced by the mace or a heavy sword or dagger. This "fish scale" design may have been the inspiration for the more densely woven mail that dates back to about 400 B.C.E. The ancient Celts of Britannia are often credited with inventing it. (It is more commonly called *chain mail,* presumably because the pieces of metal are chained together.) The Romans improved upon the basic concept of mail. The chain mail tunics of all Roman legionnaires, known as *lorica hamata* (meaning "hooked breastplate"), were made of steel and were worn under steel cuirasses. Chain mail is not impervious to the thrust of a sword, but it did provide an amazing amount of protection from slashing weapons. Obviously, rigidly formed armor afforded the most protection, but was more expensive to manufacture and more cumbersome to wear into battle. When the Greek army abandoned the heavy fabric cuirass, they replaced it with one made of bronze, and added bronze greaves as well as a bronze helmet. Roman legionnaires wore a cylindrical steel cuirass or breastplate consisting of as many as seven horizontal segmented hoops of steel that were laced together in both the front and back. These were called *lorica segmentata*, meaning "trimmed breastplate." The shoulders of the soldiers were covered with vertical steel hoops.

Over the centuries, armorers refined and refashioned body armor to give maximum protection to its wearer. The medieval knight's full-body armor is an example of this extreme craftsmanship. However, with the invention of gunpowder and artillery weaponry, armorers recognized that body armor would need to be thicker and heavier—and thus more cumbersome—in order to withstand the explosive charges of these new weapons. The bulletproof vest is an outgrowth of the original fabric or rigid metal breast coverings of ancient civilizations, although the modern version is designed to stop bullets rather than the piercing of a sword or dagger.

Helmets

For the most part, the *helmets* worn by the warriors of all ancient civilizations were crafted of leather. With the discovery of metals, leather helmets could be reinforced with bronze or iron straps, or be encrusted with metal adornments and crests of horsehair. While menacing in appearance, they afforded little protection. Later Bronze Age designs varied from a cone-shaped helmet that rested on the head above the ear line to one that covered the entire head, except for the facial area. (See Figure 11.5.) Beginning in the first millennium B.C.E., ancient Greek metalworkers began to fashion more efficient and protective headgear, eventually becoming the predominant designers of military helmets in antiquity. The Greek bronze helmet often covered the entire head and facial area, as well as part of the neck. A narrow opening in front allowed for sight and breathing. Roman helmets were varied in design, and included the round legionnaire's helmet, and the gladiator's helmet, with a broad brim and a pierced visor that gave the wearer a fair degree of protection from an opponent's sword. Helmets made in China and the Far East were constructed of leather, bronze, and horn and were lacquered for greater strength and durability. By about the sixteenth century, body armor, including helmets, was abandoned as protective gear for infantrymen. However, the helmet reappeared as a piece of military equipment during World War I. It remains an essential part of the combat gear of every soldier and sailor.

Catapults

Stories abound concerning the often mythological, highly skilled archer, from Odysseus in Homer's *Odyssey* to the medieval characters of

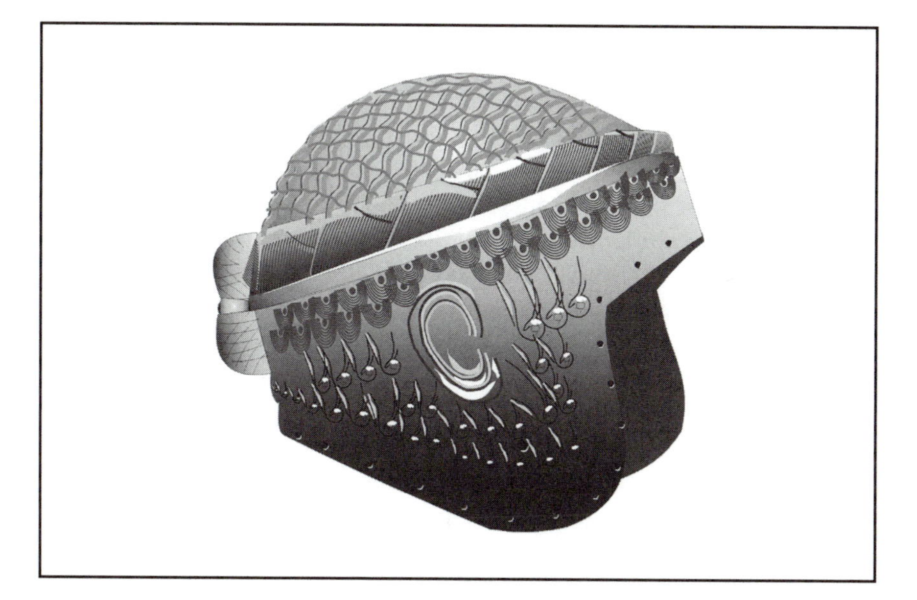

Figure 11.5 Akkadian Bronze Helmet
This *helmet* dates back to about 2500 B.C.E. Made from bronze and padded with cloth, it is believed to have belonged to an Akkadian of high rank.

Robin Hood and William Tell, who could accurately hit a target under seemingly impossible conditions. While these legends make for interesting and romantic tales, it was, nevertheless, the organized assault tactics of large bands of archers, either stationary or on horseback, that made the bow and arrow the principal weapon of antiquity. However, the human archer is limited physically by the distance a single arrow can travel, as well as by its accuracy and ultimate damage. The military objective of all armies, either ancient or modern, is to inflict the greatest amount of harm to the enemy while sustaining the least amount of harm to its own soldiers. To that end, the *catapult* and the *crossbow* were two improvements based on the prehistoric spear-thrower and bow and arrow that were engineered to provide greater distance, greater force, and greater casualties.

The basic and most primitive catapult is merely a simple mechanism that propels spears, stones, or other objects. The Greeks invented the mechanized catapult in about 400 B.C.E. This was a deliberate invention that was, in a sense, commissioned by Dionysius the Elder (ca. 430–367 B.C.E.), the despotic ruler of Syracuse, a Greek colony on the isle of Sicily. He ordered his engineers to design and build new weapons for his anticipated war with the Carthaginian Empire. (Carthage was a Phoenician city on the Mediterranean coast of Africa.) The catapult was the most successful of those weapons. Succeeding military engineers, primarily in Greece and Rome, further improved upon its design and capabilities. The catapult can be either a freestanding siege machine or a smaller, almost handheld mechanism, such as a bow-firing catapult, also known as a *gastraphĕtēs*. The *ballista* was a Roman catapult designed to hurl stones. The Roman *catapulta,* which could be wheeled from place to place during battles, hurled arrows, darts, and wooden bolts about 68.5 centimeters (27 inches) long. Archimedes, the great mathematician, is said to have designed a catapult capable of hurling a 79-kilogram stone (about 175 pounds) over a distance of 182 meters (199 yards).

Although the design evolved over the centuries, all ancient catapults worked on the same basic scientific principle—using stored or pent-up energy to achieve a specified goal. Objects were held in place on flexible wooden beams or twisted ropes, called a *bow,* made from horsehair, animal gut, tendons, or some other heavy fiber. These were cranked to produce tension and elasticity and then suddenly released, converting the potential energy to kinetic energy, and propelling the object (stone, spear, arrow) towards a directed target. A major improvement in the design of the ancient catapult was the replacement of the bow with a torsion system: a *winch* twisted a skein of animal hair and tendons to a high degree of tension, propelling objects to farther distances than those released by the flexible bow. Mathematicians, particularly the Greeks, played an important part in the design of armaments, inasmuch as their calculations dealing with angles, weights, and trajectories were essential to the design of assault weaponry. More important, all weaponry is based on the principles of the six simple machines. Siege catapults could be used either offensively or defensively and quickly shifted the advantage to whichever side possessed the weapon. The velocity and force of an over 300 pound stone could quickly destroy the walls of a city or other fortification, and the release of hundreds of spears or arrows could stop or hinder the attack of an invading army. Victory was a combination of skill, might, weapons, and luck.

With regard to the bow-firing catapult, some historians believe it was the weapon that Dionysius's engineers reportedly invented around 400 B.C.E. Others claim that the date of its actual invention is unknown but acknowledge that it was not in use until sometime after 404 B.C.E. It was called the *gastraphĕtēs,* meaning "belly-shooter" in Greek. Hero of Alexandria described and drew the plans for the original *gastraphĕtēs,*

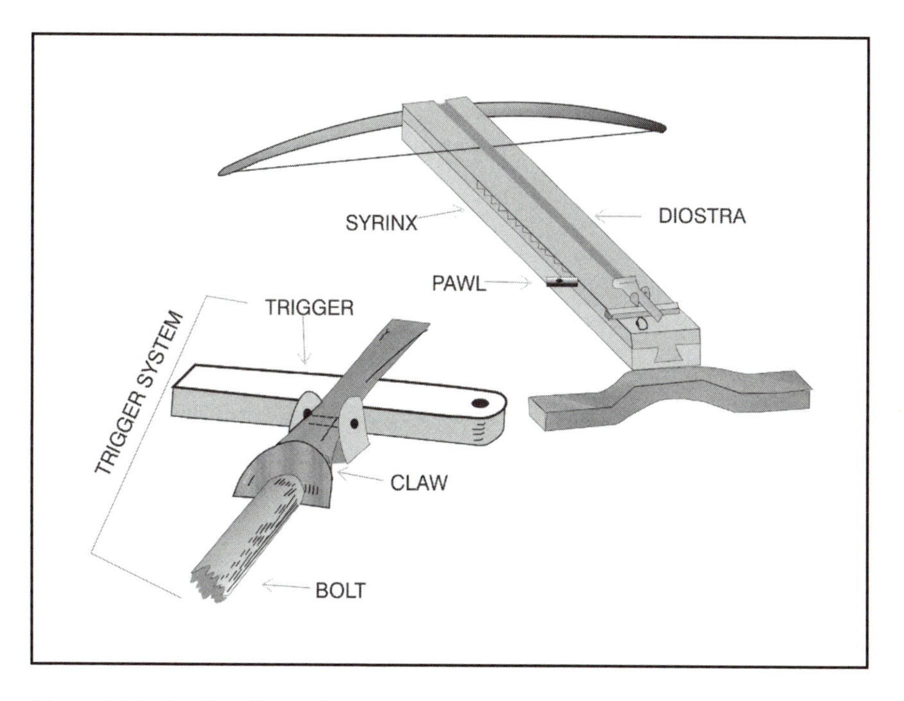

Figure 11.6 The Bow Catapult
Also called a *gastraphĕtēs* (belly-shooter) in Greek, this *bow catapult* was invented sometime between 404 and 400 B.C.E.

although his descriptions left out a number of important points and parts, including the arrow, which compelled the reader or weapon maker to assume certain facts that were either implied or not mentioned altogether. (See Figure 11.6.) However, the engineering principle of the *trigger system,* which was very similar to a crossbow, was not difficult to decipher and was performed in this manner: An *arrow* (about 183 centimeters or 6 feet long) was placed in the *syrinx* or *trough* (long, narrow depression) in the *diostra* (slider), where it was locked into place with the *claw.* The slider, with the arrow in position, was then drawn back against the tension and held in place by the *pawl* (a pivoted device that fit into the notch of the ratchet, allowing for forward motion or preventing backward motion). By lifting the claw, the trigger was moved and the arrow released. The device was called a belly-shooter because the stomach muscles, as well as those of the upper body, were needed to pull back the tightly tensed bowstring, *not* because the weapon was fired at the height of the shooter's midsection. On the contrary, the bow catapult was often placed atop a wall and aimed down at the targeted area. It was a cumbersome device that was moved after each shot. However, at the time of its invention, it was considered state-of-the-art and was quite effective and continually improved upon.

Crossbow

Historians believe that the Chinese are responsible for the design of the *crossbow,* an invention that was their armies' principal weapon until the end of the nineteenth century C.E. In simple terms, the basic difference between the bow and arrow and the crossbow is merely direction. The bow and arrow is held vertically, the crossbow horizontally. The Chinese adapted the crossbow from a primitive hunting concept used by indigenous (native) tribes, namely the "bow-trap." A bow and arrow was laid horizontally on the ground and a long piece of wood held the bow in a taut position. A long piece of vine or rope, attached to both the wood and a tree or bush, was positioned so that a prey animal would trip it, thus knocking away the wood and releasing the arrow into the target (animal). The dates for the crossbow's invention are unclear, primarily because written texts from China are unreliable or have been lost. Generally, it is believed that the crossbow's invention took place around 400 B.C.E., although it is possible that it is older by some 300 years. The justification for this is the reference to the crossbow as being a common or familiar weapon in *The Art of War* by Sun Tzu, a book that is reported to have existed since 500 B.C.E.—and certainly by about 400 B.C.E. The Chinese were clever and fierce warriors, and the crossbow became primary in their arsenal, not only because of modifications in its design that made it more deadly and accurate, but also because of the sheer numbers of the weapon. For example, the ancient Chinese crossbow's triggering system only contained three moving parts on two shafts, each of which was cast in bronze and engineered to perfect alignment. The design of the crossbow was standardized and it was mass-produced by the Chinese in numbers up to 500,000 in the second century B.C.E. This technology, however, proved difficult to replicate among other ancient tribes, who for the most part were incapable of providing the type of work environment necessary to produce such precision weaponry. While the crossbow had advantages—a lever held the string and arrow in the cocked position until the archer chose to release it—it also had its disadvantages in that it took a fair amount of time to reload, and cock, making the archer or bowman vulnerable to the opposing forces. The Chinese reengineered the crossbow so that it could fire multiple arrows or bolts, as many as 10 or 11, in rapid succession. The actual date of the crossbow's appearance on the European continent is uncertain, but historical records confirm that it was in existence in Wales in about 1000 B.C.E. Whether it was an independent Welsh invention or whether the idea was transported from China in some fashion remains unclear. The crossbow, the long bow (which was a Welsh invention), and bow catapults were standard among the armies of the Middle Ages and the Renaissance until gunpowder and artillery weapons rendered them obsolete.

The Chariot

Artifacts found in Iraq indicate that chariots were in existence from at least 3,200 B.C.E. and employed in a military capacity by about 2500 B.C.E. Archaeologists and historians believe that, at first, chariots were used solely for ceremonial purposes—for example, in the funeral processions of nobles. The framework of these early chariots was wood covered with tanned animal skins; they were pulled by yoked oxen. This rather narrow usage was most likely due to two factors: (1) the primitive design of the wheel, which in the third millennium B.C.E. was a heavy, solid piece of wood, making the ride unbearably bumpy and rough; and (2) the use of onagers (wild asses) and oxen rather than horses to pull the carts. (The use of horses as draft animals would not become popular until about 2000 B.C.E.) The wheel was a rather late invention of humans, and it took centuries before the ancients redesigned the wheel with axles and spokes, which made the ride smoother and more maneuverable. And even after horses were domesticated and bred for draft work, it took many centuries before an efficient harnessing system was designed that did not constrict their windpipes and produce irritating sores on their hides.

The earliest military chariots were cumbersome conveyances that were used as transports for spearmen. They were not themselves used in battle. In time, the design of the chariot evolved into one that was lighter in weight, thus faster. Some chariots were equipped with two rather than four wheels, but the wheels were spoked and affixed to metal axles, making them more maneuverable. Specially bred horses, in teams of two, sped the charioteer and his bowman into battle with a degree of mobility that was previously limited. Thus, it could be said chariots revolutionized the conduct of warfare of the second and first millennia B.C.E. They were used in great numbers by the armies of Egypt, Mesopotamia, India, Greece, and China in the second millennium B.C.E. The Etruscans (ancestors of the ancient Romans) ostensibly brought the chariot's design to the Celts of Britannia around 500 B.C.E.

Although the basic construction material was wood and leather, carpenters emblazoned the chariot with bronze reliefs that depicted important victories, and wheelwrights outfitted the wheels with iron rattles and knives that intimidated and ravaged the legs of the opposing force's soldiers and horses. The chariot was, however, a piece of military equipment that was relegated to an elite group of warriors who were skilled in their ability to control a fast-moving, horse-driven conveyance into battle. And it was most effective against peasant or infantry armies that did not possess this expensive item of weaponry. In a conflict where both armies commanded battalions of chariots, the charioteer and his bowman and team of horses were at a disadvantage. The impact of colliding horses, wood, metal, and knives resulted in tremendous numbers of casualties on both sides.

Over time, the cavalry (individual troops mounted on horseback) replaced the use of chariots in a military capacity. Chariot racing, however, became increasingly popular, starting in Greece in about 800 B.C.E., expanding to the Olympic games in Athens in about 500 B.C.E., and culminating in the spectacular races held in the *Circus Maximus* in Rome during the first and second centuries C.E. Racing chariots were lighter-weight, two-wheeled conveyances that were drawn by teams of up to six horses. Although the Greeks were the originators of the chariot race, the Romans celebrated the spectacle and built *hippodromes* (from the Greek word meaning "horse courses") that seated thousands of spectators. Reportedly, they were as popular as the gladiatorial matches and, by all accounts, every bit as bloody.

Mobile Assault Weapons

The fierce battles fought by armies on open fields resulted in large numbers of casualties. However, for the most part, the battles were swift and left no doubt as to the victor. On the other hand, when one side was entrenched behind the walls of a fortified city, the military tactics differed considerably and it became a deadly game between the attacker and the besieged. The first weapon of the attacker was starvation. The theory was that by preventing food supplies and water from going into a city or fortification, those within became weakened and demoralized while the advancing army waited—often for long periods—with full bellies and rested spirits until the moment of assault, when victory would be easy and assured. In antiquity, just as in modern times, the theory is not always supported by the facts. Often, it was the attacking army that suffered the pangs of starvation and defeat, particularly if the besieged had prepared for the coming onslaught. History is replete with stories

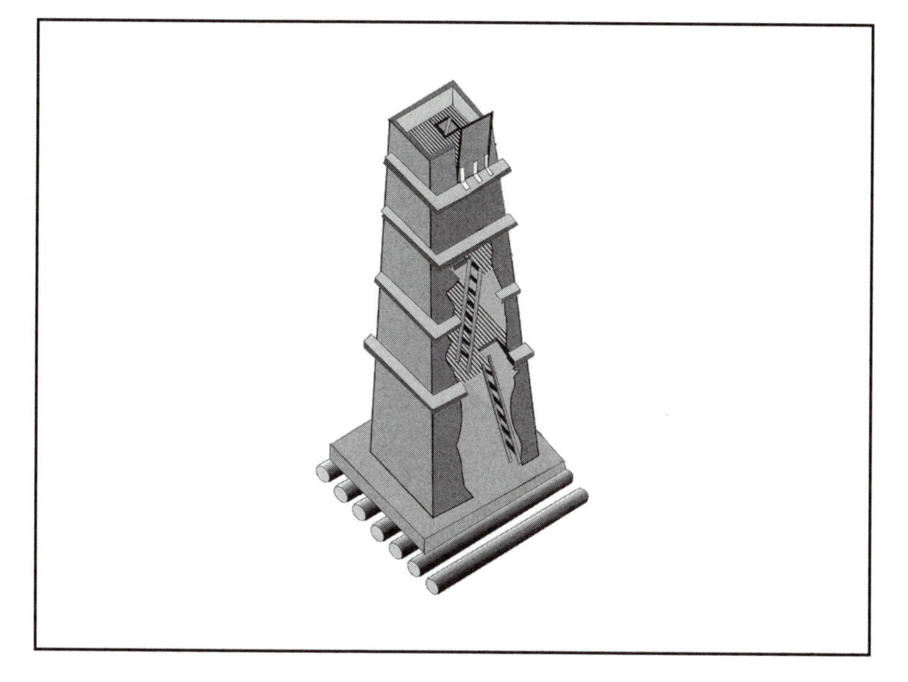

Figure 11.7 Assyrian Mobile Assault Tower
Mobile assault towers were siege weapons that were constructed on-site. They were moved by a series of rollers (logs) up to a city's walls, where the soldiers inside the tower climbed to the top, bombarded the city with various weapons, and finally scaled the walls for entrance into the city under siege.

of long and unsuccessful sieges. For example, the Roman siege of Syracuse that began in 215 B.C.E. was a showcase for the talent and ingenuity of Archimedes, who invented a number of "siege engines," including the Claws of Archimedes, that withstood the assault for nearly three years. The invention of mobile assault and siege weapons and equipment, in a sense, equalized the odds. Towers, usually constructed on-site, were moved via logs up to the walls of the city or fortress. Specially constructed platforms allowed archers sufficient access to shoot hundreds of arrows over the walls and/or soldiers to scale them altogether. (See Figure 11.7.) Assault and siege weapons have been in existence for about 1,200 years. Perhaps the most basic is the *battering ram,* which was merely a huge tree trunk that was lifted and swung by a large group of soldiers, or suspended on ropes and swung like a pendulum, and rammed into a door, breaking it down. The Romans built elaborate siege engines and assault towers, outfitted with ramps and ladders, platforms and towers, and wheels. Fire was also a siege weapon. Archers dipped the ends of their arrows into oil, then ignited and shot them into cities and fortresses where thatched roofs and other flammable items burned uncontrollably. Fires were also set around the foundation of fortresses that, in ancient civilizations, were primarily constructed from timber. The siege and assault weapons have not survived, however, as they too were constructed of timber and were often abandoned after the siege ended.

Wheelbarrow

Invented in China, the wheelbarrow was initially used for military purposes, *not* for agriculture or construction (building) purposes. Its date of origin and inventor remain in dispute. Some historians believe the wheelbarrow first appeared in the first century B.C.E., while others say it was the first century C.E., and even as late as the mid-second century C.E. As to its inventor, certain accounts name a Taoist farmer, Ko Yu, who may or may not have existed, while others credit Jugo Liang, a Chinese general. What is not in dispute, however, is that the ancient Chinese treated this invention with great secrecy. The Chinese used wheelbarrows to transport supplies and weapons, as well as the wounded and dead, over the inhospitable, hilly terrains of the Chinese provinces. During skirmishes, wheelbarrows were turned on end to act as movable barriers against incoming cavalry assaults. The Chinese also employed variations of wheelbarrow designs, including ones that placed the wheels in the center of the transport so that the weight rested entirely on the axle. Some had smaller wheels that were designed to maneuver around the boulders and potholes of the landscape. Some were constructed entirely of wood. Others had the more conventional design of molded metal and wood. Interestingly, the wheelbarrow design that is most common today in the West is not the design that was—or is—most popular in China, the country of its origin. The Chinese preferred, then as now, to fashion certain designs to fit specific tasks. The wheelbarrow did not "travel" westward into Europe, where it became more of an agricultural and building-trades implement, until the thirteenth century. The more efficient horse-drawn carts and transports eliminated the need for the wheelbarrow as a means to deliver armaments and supplies to soldiers engaged in military battles.

Summary

Anthropologists, historians, and evolutionary scholars have long agreed that one factor separating humans from other animal species is our intellectual capacity to fashion an object that has more than one purpose. In other words, our ancestors were innovators and engineers who were able to conceive and craft tools and implements to facilitate the numerous tasks and hardships they faced in their quest for survival. This innate ability, instinct, or part of our genetic makeup may be the key to our species' survival and domination. And while other animals fashion objects and/or structures, such as hives, nests, dams, dens, and so forth, humans are the only species with the creativity to deliberately modify their environment and adapt to the external changes in that environment. (Modern apes are capable of fashioning twigs from tree limbs to "fish" out termites from tree stumps or ants from anthills. Anthropologists believe this is a learned behavior. In other words, it is cultural rather than genetic, and thus they are incapable of fashioning anything more advanced.) Speech is another differential, although technically, speech is not a precondition for communication. Also, our primitive ancestors' physical makeup, with prehensile (capable of holding an object) thumbs and fingers, as well as a brain that allowed for eye-hand coordination, had anatomical advantages that were not shared by other species. Just as important, it is simply a fact that humans, as a species, are simply not powerful enough—physically—to endure the rigors of survival without tools and weapons. In our rather short tenure on the earth (hominids appeared only about 5 million years ago, while the age of the earth is believed to be about 4.5 billion years), humankind has evolved and progressed at an astounding rate. Species that are considered to be prehuman, such as *Homo erectus,* who lived 1.6 million years ago, recognized the technological potential of the rock, a concept that theretofore was unknown and beyond the grasp of other existent animal groups. Humans have taken technology from the most basic to the most intricate and have amassed a set of accomplishments that are too numerous to cite. However, it is necessary to place in proper perspective the technological accomplishments of humanity, considering the destructive nature of so many of them.

12

Transportation, Trade, and Navigation

Background and History

The existence of prehumans, as well as all early life forms on planet Earth, hinged on their ability to seek out a sustainable food supply that enabled them physically not only to endure threats but also to dominate other species. In simple terms, they migrated. It is safe to assume that on these travels they encountered other humans, as well as ferocious animals, in an unimaginably difficult environment. When the food source became depleted and the area trashed with the debris of humanity, they simply moved on to another location where the pattern was subsequently repeated. Curiosity was another factor in the development of movement or transportation, particularly after civilization took hold. The quest for food was less immediate with the advent of settled farming and animal domestication, but the compulsion to seek what lay beyond the next mountain or body of water remained every bit as strong.

The topography of the earth before the last ice age, which began about 40,000 B.C.E. and lasted some 30,000 years before climatic conditions caused the glacier sheets to melt and retreat, was quite different from present-day features. Thus, wheeled vehicles would have been useless on the treacherous ice sheets, just as they would have been on the swampy marshes that were widespread after the retreat of the glaciers about 10,000 years ago. When environmental factors are considered, it is easy to understand why the wheel is a relatively new invention that dates from only about 3500 B.C.E. Boats are considerably older than wheeled conveyances. During the last ice age, navigating icy rivers would have been difficult but not inconceivable. But navigating swamps, with their abundant food sources, was possible by merely clinging to a log that floated in the murky waters—just as other, smaller animals have always done. Archaeologists have speculated that dugouts may have been used as far back as 50,000 years ago, although no actual artifacts to support this theory have ever been recovered. In modern times, the most popular methods of transport are planes, trains, automobiles/trucks, and ships. In prehistoric times, it was the human foot, and later the *sledge* (the forerunner of the wheeled wagon) and the *dugout* (forerunner of the boat).

The invention of the wheel and the use of animals for draft purposes were not discrete discoveries. Rather, they evolved over time from other purposes into the mechanisms that drove the economies of all ancient civilizations from Rome to China. On the other hand, navigation in antiquity had as much to do with adventure and warfare as with commerce. For those ancient civilizations located around coastal areas, the sea brought great bounties and great challenges. The climate and the food source were desirable, yet the threat of invasion from opportunistic states and tribes was always present. Ancient harbors were filled with fleets of fishing boats, cargo ships, and naval vessels, and almost without exception, ancient towns located along the coastline of the Mediterranean became important seaports in antiquity and beyond. This chapter deals with the ancient discoveries and inventions of land and sea transportation and their effects upon the ancient peoples of the Middle East, Far East, and the European continent.

Transportation

The Wheel

The wheel was believed to have been invented in Mesopotamia around 3500 B.C.E. for use in the making of pottery and as part of a pulley system (windlass) to raise water from wells. At some point, humans realized its potential as a labor-saving device, but exactly who or when will never be known. It is probably a safe assumption that the first wheels used in a transport capacity were made of solid wood, and humans—not draft animals—either pulled or pushed the crude carts. Animals, particularly oxen, horses, camels, and onagers (wild asses), had been domesticated at least several thousand years before the invention of the wheel. Oxen were outfitted with yokes to pull plows, and horses, onagers, camels, and even elephants were used as riding and pack animals. When, at some point, another human realized that an animal could also pull a sledge or a cart, as well as a plow, the history of transportation began in earnest. This concept is not as self-evident as it appears today. First, the earliest domesticated animals were difficult to control. They were larger and more aggressive than present-day domesticated animals. Second, there were few, if any, effective harnessing systems. The yoke, in its crude form for oxen, had been invented almost simultaneously with the domestication of the ox as a draft animal. The first yokes were difficult-to-control wooden frameworks that were lashed together by leather straps and were used only for farming purposes. Various civilizations, including the Greeks and Romans, experimented continuously with harnessing systems that would not constrict the windpipes of draft animals. While the harnessing systems, including those used by charioteers in races and in battle, were acceptable, it was not until 500 C.E. that the Chinese developed a sophisticated and effective harness. Third, the wheel itself was a hindrance to an efficient and comfortable transport system. As a matter of fact, in the beginning, comfort was not a consideration at all.

The first wheels were heavy, *solid* pieces of wood cut from tree trunks. Size was uneven and the wheel itself was a plodding, unyielding, and bumpy device. Later, carpenters and wheelwrights fashioned wheels from several pieces of lumber that were pegged together and then rounded into shape. While the size was more uniform, it was still a heavy wheel. At some point after 2000 B.C.E., someone realized that cutting pieces out of this solid wheel would not compromise the wheel's mechanical ability, yet the wheel would be lighter in weight and would absorb the natural shocks of the road more effectively. In essence, carving *spokes* into the wheels had a twofold benefit: a heavier load could be placed on the wagon or cart, since the weight would not be in the wheels themselves, and it would be easier on the draft animal. The *crossbar* wheel was also built in antiquity. It was less popular primarily because it was less durable and broke more frequently than either the solid or spoked wheels. (See Figure 12.1.) Around 1400 B.C.E. the ancient Egyptians redesigned the wheel further by assembling it in three parts—the rim, spokes, and hub—making it lighter and more balanced. Copper or bronze strips were affixed to the outside of the wheel to reduce wear. The availability of these high-grade wheels, however, was limited to only the wealthy or to charioteers who used them in racing or in battle.

Aside from retooling the basic construction of the wheel itself, the axle proved another daunting challenge for the ancients. An *axle* is defined as a supporting shaft on which a pair of wheels revolves. Without an efficient axle, carts or wagons are difficult to maneuver, particularly when turning, since the load would shift. This was a problem in

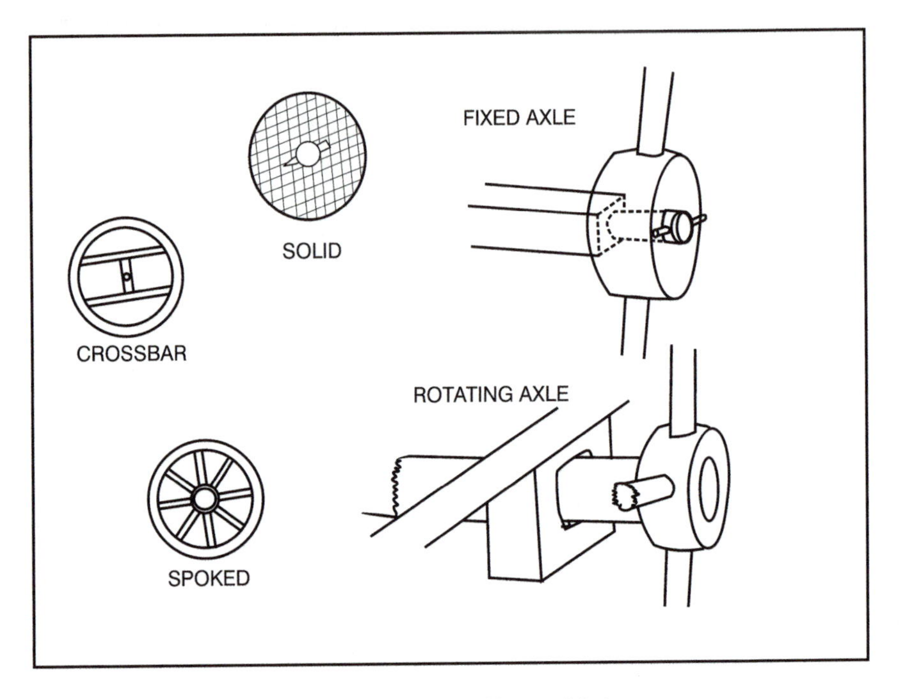

Figure 12.1 Three Types of Wheels and Two Types of Axles
The original *wheel* for transportation purposes was a solid piece of wood that was heavy and unstable. The addition of a crossbar and, finally, spokes made the wheel lighter in weight and gave it a greater ability to absorb the shocks of the road. *Axles* gave the wheel system greater maneuverability.

ancient towns where narrow lanes and alleys were the norm. There were two basics types of axles in antiquity, the *fixed axle* and the *rotating axle.* As the name implies, a fixed axle does not move. Rather, the wheel rotates freely at the end of the axle. The pole (axle) extends through a hole cut into the center of the wheel and is held there by a wooden peg, sometimes called a lynchpin. *Rotating* axles allowed both the wheel and axle to rotate simultaneously as a unit. The advantage of the fixed axle is greater maneuverability. Just like the wheel, it took hundreds of years to perfect the design of the axle. *Bearings,* devices that both supported and guided the weight on the axle/wheel unit, and *lubricants,* which reduced the friction on the axle, were improvements made to the basic design. Bearings were more popular than lubricants, which were used mainly by charioteers in racing venues. In antiquity, lighter vehicles were equipped with fixed axles even though tilting of the wheels occurred occasionally. Wagons used for carrying heavier loads operated with rotating axles. Despite the added wear on the bearings of the axle, there was less tilting of the wheels, thus more stability. In present-day usage, the fixed axle is standard on automobiles and trucks, while the rotating axle is used on railway engines and cars.

Draft Animals

For the ancients, oxen were the preferred draft animals, primarily because oxen were less expensive to maintain. While cave pictographs of harnessed horses dating back about 17,000 years have been uncovered, as well as Stone Age pictures of hunters on horseback, horses were not used initially as draft animals. At first, they were a prey animal, that is, food. But, over time, this changed, and horses were considered prestigious and valuable possessions rather than disagreeable but necessary beasts of burden. Although the ancients were unaware of it, there is a scientific reason why oxen rather than horses are more efficient draft animals. Oxen (bovines) have a rumen—the first part of the stomach where food is partially digested before being regurgitated and chewed further; horses (equines) do not. Therefore, all things being equal, if an ox and a horse consume equal amounts of food, the ox will gain more weight, nutrition, and strength—even when feeding on inferior-grade fodder—while the horse will actually lose weight and muscle mass. The digestive system of equines uses up essential proteins that are eliminated in fecal matter. They cannot survive on hay or straw alone and must have other food sources to replenish those lost proteins. Farmers and tradesmen learned early on that it was cheaper to maintain an ox than a horse while benefiting from the animal's hard labors. Horses were limited to the wealthy, who could afford their feed. The horse can be temperamental, difficult to control, and prone to more illness. In addition, the domesticated horse needed to be shod routinely to compensate for the crude and rocky roads that sprung up with the population centers of civilization. For these reasons, the horse was simply not an efficient draft animal prior to the fall of Rome. After the collapse of the Roman Empire, however, the living and working conditions of the horse changed markedly, just as they did for humanity. In other words, life was decidedly more harsh and difficult.

Trade

Bartering and Currency

Neolithic humans established the first communities when they embarked on the risky enterprise of settled farming. Before that, humans continually moved in an effort to find food sources and adequate shelter. Once agriculture took hold and the planting and harvesting of crops became more predictable, the bounty from those crops was able to feed not only the farmer and his family but others as well. In ancient times, the source and supply of food was the primary consideration. Thus, a successful farmer had something of value, and **bartering** became a mechanism by which people were able to exchange crops, services, and goods with other members of the extended community. For thousands of years bartering was the accepted way of acquiring that which could not be raised, produced, or created by the individual. Wheat was traded for goats, linen for pottery, chickens for beads, and so on. As civilization and cities grew, bartering became less efficient and more burdensome, and the necessity for a different means of exchange became apparent. Five-thousand-year-old cuneiform tablets found in Mesopotamia attest to this fact, inasmuch as cuneiform, the earliest form of writing, was invented to maintain inventories of grain, and later other goods.

The ancients grappled with the concept of how to record inventories as well as how to reward the farmer or merchant. The ancient Chinese of the Shang (Yin) Dynasty, which reigned from about 1520 B.C.E. to about 1030 B.C.E., established a currency that used *cowry* (also spelled cowrie) shells. These are glossy, brightly marked shells from Cypraeidae mollusks. For example, in ancient China a string of 5 or 10 cowry shells would have been similar to a modern-day U.S. nickel or dime. An accumulation of thousands of shells indicated great wealth.

The Chinese also cast, rather than stamped, knife-shaped metal coins that reportedly date back to the ninth century B.C.E. In fact, all Chinese coins from ancient to modern times have been *cast* rather than *stamped.* (Casting means that the metal, in whatever shape chosen for the coin, is poured into a mold rather than being punched or "stamped" with an identifying mark of the city, state, merchant, or moneylender.) During the second millennium B.C.E. the Mesopotamians and Egyptians used bronze rings attached to a larger bronze ring that could be weighed to verify their worth. The Babylonians had invented metal weights during this period for precisely this purpose. And in the latter part of the eighth century B.C.E., the Assyrians are believed to have used pieces of silver, known as Ishtar heads. These were stamped with the image of and named after Ishtar, the Babylonian goddess of fertility. India also had silver currency in the shape of bars as well as coins. Metal coinage, however, in its true form was not invented until about the seventh century B.C.E., when King Gyges of Lydia ordered that pieces of *electrum* be stamped with his insignia, the lion, on one side and a certification of weight and genuineness on the other. (Lydia was an ancient Aegean country located in present-day Turkey. Electrum is a naturally occurring amber-colored alloy of silver and gold, found in the rivers of this region.)

The popularity and minting of stamped metal coins grew rapidly in both the East and West. The challenge was to mint these coins in the proper ratios of whatever metals were chosen. In antiquity, this was either silver or gold. Reportedly, Alexander the Great decreed a silver-to-gold ratio of 10 to 1 in the countries of his empire. Counterfeiting was not unheard of, and silver-plated, copper-cored coins were minted that, today, are many thousands of times more valuable than when first produced. It was, however, the Romans who created a single currency system that was the longest-lasting and most successful, if not the most stable. It consisted of the *denarius,* a silver coin worth approximately 25 cents, and the *solidus,* a solid gold coin. (The English word "pecuniary," meaning "related to money matters," is derived from the Latin word for money, *pecunia.)* The efforts of the Roman government to hyperinflate the solidus, a coin that only the wealthy and elite could afford, made the denarius nearly worthless, particularly toward the middle of the fifth century B.C.E.—the time of the Roman Empire's collapse. The common people were saddled with currency that had little or no value, and historians believe that this was one of the contributory factors to the Empire's decline and fall.

From about the second century B.C.E. to about the ninth century C.E., the Chinese had experimented with the idea of a currency other than cowry shells or coins. Despite having invented paper in the first century C.E., the Chinese at first used the hide of a white stag (a castrated deer), cut into a 12-inch square and imprinted with a pattern. Each note was worth 400,000 copper coins. It was not a successful undertaking, primarily because the rarity of the white stag limited the printing of these notes, likewise called *white stags.* The printing press had not yet been invented, and Chinese paper money was lightweight and physically unstable, as well as being difficult to produce. Paper money did not become popular until the about the seventeenth century in Europe. However, from that time up until about the middle of the twentieth century, paper money was based on the **gold standard.**

Caravans and the Silk Road

Until the fourteenth and fifteenth centuries C.E., when the Europeans actively sought sea routes between their port cities and the Far East, virtually the only trade route available was the famous *Silk Road* that was first established in the latter part of the third century B.C.E. Local trade routes, both on land and sea, between countries in the Middle East and those on the Aegean had long been established. However, trade between those countries and China, which began in the eighth century B.C.E., though arduous, was nevertheless undertaken by hearty traders who brought back from the Far East the precious and desired commodity of silk.

The Chinese jealously guarded the secret of and monetarily exploited the production of silk for over 3,500 years. Silkworms that feed exclusively on the leaves of the white mulberry tree produce a silky filament that they wrap around their larval cocoon. Chinese farmers cultivated the mulberry trees and bred the silkworms for just this purpose—to produce the silky threads from which the lustrous fabric is woven. For millennia, the outside world believed that the Chinese alone were able to produce this material, and traders made the long and dangerous journey to purchase yards of silk at exorbitant prices, which they sold at equally exorbitant prices in the Middle East, Egypt, Greece, India, and Rome.

Just as silk was not the only commodity purchased from the Chinese, the caravans from the West arrived with goods that the Chinese coveted and for which they paid handsomely. Western and Asian traders brought glassware, amber, wool, rugs, linen, wine, furs, fruits, animals (elephants, lions, exotic birds, horses), and even human slaves. Their return caravans were then stocked with bronze, porcelain, jade, spices, paper, and of course, silk. But perhaps the most important commodity carried over the Silk Road was the exchange of ideas. Silk itself became less exotic in about 555 C.E., when Catholic monks brought cocoons from China back to Byzantium, and the secret was revealed and later cultivated in the countries along the Mediterranean and into France.

Although trade with the Far East had flourished for four or five centuries, the Silk Road, as a specific route, was really established around 200 B.C.E. It was not a road in the conventional sense of a continuous unbroken passage. Rather, it was a series of vast and harsh geographical routes that began in Xi'an in central China (near present-day Beijing) and then progressed over selected roads that depended on weather and accessibility. Caravans could follow the road through the Afghan valleys that led to the Caspian Sea or cut through the Karakorum Mountains in Kashmir and China into Persia (present-day Iran) and arrive in Anatolia (present-day Turkey). From that point, caravans could choose to go by sea to the Mediterranean ports or continue overland through Thrace (the region on the present-day Balkan peninsula). The entire journey was about 7,000 kilometers (nearly 4,400 miles). It was rare that a single individual made the entire journey. Usually, traders went part of the way, trading their goods with merchants in the towns along the way or with other caravans. Supposedly, the tandem hitching of animals (one behind the other) was invented on the narrow, winding passages of the Silk Road. There is probably some truth in this assertion, although horses were used only part of the way in caravans. Generally, the horse is ill-suited to carrying or pulling heavy loads over desert regions.

Just as it had locally for thousands of years, the camel became the ultimate beast of burden on the caravan routes over the Silk Road. First domesticated between 6000 and 5000 B.C.E., the camel is an ill-natured animal with body odor and a disagreeable temperament. However, it has several attributes that were invaluable to ancient transportation. Camels can tolerate extremely hot weather. They can go without water for a week or more. And they can carry weights of up to 150 kilograms ($331^{1}/_{2}$ pounds) over an 8- to 10-hour day. For short distances, camels have been known to carry as much as 450 kilograms ($994^{1}/_{2}$ pounds). Pull weights for camels are 750 kilograms ($1,657^{1}/_{2}$ pounds). Their durability in the impenetrable deserts of Asia was an important factor in the successful commerce between the East and West.

In the desert regions along the Silk Road, an oasis town gave respite to the travelers. In the Middle East especially, provision for these large convoys of animals and men was found in a public building called a *caravansary*. Built outside the town, the caravansary was a high-walled, wooden quadrangle with a single, tall entranceway that was wide enough to admit entrance to camels bearing full loads. The perimeter of the bottom part of the quadrangle consisted of stables for the camels and other animals, storerooms, quarters for cooking, and a central fountain in the courtyard that was open to the sky. Stairways led to rooms for lodging that contained the only true windows in the structure. The bottom portion of the building contained narrow air holes, but the main source of ventilation was the structure's open-air courtyard. The purpose of the single entranceway was to secure the cargo of the caravans through chain-locking the structure until one caravan departed or another caravan arrived. Guards or porters were employed for this purpose, although thievery was not uncommon.

Caravans remained an important vehicle of transportation between East and West until the early Middle Ages, when the Silk Road became increasingly unsafe and thus less traveled. Marco Polo (ca. 1254–ca. 1325 C.E.) reopened the route in the late thirteenth and early fourteenth centuries, when he traveled to Cathay in China. In the fifteenth and sixteenth centuries European oceangoing vessels supplanted the camel as the most efficient carrier of commodities. Now a popular tourist attraction for the physically hearty adventurer, the routes of the ancient Silk Road still exist, although the only paved portion is a highway between Pakistan and China.

Navigation and Exploration

Cartography

Many archaeologists and historians believe that maps in one form or another were in existence since the emergence of *Homo sapiens sapiens* about 40,000 years ago. Artifacts from the Stone Age and early Neolithic Period indicate that this belief is more than mere speculation. In addition, archaeologists have uncovered thousands of examples of ancient maps in all parts of the world, indicating that mapmaking was a universal human undertaking and evolved independently in all of the earth's regions. Carvings on a 12,000-year-old mammoth tusk that represent a stream and an early Neolithic community were found in the Ukraine in 1966. A cave-painting map, about 8,200 years old, found in Çatal Hüyük in present-day Turkey, depicts houses and streets as well as a volcanic eruption. However, the oldest known regional and topographical maps were found in Mesopotamia. A 2,500-year-old clay map illustrates the Akkadian region of Mesopotamia (present-day northern Iraq) with the pivotal north, south, east, and west points.

The Babylonians also produced clay tablet maps, one of which diagrams, to scale, the ancient city of Nippur on the Euphrates River, complete with lines that indicate canals, temples, walls, gates, storehouses, and so on. The Babylonians also produced on a clay tablet what is believed to be the oldest extant map of the world, dating to about the sixth century B.C.E. However, this particular map is speculation and myth, based on Babylonian astrology and religion.

The Egyptians also engaged in cartography, using their invention, the papyrus scroll. However, most of their maps were religious or astrological in nature and contained elaborate routes to and diagrams of the netherworld and the universe. One ancient Egyptian map, however, still exists. Referred to as the "Turin Papyrus," this is a 1,350-year-old map of gold mines in the region of the Red Sea.

The ancient Chinese also produced maps, as early as the seventh century B.C.E., that reportedly were more accurate and detailed than their

Western counterparts. Mesoamerican maps used footprints to indicate roads.

The age of classical Greek thought and science really began in the seventh century B.C.E. Many of the theories constructed by the Ionians and the Miletians were based on Egyptian and Mesopotamian philosophy and texts. Some historians credit the Greek philosopher and astrologer, Anaximander of Miletus, with creating, in the sixth century B.C.E., the first map of the world, at least as he perceived it. And for his time, his perceptions were partially valid. He believed that the earth's surface was curved, but only in the north-south direction. However, none of his actual writings survive. Anaximander's map of the earth was reportedly redrawn by Hecateus of Miletus in about 500 B.C.E. and is known only through the works of Aristotle and Theophrastus.

In the third century B.C.E. Eratosthenes measured the circumference of the Earth, arriving at a figure that is remarkably close to today's accepted figure. A true measurement of the Earth's size is essential in the development of any map, but most especially in ancient times when so little was known of the physical world beyond accessible points. The Greeks, however, believed that answers to the natural earth and the universe resided less in mythological and religious concepts than in the physical realities of the world. And as time went on, the Greeks approached cartography in this same manner. In the fifth century B.C.E. Herodotus, the Greek historian, is believed to be the first ancient who actually traveled to distant locations *before* producing a map of the then-known world; he also wrote a treatise on the geography of those regions. His map still exists. However, Greek mapmaking was slow to evolve into a precise and accurate endeavor. Even though Greek sailors returned from their voyages with correct topographical information, many of the Greek cartographers of this period were reluctant to transform this knowledge into more accurate maps and insisted, for whatever reason, on copying old errors onto new maps.

The growing interest in and influence of Greek astrology (astronomy) eventually transformed mapmaking, specifically the discovery of the concepts of geographic latitude and longitude over a period of several hundred years. (There are two other types of latitude—astronomical and geocentric.) *Geographic latitude* is defined as the angular north-south distance from a primary great circle or plane—that is, the equator—measured in horizontal bands or lines from the equator, and expressed in degrees from the equator, which is zero degrees. Greek mathematicians/astronomers first divided the globe into these lines in the fourth century B.C.E. when they were constructing celestial maps or catalogs of stars. The concept of longitude evolved over several centuries. The Phoenician mathematician/geographer, Marinus of Tyre (present-day Lebanon), created a map that showed vertical as well as horizontal bands or lines, in essence assigning what he mistakenly believed were the correct latitudes and longitudes for each location on his map.

In present-day terms, *longitude* is defined as the angular distance on the Earth, map, or globe east or west of the prime meridian at Greenwich, England. It is measured from the prime meridian to the point at which the longitude (meridian) is being determined and is expressed in degrees west of the prime meridian, which is at zero degrees. In antiquity, the prime meridian was placed at the Fortunate Isles (present-day Canary Islands), not at Greenwich. However, until the eighteenth century, when John Harrison (1693–1776) invented an accurate clock that could withstand the rigors of ocean travel, determining one's exact longitude was almost always an inexact and often dangerous calculation. Marinus (ca. 70–130 C.E.) was also the first to use map projections—placing the topographical or geometrical features of the globe onto a flat or plane surface, along with the lines of latitude and longitude.

The Romans expanded the art of mapmaking by utilizing information received not only from Greek sailors but from their own military personnel, who had traversed great distances over the ever-expanding Roman Empire. The Romans wanted accurate and practical maps that would aid them in their quest for territory and economic superiority. They were less concerned with perceptions and theories and more concerned with precise depictions of roads, mountains, rivers, distances, and obstacles. Roman commerce and Roman armies depended on the accuracy of their maps, and the Roman government actively supported the study of geography and its related professions of surveying, road building, and, most notably, cartography.

The second-century mathematician, astronomer, and geographer, Ptolemy of Alexandria, is considered the most influential geographer and cartographer of antiquity. Also known as Claudius Ptolomaeus, little is known about his actual life. This is unusual, considering that he exerted considerable influence in the fields of astronomy and geography well into the Renaissance. He is generally believed to have been a Roman citizen of Greek birth who lived in Egypt but who did *not* travel extensively. Rather, he based his maps on the information he received from returning Roman legionnaires and commanders, as well as the writings of other geographers. He is the author of four major works: the *Almagest,* the *Tetrabiblos,* the *Optics,* and the *Geography.* And although Ptolemy criticized Marinus's geographical findings and calculations, much of what is contained in *Geography* is based on the writings of Marinus. *Geography* was an eight-volume text, six volumes of which contain latitudinal and longitudinal tables.

While Ptolemy's map projections were more advanced than Marinus's drawings, he made many serious errors in geographical calculations and measurements as well as topographical locations. For example, Ptolemy used Marinus's prime meridian (zero longitude) at the Fortunate Isles, but placed those islands too far eastward toward the European continent. This was a seven-degree error that resulted in a misaligned map. Also, contrary to Herodotus's writings of the *oikoumenē,* the Greek word meaning "the known world," Ptolemy believed that, at its southern tip, the African continent was joined to the eastern part of the Asian continent by a narrow strip of land that he called *Terra Incognita* or the "unknown land." Ptolemy also disregarded Eratosthenes' calculation of the earth's circumference—which at 24,700 miles was remarkably accurate to the actual figure of 24,902—and used his own calculation of 17,800 miles, an error of over 7,000 miles. This last error had serious but, in the end, serendipitous repercussions, primarily because Ptolemy's maps and calculations were used almost without question well into the fifteenth century.

Developing his own navigational charts but using Ptolemy's measurement of the Earth's circumference, Christopher Columbus (ca. 1451–1506 C.E.) undertook his famous voyage to find a new trade route to India. Historians believe that if he had had a correct figure of the sailing distance to the Far East, Columbus may have reconsidered this undertaking. The discovery and naming of the West Indies, the Caribbean Islands where Columbus first landed, and the subsequent voyages of other explorers to the Americas, can be attributed to Ptolemy's erroneous idea that Asia lay more westward, closer to the European continent.

In fairness to Ptolemy, ancient mapmakers were forced to rely on the geographic information gathered by their predecessors—although Ptolemy did choose to ignore Eratosthenes' more correct figure. Traveling to the many locales of the then-known world was beyond the ability of the ancients, and, of necessity, they relied on dubious anecdotal information given them by sailors, soldiers, and other adventurers. Distances, particularly those over the oceans, were simply guesses. Ptolemy's maps, written in the original Greek, were not translated into

Latin until about 1405 C.E., which was the beginning of the age of European exploration. Maps drawn in the early Middle Ages essentially followed Ptolemy's basic design, but they were grossly incorrect. Influenced by the Roman Church, the maps were known as *T-maps* because they contained only three continents: Europe, Asia, and Africa. Jerusalem was a central location, and the East was at the top. The twelfth-century C.E. invention of the magnetic compass and the drawing of sailors' charts in the fourteenth century led to the production of more accurate navigational, as well as territorial, maps.

Compass

The scope of these early migrations, as well as the courage of early humans, is almost imponderable. They literally did not know where they were going or what fate would befall them along the way. There were no navigational tools beyond the celestial bodies of the Sun, Moon, and stars that guided their time and direction until the ancient Chinese experimented with the concept of the compass perhaps as early as 2,400 years ago. The secrecy of the ancient Chinese and the reliability of ancient Chinese writings make it difficult to provide an exact date. However, the *sinan,* a navigational device, was written about in a first-century C.E. Chinese text. There are no actual extant artifacts of the *sinan,* thus all the scientific community has, at this point, are reconstructions of the "first compass."

The *sinan* has two parts. The first is a square bronze or polished stone (e.g., jade) plate, sometimes elaborately carved with Chinese characters, and indicating the four directions—north, south, east, and west. Each of these compass points played a significant role in the lives of the ancient Chinese. Their god, the Yellow Emperor, was at the center of the world, and four other gods ruled the other directions, the most important of which was south. The second part of the *sinan* was a lodestone, carved into the shape of a ladle or spoon, that rested in the center of the plate. *Lodestone* is a form of magnetite. *Magnetite,* the mineral form of black iron oxide (iron ore), is a natural magnet that has polarity. The word "lodestone" is derived from the Old English word *lode,* which means "way" or "journey." The word "compass" derives from the Old French word *compasser,* meaning "to measure."

Chinese alchemists had long realized that lodestones could attract ferrious metals. We can only assume that after some experimentation, they found that the lodestone of the *sinan* always pointed in the same direction. (Chinese compasses always had the handle of the ladle pointing southward, the most important direction in their culture. Today, the needle of the compass always points north to south.) The shape of the ladle was chosen to represent the Chinese constellation of the Great Bear, known today as the Big Dipper. It is believed that tradesmen involved in the mining of jade may have been instrumental in the design of the *sinan*. It was necessary to travel vast distances to find the green minerals (nephrite and jadeite) that the wealthy Chinese coveted for jewelry and carvings. For the jade miners and merchants, the *sinan* provided a measure of accuracy with regard to angular direction, although as a result of the extensive carving process, the lodestone ladle of the *sinan* probably did not retain much of its original magnetism. Sometime in either the late seventh century or early eighth century C.E., the Chinese discovered that rubbing a steel needle on a lodestone would produce magnetism in the needle. Thus, they began to produce compasses that were less cumbersome than the two-part *sinan*.

The furtiveness of the ancient Chinese, as well as their isolation from much of the civilized world, probably delayed the use of the compass in the West until the twelfth century. Historians believe it was independently invented in Europe at that time. However, trade with the Far East had been ongoing for centuries, and more than likely, the invention of the compass traveled east to west. In any event, European astronomers and mariners recognized and exploited the location at the north celestial pole of the polestar (Polaris) in the northern hemisphere. The compass enabled them to ascertain their position, particularly at sea, even when Polaris was not visible. The magnetic compass was the essential, indispensable, and only navigational and directional tool available until the twentieth-century discovery of GPS (Global Positioning System), which relies on space satellite technology.

History of Shipbuilding

No matter their size, in order to float, all boats and ships must follow the scientific principles of buoyancy and displacement, often known as *Archimedes' principle*. (Actually, this is a bit of a misnomer, since Archimedes did not actually state the principle of buoyancy and displacement in exact terms, although he certainly understood buoyancy in practical terms long before his famous experiment. His discovery concerning the concepts of relative density and specific gravity came about when he was asked to verify the amount of gold in the crown of King Hiero.) In any event, since boats and ships were successfully built and sailed, probably for thousands of years, it is obvious that ancient humans grasped the "physics" of the boat long before Archimedes explained the "why." Simply stated, a boat floats because the weight of the boat is less than the weight of the water that is displaced by the boat. If the boat was in the form of a solid stone it would sink, since the weight of the volume of the water displaced is less than the weight of the stone.

There is evidence in the form of fossils and tools that establishes that early humans migrated from continent to continent across bodies of water. If it is true, as scientists generally believe, that human life began on the African continent, and the presence of humanity has been found on all continents and on most major island groups, then we must acknowledge that our earliest ancestors trekked slowly, steadily, and over great distances, both on land and sea, in search of food and shelter for hundreds of thousands of years. Indeed, if the existence of humanity on the continent of Australia, in the countries of Japan and Korea, and on a number of Pacific Island groups can be traced back nearly 50,000 years, there must have been some early conveyance that transported these hearty individuals across the open ocean—the boat. Anthropological studies indicate that this migration, possibly in three different waves, began about 1.8 million years ago.

Humanity has looked to the oceans for food, war, and adventure, and the progression of civilization is evident in the history of ancient boats and ships. A detailed account of all the boats, ships, and notable seafarers of antiquity is beyond the scope of this book. And much of what present-day historians know about ancient sailing and shipbuilding has been found in research of ancient writings or other artifacts, such as paintings, sculptures, and reliefs. They have not had the luxury of examining the remains of actual sailing vessels, thus much is speculative. The following represents the most important types of boats and ships designed and constructed in the ancient world, and those civilizations who made significant contributions in ancient navigation.

Dugouts, Rafts, and Canoes

The earliest scientific evidence for the existence of manmade boats is only about 10,500 years old, but archaeologists and historians speculate that prehistoric humans constructed boats out of the natural materials of the environment long before the Neolithic Period. Due to the organic decomposition of these materials, however, there are no physical artifacts to support this contention beyond the 4-meter (13-foot) dugout, known as the Pesse canoe, that was uncovered in the Netherlands and dates back to about 7400 B.C.E., and a wooden paddle, dating back to 8500 B.C.E., found in Yorkshire, England. The empirical evidence for boats and navigation before the Neolithic age is based on writings, paintings, reliefs, fossils, tools, and so forth. The first *raft* was, most likely, a series of logs that were lashed together by vines. Archaic humans would have noticed that many animals "hitched a ride" on logs that floated down streams or rivers. The first dugouts, which are similar to canoes, were constructed after humans discovered tools, such as the axe and the adz, which would have enabled them to carve out the wood and pulpy centers of the log—hence the name dugout. Archaeologists also believe that ancient humans used fire to "burn out" the center seating area of dugouts. The invention of the *oar* or *paddle* is merely an extension of the human arm. *Canoes* made from bark are also believed to be among the earliest boats designed by Stone Age humans, although no physical artifact to support this has yet been found.

Leather Boats

As early as the seventh century B.C.E., the Assyrians constructed boats, called *keleks,* using tanned sheepskins. *Keleks* were inflatable boats similar to present-day rubber rafts that used up to 20 sheepskins sewn tightly together and inflated like balloons. The frame was built with willow rods, and the platform was made from river grasses, reeds, bulrushes, and moss. Historians believe that the Assyrians borrowed the concept from boats or rafts built by primitive humans who used animal skins stretched over tree limbs. The Assyrians also built a round transport boat, also using sheepskins and willow-rod frames. The size varied from a two-person vessel to larger ones capable of transporting a small animal and/or carrying up to 135 kilograms (300 pounds). An oarsman propelled both the *kelek* and the round transport boat.

The Celts of Britain and Ireland were reputed to be master boat builders and sailors. One of their most famous is the leather *coracle.* Evidence for its existence is based on the writings of Julius Caesar, who first saw one when he invaded Britannia in 55 B.C.E., although it is probable that coracles had been in existence long before the first century B.C.E. The Celts used coracles for river fishing, but Caesar reportedly witnessed Celts sailing them in the rough seas of the Atlantic Ocean. Caesar also wrote about the Celts' use of *leather sails* on wooden ships built with planks, crossbeams, and iron bolts. Apparently, the Roman galleys were no match for these huge oceangoing vessels that were constructed from solid oak. The *curragh* was a longboat made from leather that the Irish monks used to navigate around the islands and coastline of the Atlantic. Carvings and drawings that date back to at least 5000 B.C.E. depict Bronze Age boats of similar design, giving further credence to the belief that leather boats were invented simultaneously and independently in many parts of the colonized world.

Egyptian Wooden-Plank Ships

Interestingly, the oldest known wooden-plank ship in existence, some 43 meters long and 5.7 meters wide (141 feet long, 19 feet wide), never saw water. It was discovered in Egypt in 1952 at the site of the Great Pyramid of Khufu (also known as Giza). Believed to date back to about 2600 B.C.E., its construction included 1,224 pieces of sycamore and cedar timbers and was intended to provide transportation for the Pharaoh Khufu (sometimes called Cheops) after his death. Historians, however, generally accept that the Egyptians constructed similar wooden-plank ships centuries earlier. And at least 1,000 years before, the Egyptians used papyrus reeds that grew exclusively along the banks of the Nile River to build boats to transport themselves and goods along the river. It was on these *reed boats* that the use of multiple oarsmen was first tried, with a single individual using a steering paddle. Square sails that could be raised or lowered depending on wind and current were also attached. Reed boats were sailed only on the Nile River.

With the construction of *wooden boats,* the Egyptians expanded their trade to the Mediterranean by about 3000 B.C.E. The design of these wooden ships was based on the earlier reed boats— they had no keel or ribs. This proved to be a design flaw, at least in wooden construction. For while the water below exerted pressure on the planks and pressed them together, the heavy cargo also exerted pressure on the timbers, whereupon the planks would spread, causing leaks. The Egyptians quickly learned to accommodate this problem and built cargo decks mounted on transverse (crosswise) planks. Rope trusses that were tightened by a windlass provided stability from the bow to the stern. Mortises and tenons and leather thongs supported the timbers and the overlapping joints. Linen was used for sails. Depending upon the size of the ship, as few as 10 and as many as 30 oarsmen provided the power, while 2 oarsmen at the stern were responsible for steering. Egyptian ships, up to 55 meters long and 18 meters wide (180 by 59 feet), were used primarily as merchant vessels. The Egyptians also built naval warships to protect their merchant fleet.

Egyptian wooden-plank ships performed at least two remarkable navigational feats in antiquity. First, it is believed that they used large wooden boats to transport some of the enormous stones found in the construction of the Great Pyramids. Second, one of the few women pharaohs of Egypt, Queen Hatsheput (fl. 1500 B.C.E.), reportedly sent a fleet of five large vessels to Punt (present-day Somalia) to retrieve whole frankincense trees from this remote east African country and bring them back to Egypt. The ships ostensibly voyaged through a canal, built by the Egyptians some 300 years earlier, that connected the Red Sea to the Mediterranean, the site of the present-day Suez Canal.

The Phoenicians

The influence of the Phoenicians, who are considered to be the most accomplished mariners of antiquity, remains to the present day. Phoenicia, a chain of city-states headed by the city of Tyre, was located on the coast of present-day Lebanon and Syria. All Phoenician cities, including those they conquered in battles, were located on the coast, among them Byblos, Sidon, and Carthage.

Phoenicia was formed early in the first millennium B.C.E. after being repeatedly attacked and eventually conquered by sea marauders of indeterminate origin. Phoenicians descended from a variety of ancient tribes from the Middle East, among them the Canaanites and Semites, as well as the Aegeans and Cypriots of the Mediterranean, and their "unknown" vanquishers. Because of the aggregation of cultures, the Phoenicians were fierce sailors as well as merchants skilled and adept at all trades. They successfully exported the products of these trades to their own colonies and other countries via the sea.

Geographically, Phoenicia was located in a region of dense forestation. Thus, Phoenicians had the best timbers available for both their merchant ships and war vessels, which were among the fastest and best built in antiquity. Phoenician sailors also were among the earliest of the ancients to recognize that the wind was a more efficient source of power than the arms of oarsmen. The Phoenician *gauli,* a ship designed for long ocean voyages and equipped with both sails and oarsmen, relied primarily on sails to propel it through rough waters and high wind. Phoenicians possessed a highly developed knowledge of astronomy that enabled them to navigate by the stars. They were skilled merchants who exported jewelry and statuary of gold, ivory, woodcarvings, dyes, textiles, and glass. And they were the most skilled glassmakers of antiquity.

The Phoenicians, who used cuneiform writing, also devised a script of their own that progressed from them to the Greek, Etruscan, and Latin alphabets, and finally to the Modern English alphabet. This progression can be directly attributed to Phoenician sea trade and its export of not only goods but also ideas to the many colonies along the Mediterranean trade route. Phoenician sailors voyaged beyond the Pillars of Hercules (Straits of Gibraltar) to Cornwall in Britain for tin to alloy with copper to make bronze, and to the Azores.

Though it has never been proven in fact, the Greek historian Herodotus reported, somewhat skeptically, that the Phoenicians successfully circumnavigated the African continent from the Arabian Sea, back through the Pillars of Hercules and along the northern coast of Africa, and then home. This feat could have been possible because sometime around 1850 B.C.E. the ancient Egyptians began constructing a ship canal that connected the Red Sea and the Mediterranean, more or less at the site of the present-day Suez Canal. Thus, if the Phoenicians were able to sail from the Mediterranean to the Red Sea, they would have been able to sail eastward to the Arabian Sea and begin their long journey around Africa.

For most of its short history, Phoenicia was subject to a series of Assyrian invasions and control. From time to time the individual city-states of Phoenicia would form alliances—some successful—to thwart these invasions. However, in 586 B.C.E. the Babylonians conquered all of Phoenicia, except for the city of Tyre, which withstood the Babylonian siege for 13 years before finally capitulating. In about 523 B.C.E. the Phoenician culture was absorbed into the Persian civilization when the Babylonians fell to the Persian Empire, led by Cyrus the Great.

Greek Penteconters, Biremes, Triremes, and Quinqueremes

After the slow decline of the Phoenician civilization, the ancient Greeks, who, primarily because of geography, relied on sailing to import and export goods, were the dominant seafaring presence in the Mediterranean region. At the time, the Greek city-states were located either amidst impassable mountain ranges that made trade almost impossible or on the various islands that were cut off from each other—except by boat. Thus, the Greeks needed an effective method of trade to compensate for these geographical obstacles. Unlike the Phoenicians, they limited their trading to ports on the Mediterranean Sea and on the Black Sea, rather than venturing out into the Atlantic on voyages of exploration. The almost constant threat of invasion and war was another factor in the Greeks' construction of large and powerful seagoing vessels. The ancient Greeks maintained fleets of three types of ships: (1) merchant or cargo ships; (2) naval warships; and (3) small passenger boats. The cargo or merchant ships were what made the Greeks wealthy. These 22-meter (72-foot) wooden ships were deep-hulled and broad in the beam and were equipped with sails and rowers. Averaging only about five knots (less than six miles per hour), they sailed in a zigzag pattern to compensate for the necessity of sailing or tacking into the wind. Thus, their trading voyages were arduous and lengthy.

The original Greek naval ships were called *penteconters* ("fifty-oarers"). (See Figure 12.2.) *Penteconters* were equipped with massive sails with a single row of 24 oarsmen on each side. Steering paddles made up the remaining pair of oars. *Penteconters* were up to 30.5 meters (100 feet) in length and about 3 meters (10 feet) across at the widest part. The Greeks soon realized that although the design of the *penteconter* was satisfactory for cargo or merchant purposes, it was inadequate for the rigors of naval warfare, where speed and agility during sea battles was paramount. Thus, the Greeks developed *biremes* and *triremes.*

Biremes were ships with two banks of oars; *triremes,* as the name implies, had three banks of oars. "Bireme" was a merely generic term for a ship with two tiers of oars, the design used by Phoenician sailors. There was, however, a standard design for warships, the trireme. The Greeks borrowed heavily from the Phoenicians in the design of their triremes, first built in about 500 B.C.E. These were two-masted sailing ships whose main weapon was a 3-meter (10-foot) bronze-covered battering ram. The ship was about 43 meters (141 feet) long and about 6 meters (20 feet) at its widest point, called the beam. Reaching speeds of up to 14 knots, or about 16 miles per hour, in good weather, they were propelled by the power of 170 oarsmen, with an additional crew of 30. (See Figure 12.2.) The three levels of oarsmen on a *trireme* were: (1) the *thalamites,* the lowest level where the elevation of the oars above the water was a mere half-meter, or less than 2 feet; (2) the *zygites,* the middle level; and (3) the *thranites,* the highest. The design of the triremes did not remain static, and shipbuilders, engineers, and sailors continually realigned the positioning of the banks of oars in an effort to effectuate a more efficient arrangement. The oarsmen's job was the most difficult and dangerous aboard ship. They bore the brunt of the damage from enemy ships, often losing limbs, sustaining serious injury, drowning, or being captured. The arrangement of the banks of oars meant that the lowest levels *(thalamites)* were continually dealing with water at their feet or in their faces. In addition, they and the middle level *(zygites)* were positioned closely to the bottoms of the oarsmen above. The highest level *(thranites)* were directly below the upper deck. None of the oarsmen had sufficient headroom. The trireme, however, was light and easily blew off course during periods of high winds.

The Greeks then built *quinqueremes* in about 400 B.C.E. These were ships with four tiers of oars, with two men to each oar, although the total number of oarsmen was less than on a trireme—probably because of the weight factor. They were constructed with fir, cedar, and pine timbers, metal spikes (nails), and lead-sheathed bottoms. The result was a

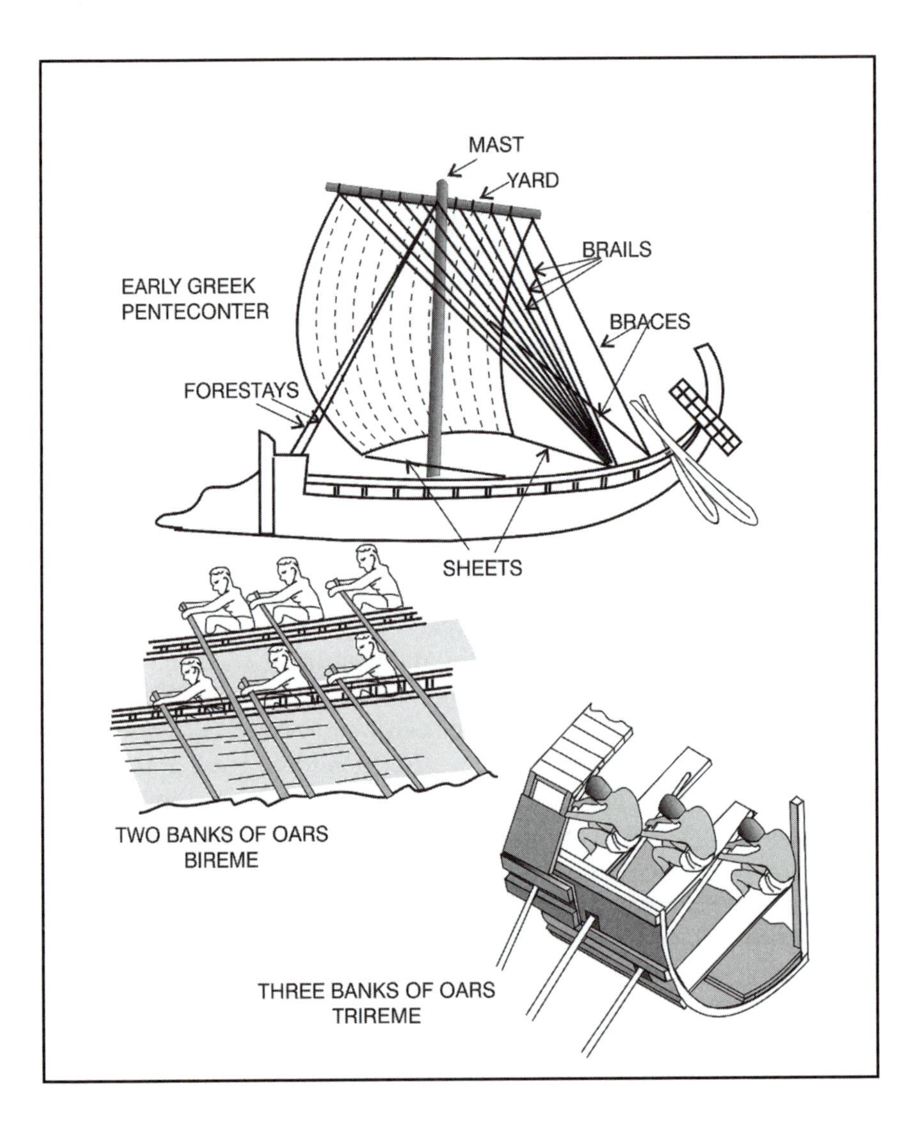

Figure 12.2 Penteconter, Bireme, and Trireme
The original ancient Greek naval ships were called *penteconters* ("fifty-oarers"). Using the fast-moving ships of the Phoenician navy as their models, the Greeks later redesigned their fleet and used *biremes* (ships with two banks of oarsmen) and later *triremes* (ships with three banks of oarsmen). All had their design shortcomings, but the trireme was the main warship of the Greek navy.

heavier ship less likely to blow off course, but also with decreased speed and agility. Despite its shortcomings, the *trireme* was the main warship of the Greek navy. Triremes were used in the famous Battle of Salamis in 480 B.C.E. when the Greeks defeated the Persians, who actually had a larger fleet of warships. Historians believe that victory may have been the result of some trickery and luck as well as the effectiveness of the Greek triremes.

The Roman Navy

By nature the Romans were soldiers—not sailors—and the Roman navy was always held in lower esteem than the generals and legionnaires of the Roman army. Nevertheless, the Romans recognized that developing a strong navy was essential for them, given the seafaring superiority of neighboring countries. The expansion of the Roman Empire, however, did not begin in earnest until after the end of the Second Punic War in 201 B.C.E. At this time, trade and transportation in the Mediterranean region became increasingly important as a source of revenue for the Romans.

From about the eighth century B.C.E., Rome had been in an almost constant state of war in its quest to control Italy and expand territorial control over neighboring countries. A series of battles between Rome and the Phoenicians that took place over a period of almost 120 years became known as the First, Second, and Third Punic Wars. ("Punic" is the Latin word for "Phoenician.") In 146 B.C.E. the Romans defeated the powerful Phoenician city of Carthage on the northern coast of Africa. Very shortly thereafter, the Romans realized they needed to develop a strong merchant marine in the region in order to exploit the riches of their empire. They also knew that a strong navy could defend conquered territories from outside threats or invasion and quell rebellions, as well as transport Roman armies across seas on their way to new conquests. An increasing, and eventually dominant, Roman presence on the seas had profound effects on the empire. Increased sea trade meant that goods and ideas traveled among countries. The Greek, Roman, and even the Egyptian aristocracy that had become Hellenized over the centuries achieved even greater wealth. Slavery increased in their households as captives from conquered territories were now transported via the merchant ships into the households of the elite. Slaves were now exclusively used as the rowers on all Roman vessels. Roman farm life, however, suffered. Roman peasant farmers were conscripted into the expanding Roman armies and navies. The slave labor that replaced them was not altogether satisfactory, and poverty among Roman peasants was pervasive. The chasm between the common people and the wealthy grew wider, sowing the seeds of sociological problems that hastened the collapse of the empire in the mid-fifth century C.E.

The design of Roman naval warships borrowed heavily from captured Carthaginian vessels, as well as from the Grecian triremes. However, the Romans, who were not a seafaring race, believed that heavier ships would be superior to the fairly lightweight triremes of Greece. This heavy construction had both benefits and drawbacks. Roman warships were designed primarily to carry larger numbers of soldiers, who would board the enemy's ships during battle, rather than employ the traditional battle tactic—use of the battering ram. Roman ships were equipped with a *corvus* (the Latin word for "raven"), which was a large, heavy boarding timber with a huge spike on the end. Like a drawbridge, the *corvus* was raised before battle and then dropped onto the opponent's ship. The spiked end of the *corvus* became embedded in the deck, stabilizing the plank, whereupon Roman soldiers would board and engage in pitched battles. The invention was quite successful and credited with at least five of the major Roman victories at sea. But these heavier ships that carried large numbers of troops, as well as the heavy *corvus,* were also unstable in rough seas, often capsizing, the result of which was large numbers of casualties. The one exception was the *Liburnae*—lightweight galleys with two banks of oarsmen that were used as a courier ship and transport of the Roman elite. Liburnian galleys were the most common Roman ship after the Battle of Actium in 31 B.C.E. Roman merchant ships were similar to those popular in the region for centuries before Rome's domination. Two-masted, with wide and deep hulls, they were about 55 meters (180 feet) long and 14 meters (46 feet) wide. These were the ships that carried the wealth of the empire back to the city of Rome.

The Chinese

Little is known about Chinese ships and shipbuilding prior to the sixth century B.C.E., when, reportedly, one of the southern states of China amassed a navy of some 10,000 men. Known as the Period of the Warring States, this was a time of intrigue and conflict among China's numerous warlords. The many rivers of China played important roles in the history of its civilization and culture. Rivers are where Chinese civilization began at least 7,000 years ago. The rivers provided drinking water, food, irrigation for crops, and, in time, were the conduits of transportation and trade. However, the rivers usually did not flow into each other. Thus, sometime from about 600 B.C.E. and for the next 400 years or so, the Chinese began building a series of canals to connect many of their rivers, creating a system whereby large numbers of Chinese troops could be transported from one state to another to wage war on each other. The by-product of these canals, the most famous of which are the Grand and the Wild Geese canals, was trade between the states that eventually learned to live more or less in peace—but not until about 220 B.C.E., the time of the first unified Chinese empire. In addition to the compass, the Chinese made several important contributions to the maritime industry. The most important are the *rudder,* the *watertight bulkhead,* and *sails.*

A *rudder* is a vertically hinged piece of material (wood or metal), mounted at the stern (rear or aft) of a ship, which steers or directs its course. The Chinese invented the rudder in the first century C.E. Before that time, all ships, from the smallest boat to the largest cargo vessel and warship, were steered by the paddles of oarsmen. In fact, there is no evidence indicating that Western vessels used rudders until at least the eleventh century C.E. Figure 12.3 illustrates a Chinese first-century C.E. *axial* or *slung rudder,* designed to be lowered and raised by a windlass. Ships entering shallow waters would raise the rudder so as to prevent its breaking off, and then lower it when returning to deeper water. Over the centuries, the Chinese continually redesigned, reengineered, and improved their original invention, making rudders larger and larger to accommodate the heavier and longer seagoing vessels. Two of these variations are the *fenestrated rudder,* which is a rudder with a series of symmetrical holes that eases the task of turning the vessel, and the *balanced rudder,* which has a blade that juts out from the center post and makes it easier to turn. Interestingly, the first Western ship to use the balance rudder was a British warship—but not until the nineteenth century.

The Chinese invented *watertight bulkheads* (or compartments) in the second century C.E. A *bulkhead* is a vertical or upright partition that divides a ship into compartments and prevents water leakage from one section to another. (In present-day vessels, bulkheads also prevent the spread of fire.) Bulkheads strengthen (reinforce) the ship's hull. Again, this seems a rather self-evident concept, but it was not brought to the West until about the eighteenth century—although Westerners trading with Chinese merchant seamen must have been exposed to this design. Nevertheless, the Chinese knew that if one compartment flooded, it was possible to seal it off and prevent leakage into neighboring compart-

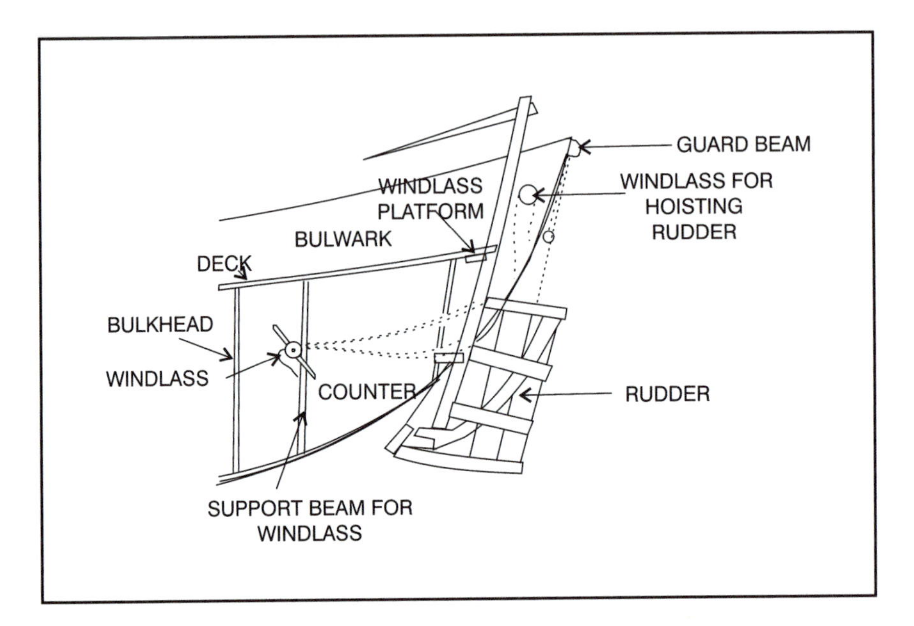

Figure 12.3 Chinese Rudder
The *rudder,* a vertically hinged piece of wood or metal mounted at the stern of a ship, directs its course. Prior to its invention by the Chinese in the first century C.E., all boats and ships used oarsmen equipped with paddles at the ship's stern to steer the vessel.

ments until such time as the ship could be repaired. The obvious benefit was that fewer Chinese ships sank after sustaining minor damage to their hulls.

The Chinese were also master sail makers. In the West, leather sails were replaced by linen and other heavy fabrics, and sailors were forced to climb up tall yardarms to roll up (furl) and roll down (unfurl) the large sails. The Chinese designed sails with *battens,* narrow strips of bamboo. Covered with a bamboo matting or linen that was stretched between the battens, the sail functioned similarly to a Venetian blind. Chinese sailors furled or unfurled the sails using windlasses and ropes, also called *halyards.* Sails made with battens are aerodynamically superior; the battens maintain the sail's rigidity, even if it is torn or ripped. When sails lose their tautness and "belly," they are more susceptible to wind turbulence. Hence, the ship is more difficult to control and steer, while speed is reduced. Although they are superior, with the exception of modern racing yachts that are partially equipped with them, sails with battens have not been adopted in the West. Reportedly, in the second century C.E. the Chinese replaced the single square sail with *fore and aft rigs (sails)* that are aligned along the length of ship. This enabled Chinese ships to sail into the wind, and ocean or sea voyages were not dependent on oarsmen or weather patterns. On the other hand, Western or European ships as recently as the nineteenth century often had to wait in ports for months on end until the prevailing winds were favorable.

The most famous Chinese ship is the *junk*, believed to have been in existence for about 10,000 years. In China junks were known as *fangzhous* (pronounced "fang ju"). The term "junk" originated with the Portuguese, one of China's most important Western trading partners. The Portuguese referred to a reedy plant used in the rigging of these boats as *junco*. (The Latin word for reed is *juncus*.) Although at first glance they appear ungainly and slow, their design actually has the opposite effect. Junks have a flat bottom that aids in sailing shallow waters and driving ashore on beaches for removal of cargo. Their high stern prevented water from high waves from crashing onto the decks. Thus, they remained drier. And, when anchored in port, the high stern assured that the ship would naturally turn with its bow to the wind. And from the early part of the first millennium C.E., rudders, sails with battens, and watertight bulkheads were standard on Chinese junks. This meant that these merchant ships were easier to steer, faster, and more capable of delivering their cargo intact. The junk is still a popular boat in present-day China, although its sail power is often augmented by small motors.

The Chinese navy continued to build up its strength, and in the Middle Ages reportedly had the largest complement of sailors, some 28,000 men. By the fifteenth century, it had 250 galleons, 400 warships, 400 freighters, and nearly 1,400 patrol boats and combat vessels. The Chinese, however, limited their trade routes and naval warfare to the Asian continent, rather than moving beyond to far and distant ports. Their reluctance to venture beyond a certain point on the seas may have had more to do with superstition and myth than their navigational abilities, which were probably far superior to those of their Western counterparts.

Summary

Dinosaurs existed and thrived for nearly 100 million years before their extinction. By contrast, humans, including our earliest hominid ancestors, have been around only for a mere 5 million years. Nonetheless, it has been humanity that has brought movement (migration) to near unimaginable dimensions. Of necessity, other species travel beyond the confines of the nest or den in search of food. However, humans are the only species that grasped the potential of other objects and other animals to aid them in that search and had the instinctive or genetic ability to engineer and construct conveyances that made that search more efficient. For millions of years, archaic humans traveled from place to place on foot searching for food and places of shelter. Civilization began with agriculture. Neolithic humans who chose to work the land rather than continue the harsh and continual search for food established the first villages that, in turn, became cities and cultures with individual characteristics. The concepts of trade and commerce soon followed: someone raised something that someone else wanted and was willing to barter for. The discovery of metals led to the establishment of trades. Trades brought commerce, and commerce meant currency. Currency meant wealth, and wealth, then as now, was inextricably identified with the accumulation of things. Getting those "things" from place to place was a challenge that required initiative, ingenuity, and a bit of serendipity.

The economy of the modern world depends upon the ancient concepts of trade and the exchange of goods, services, and ideas. Today, we are limited only by those technological advances that have yet to be made—not by fear of the unknown. The ancient Phoenician sailors who refused to venture very far beyond the Pillars of Hercules (Straits of Gibraltar) believed they would drop off the end of the earth. Chinese sailors, who built ships capable of sailing great distances, nevertheless refused to venture beyond their own continent because of superstition and myth. Using the principles, concepts, inventions, and discoveries made by the ancients, modern-day scientists continue to explore the boundaries of a world that is far wider and more complex than the great philosophers of antiquity ever dreamed.

Glossary

abortifacient
A substance, usually a plant derivative, that induces abortion.

acids
Substances that release hydrogen ions when added to water. HCl is an example of a strong acid. Acids are sour to the taste, turn litmus paper red, react with some metals to release hydrogen gas, and register below 7 on the pH scale.

alidade
A sighting device that is used for angular measurement. Alidades were part of ancient astronomical instruments, such as the astrolabon.

alloy
A mixture of two or more metals, or a mixture of metals and nonmetals.

alluvial
Characteristic of soil rich in minerals, particularly alluvium, that is deposited as a result of the flow of water. The Nile River's alluvial plain was formed by the river's annual flooding.

anthropomorphic
Describes the assigning of human qualities (feelings, sensibilities, etc.) to nonhumans (animals, objects, etc.).

axioms
Self-evident principles that are accepted. Synonymous with presuppositions and assumptions.

bartering
The trading of goods or services without the exchange of money.

base
An alkali substance that reacts with (neutralizes) an acid to form a salt. A base in a water solution tastes bitter, feels slippery or soapy, turns litmus paper blue, and registers above 7 on the pH scale.

biorhythmic
Describes the inherent cyclical, ordered biological functions or processes of living organisms.

conundrum
A problem without a satisfactory solution.

corbelled
Refers to an L-shaped stone or brick structure projecting from the face of a building or wall that supports an arch or a cornice.

cupping
An ancient medical practice. Heated glass cups were applied to the patient's skin in an effort to draw blood to the surface.

deduction
A method for gaining knowledge; for example, a deduction is inferred in the statement "if-then" (from the general to the specific).

distillation
A process used to purify and separate liquids. A liquid is said to have been distilled when it has been heated to a vapor and then condensed back into a liquid.

DNA
Deoxyribonucleic acid. The complex, ladderlike, double-stranded nucleic acid molecule, present in the chromosomes, that forms a double helix of repetitive building blocks. DNA shapes the inherited characteristics of all living organisms, with the exception of a small number of viruses.

domestication
Refers to the process whereby plants and/or animals are either trained or made to adapt to a human environment. It can also mean to naturalize—that is, to introduce to another region where the plant or animal is not indigenous.

ductile
Easily shaped.

ecliptic
The apparent annual path of the Sun among the stars. The ancients projected the star patterns of the zodiac within this zone.

elliptical
Having the form of a two-dimensional, oval plane curve.

eon
The longest unit of geological time, having two or more eras.

epoch
An interval of geological time that is longer than an age, shorter than a period, and characterized by a notable or extraordinary event.

eras
Periods of geological time that comprise one or more periods. For example, the Mesozoic Era comprises the Triassic, Jurassic, and Cretaceous Periods.

extant
Still in existence.

feudalism
Refers to a system in Europe during the Middle Ages (ninth to fifteenth centuries) where the king awarded land grants to his loyal nobility and clergy in return for their contribution of soldiers for the king's army. The lowest echelon in this system was the serfs, who in exchange for living on and working the land were given the king's protection. Under this system, there was a widespread social and economic gulf between the landed gentry and the impoverished peasants who supported them with their labors.

flux
A chemical that assists in the fusing of metals while preventing oxidation of the metals, such as in welding and soldering. Also, lime added to molten metal in a furnace to absorb impurities and form a slag that can be removed.

forage
Refers to a type of plant (e.g., alfalfa) that provided a food source (fodder) for early humans as well as a number of animals during the Neolithic Period, at the advent of settled farming.

formalism
A strict adherence to accepted practices, particularly religious.

fossil
The remains, trace, or imprint of an animal or plant that has been preserved in the earth's crust since some past geologic age.

gold standard
The use of gold as a standard value for the currency of a country.

gymnasium
Established by the health-conscious ancient Greeks, gymnasiums were places, exclusively for males, where physical exercise was encouraged in the form of games and sports.

hydraulic
Something that is operated or affected by the action of a pressurized fluid of low viscosity, usually water.

intercalary
Refers to the insertion of a day or a month into the annual calendar to make it correspond to the solar year.

lunation
The elapsed time period between two successive new moons, averaging 29 days, 12 hours, 44 minutes, 28 seconds.

Medieval Warm Epoch
A period of time from the ninth through the fourteenth centuries C.E. when global temperatures reached levels 0.5–10 degrees Centigrade (33–50 degrees Fahrenheit) warmer than they are presently. The degrees of warmth varied from region to region.

mordant
An agent, such as tannic acid, that "fixes" color during the dyeing process.

morphology
The study of the structure and form—but not the functions—of living organisms.

necropolis
In antiquity, a large and elaborate burial ground that was part of a larger city or complex, such as one of the Egyptian pyramidal structures.

oviparous
Refers to those animals who produce eggs that hatch outside the body of the mother.

paradigm
In geometry, the general plan for the development of the logical statement. In science, it is referred to as a "ruling theory" or "dominant hypothesis."

parallax
The apparent change in direction and/or position of an object viewed through an optical instrument (e.g., sighting tube or modern telescope), which occurs due to the shifting position of the observer's line of sight.

pastoralism
The practice of herding and breeding domesticated animals.

perturbations
Variations of a celestial body, such as Earth, from its orbit as a result of the influence of one or more external bodies or forces acting upon it. In simple terms, it is the "wobble" of the planet on its axis.

physiology
The study of the vital activities, functions, and processes of living organisms that occur in their organs, tissues, and cells.

pleistocene
Refers to the period beginning two or more million years ago to just 8,000 to 10,000 years ago during which time glaciers periodically advanced and retreated, and during which time *Homo sapiens* migrated out of Africa.

postulate
To make a claim as a basis of an argument, or to assume something without proof as being self-evident or apparent.

pulses
The edible seeds of pod-bearing plants, such as beans or peas.

relief
The raised projection of figures or objects from a flat surface. In antiquity, large stones and marble were the preferred materials on which to carve reliefs. Reliefs were also part of ancient mapmaking.

sedentism
Living permanently in one place—that is, replacing the hunter/gatherer lifestyle with that of settled farming.

solstitial
Refers to the Sun's positions during the two annual solstices: *summer solstice,* when the Sun is at its highest point over the Tropic of Cancer (about June 21), and *winter solstice,* when the Sun is at its lowest point over the Tropic of Capricorn (about December 21).

stele
A stone or slab of marble, usually part of a monument or building, inscribed or sculpted with commemorative writing.

surfactant
Also called a surface-active agent. A soluble compound that reduces the surface tension of a liquid into which it is dissolved. A necessary ingredient of all soaps and detergents.

tallow
Animal fat obtained from cattle, sheep, and horses that, in antiquity, was a fuel for lamps. Also used in candles and soaps.

tempered
In metallurgy, refers to the degree of hardness achieved by heating metals to a specified temperature. Tempered steel is an example.

theorem
In mathematics, a proven proposition.

vectors
Quantities specified by magnitude *and* direction whose components convert from one coordinate system to another in the same manner as the components of a displacement. Vector quantities may be added and subtracted. For example, in the equation $F = ma$, the F stands for force, which is the vector quality that has both magnitude and direction and is calculated by the product of the m (mass) and a (acceleration) related to the "push" or "pull" on an object.

vermiparous
Producing or breeding worms.

viviparous
Refers to animals that give live birth after the offspring develops inside the body of the mother.

waned
Synonymous with declined. It refers to a gradual *decrease* in the illuminated visible surface of the Earth's Moon.

waxed
Opposite of waned. Refers to the gradual *increase* in the illuminated visible surface area of the Earth's Moon.

ziggurat
An ancient Mesopotamian (Assyrian, Babylonian) tower constructed with clay bricks and built in the form of a stepped pyramid. Ancient astrologers would climb to the top to view celestial events.

SELECTED BIBLIOGRAPHY

Ackerknecht, Erwin H., M.D. *A Short History of Medicine.* Baltimore: Johns Hopkins University Press, 1982.

Aristotle. *The Complete Works of Aristotle.* 2 vols. Jonathan Barnes, ed. Princeton, NJ: Princeton University Press, 1984.

Asimov, Isaac. *A Short History of Biology.* Westport, CT: Greenwood Press, 1980.

Balter, Michael. "From a Modern Human's Brow—or Doodling?" *Science* (January 11, 2002): n.p.

———. "What—or Who—Did in the Neanderthals?" *Science* (September 14, 2001): n.p.

Barlow, Connie. "Ghost Stories From the Ice Age." *Natural History* (September 2001).

Beckmann, Petr. *A History of π (PI).* New York: St. Martin's Press, 1971.

Bendick, Jeanne. *Archimedes and the Door of Science.* Bathgate, ND: Bethlehem Books, 1995.

Biemer, John. "Ancient Camp Found in Far North." Associated Press, September 6, 2001.

Boorstin, Daniel. *The Discoverers: A History of Man's Search to Know His World and Himself.* New York: Vintage Books, 1985.

Brown, Kathryn. "New Trips through the Back Alleys of Agriculture." *Science* (April 27, 2001): n.p.

Bunch, Bryan. *Handbook of Current Science and Technology.* Detroit: Gale, 1996.

Cambridge Illustrated History of Medicine. Roy Porter, ed. Cambridge, UK: Cambridge University Press, 1998.

Cambridge Paperback Encyclopedia. 2nd ed. David Crystal, ed. Cambridge, UK: Cambridge University Press, 1995.

Cassirer, Ernst. *Language and Myth.* Trans. Susanne K. Langer. New York: Harper and Brothers, 1946.

Cobb, Cathy, and Harold Goldwhite. *Creations of Fire: Chemistry's Lively History from Alchemy to the Atomic Age.* New York: Plenum Press, 1995.

Crombie, A. C. *The History of Science: From Augustine to Galileo.* New York: Dover, 1995.

Culotta, Elizabeth, Andrew Sugden, and Brooks Hanson. "Humans on the Move." *Science* (March 2, 2001): n.p.

Daintith, John, Sarah Mitchell, Elizabeth Tootill, and Derek Gjertsen. *Biographical Encyclopedia of Scientists.* 2nd ed. 2 vols. Bristol, UK: Institute of Physics Publishing, 1981–1994.

Davies, Paul. *About Time: Einstein's Unfinished Revolution.* New York: Simon & Schuster, 1995.

———. "That Mysterious Flow." *Scientific American* (September 2002): n.p.

De Camp, L. Sprague. *The Ancient Engineers.* New York: Ballantine, 1974.

Derry, T. K., and Trevor I. Williams. *A Short History of Technology: From the Earliest Times to A.D. 1900.* New York: Oxford University Press, 1960.

Dyson, James. *A History of Great Inventions.* New York: Carroll & Graf, 2001.

Ehrlich, Paul R. *Human Natures: Genes, Cultures and the Human Prospect.* Washington: Island Press, 2000.

Encyclopedia Britannica CD. Chicago: n.p., 2002.

Ferris, Timothy. *Coming of Age in the Milky Way.* New York: Doubleday, 1988.

Foster, Karen Polinger. "The Earliest Zoos and Gardens." *Scientific American* (July 1999): n.p.

Gibbons, Ann. "First Member of Human Family Uncovered." *Science* (July 12, 2002): n.p.

———. "Tools Show Humans Reached Asia Early." *Science* (September 28, 2001): n.p.

Glausiusz, Josie. "Caravaggio of the Caves." *Discover* (February 2002).

Hamey, L. A. and J. A. Hamey. *The Roman Engineers.* Cambridge, UK: Cambridge University Press, 1981.

Hawking, Stephen W. *A Brief History of Time: From the Big Bang to Black Holes.* New York: Bantam Books, 1988.

Hogben, Lancelot. *Mathematics for the Millions. How to Master the Magic of Numbers.* Rev. ed. New York: W. W. Norton & Company, 1993.

Humboldt, Wilhelm von. *On Language: On the Diversity of Human Language Construction and Its Influence on the Mental Development of the Human Species.* Michael Losonsky, ed. Cambridge, UK: Cambridge University Press, 1999.

James, Peter, and Nick Thorpe. *Ancient Inventions.* New York: Ballantine, 1994.

Kerr, Richard A. "Paleoclimate: A Variable Sun and the May Collapse." *Science* (May 18, 2001): n.p.

Krebs, Robert E. *Basics of Earth Science: A Reference Guide.* Westport, CT: Greenwood Press, 2003.

———. *History and Use of the Earth's Chemical Elements: A Reference Guide.* Wesport, CT: Greenwood Press, 1998.

———. *Scientific Development and Misconceptions through the Ages: A Reference Guide.* Westport, CT: Greenwood, 1999.

———. *Scientific Laws, Principles, and Theories: A Reference Guide.* Westport, CT: Greenwood, 2001.

Kunzig, Robert. "The Earliest Odyssey." *U.S. News and World Report* (April 8, 2002): n.p.

Landels, J. G. *Engineering in the Ancient World.* Berkeley: University of California Press, 1978.

Larson, Richard B., and Volker Bromm. "The First Stars in the Universe." *Scientific American* (December 2001): n.p.

Lawler, Andrew. "Report of Oldest Boat Hints at Early Trade Routes." *Science* (June 7, 2002): n.p.

———. "Writing Gets a Rewrite." *Science* (June 29, 2001): n.p.

Lennox, James G. *Aristotle's Philosophy of Biology: Studies in the Origins of Life Science.* Cambridge, UK: Cambridge University Press, 2001.

Leonard, Jennifer A., Robert K. Wayne, Jane Wheeler, Raül Valadez, Sonia Guillén, and Carlos Vilà. "Ancient DNA Evidence for Old World Origin of New World Dogs." *Science* (November 22, 2002): n.p.

Leonard, William R. "Food for Thought: Dietary Change Was a Driving Force in Human Evolution." *Scientific American* (December 2002): n.p.

Lewis, Richard J., Sr. *Hawley's Condensed Chemical Dictionary,.* 12th ed. New York: Van Nostrand Reinhold Co., 1993.

Lindberg, David C. *The Beginnings of Western Science: The European Scientific Tradition in Philosophical, Religious, and Institutional Context, 600 B.C. to A.D. 1450.* Chicago: University of Chicago Press, 1992.

Lives and Legacies: An Encyclopedia of People Who Changed the World. Doris Simonis, ed. Phoenix, AZ: Oryx Press, 1999.

Lloyd, G.E.R. *Science, Folklore and Ideology: Studies in the Life Sciences in Ancient Greece.* Cambridge, UK: Cambridge University Press, 1983.

Margotta, Roberto. *The History of Medicine.* Paul Lewis, ed. New York: Smithmark, 1996.

Margulis, Lynn, and Karlene V. Schwartz. *Five Kingdoms: An Illustrated Guide to the Phyla of Life on Earth.* 3rd ed. New York: W. H. Freeman and Company, 1982–1998.

Mason, Stephen F. *A History of the Sciences.* New York: Macmillan, 1962.

McClellan, James E., and Harold Dorn. *Science and Technology in World History: An Introduction.* Baltimore: The Johns Hopkins University Press, 1999.

Motz, Lloyd, and Jefferson Hane Weaver. *Conquering Mathematics: From Arithmetic to Calculus.* New York: Plenum Press, 1991.

Neanderthal. A Discovery Channel Production. 2001. Videocassette.

Nemecek, Sasha. "Who Were the First Americans?" *Scientific American* (September 2000): n.p.

Nordenskiöld, Erik. *The History of Biology: A Survey.* Trans. Leonard Bucknall Byre. New York: Alfred A. Knopf, 1928.

North, John. *Astronomy and Cosmology.* New York: Norton, 1995.

Oliphant, Margaret. *The Atlas of the Ancient World: Charting the Great Civilizations of the Past.* New York: Simon & Schuster, 1992.

Otto, Sarah P., and Philippe Jarne. "Haploids—Hapless or Happening?" *Science* (June 29, 2001): n.p.

Pannekoek, A. *A History of Astronomy.* New York: Dover, 1961.

Pinker, Steven. *The Blank Slate: The Modern Denial of Human Nature.* New York: Viking, 2002.

Pope, K. O., M.E.D. Pohl, J. G. Jones, D. L. Lentz, C. von Nagy, F. J. Vega, I. R. Quitmeyer. "Origin and Environmental Setting of Ancient Agriculture in the Lowlands of Mesoamerica." *Science* (May 18, 2001): n.p.

Roberts, J. M. *History of the World.* New York: Oxford University Press, 1993.

Romm, James S. *The Edges of the Earth in Ancient Thought: Geography, Exploration, and Fiction.* Princeton, NJ: Princeton University Press, 1992.

Russelle, Michael P. "Alfalfa." *American Scientist* (Volume 89, May/June 2001): n.p.

Sarton, George. *Hellenistic Science and Culture: In the Last Three Centuries B.C.* New York: Dover, 1987.

Savolainen, Peter, Ya-ping Zhang, Jing Luo, Joakim Lundeberg, and Thomas Leitner. "Genetic Evidence for an East Asian Origin of Domestic Dogs." *Science* (November 22, 2002): n.p.

Science and Technology Desk Reference. Compiled by the Science and Technology Department of the Carnegie Library of Pittsburgh. Detroit: Gale, 1996.

Singer, Charles. *A History of Scientific Ideas.* New York: Barnes & Noble, 1966.

Sobel, Dava. *Longitude: The True Story of a Lone Genius Who Solved the Greatest Scientific Problem of His Time.* New York: Penguin Books, 1995.

Stengel, Marc K. "The Diffusionists Have Landed." *The Atlantic Monthly* (January 2000): n.p.

Struik, Dirk J. *A Concise History of Mathematics.* 4th ed. New York: Dover, 1987.

Sun Tzu. *The Art of War.* Trans. Samuel B. Griffith. London: Oxford University Press, 1963.

Temple, Robert. *The Genius of China: 3000 Years of Science, Discovery & Invention.* London: Prion, 1986.

Tudge, Colin. *The Time Before History: 5 Million Years of Human Impact.* New York: Simon & Schuster, 1996.

Usher, Abbott Payson. *A History of Mechanical Inventions.* Rev. ed. New York: Dover, 1988.

Vergano, Dan. "Ancient Writing System Gets Internet Update." *USA Today* (May 21, 2002).

Vogel, Steven. "A Short History of Muscle-Powered Machines." *Natural History* (March 2002): n.p.

Williams, Suzanne. *Made in China: Ideas and Inventions from Ancient China.* Berkeley, CA: Pacific View Press, 1996.

Williams, Trevor I. *A History of Invention: From Stone Axes to Silicon Chips.* Rev. William E. Schaaf, Jr. New York: Checkmark Books, 2000.

Woods, Michael, and Mary B. Woods. *Ancient Construction: From Tents to Towers.* Minneapolis: Runestone Press, 2000.

—— *Ancient Machines: From Wedges to Waterwheels.* Minneapolis: Runestone Press, 2000.

—— *Ancient Transportation: From Camels to Canals.* Minneapolis: Runestone Press, 2000.

Wright, Karen. "Neanderthals Like Us." *Discover* (March 2002): n.p.

Online Sources

Acupuncture: A History: <http://www.hrc.org>

Ancient Greek Methods of Boating and Shipping: <http://www-adm.pdx.edu>

Ancient Roman Transportation: <http://www.crystalinks.com>

Antiqua Medicina:<http://www.med.Virginia.edu>
Aristotle and His Study of Animals: <http://www.library.scar.utoronto.ca>
Aristotle and the Great Chain: <http://www.cod.edu>
Brief History of Ancient Greek Medicine: <http://www.historyforkids.org>
Calendars through the Ages: <http://webexhibits.org/calendars>
David Joyce's History of Mathematics: <http://alephO.clark.edu>
Development of Mathematics in Ancient China: <http://saxakali.com>
Doctor in Roman Society: <http://www.med.virginia.edu>
Domestication of Animals: <http://www.le.ac.uk>
Egypt Web Search: <http://www.egyptmonth.com>
Encyclopedia of Philosophy:
<http://www.utm.edu/research/eip/g/greekphi.htm>
Etruscan and Roman Medicine: <http://www.med.virginia.edu>
Foundations of Hippocratic Medicine: <http://www.Indiana.edu>
Glass History: <http://www.cornucopia-of-colors.com>
Great Harness Controversy: <http://scholar.chem.nyu.edu>
Great Pyramid of Giza: <http://ce.eng.usf.edu>
History and Use of Roman Numerals: <http://www.deadline.demon.co.uk>
History of Alchemy: <http://www.alchemylab.com/history>
History of Plumbing: <http://www.theplumber.com>
History of Science: <http://biology.clc.uc.edu>
History of Soapmaking: <http://www.simplysoap.com>
History of the Abacus: <http://www.gol.com>
History of Zero: <http://www-groups.dcs.st-and.ac.uk>
History Topics Index: <http://www-groups.dcs.pt-and.ac.uk>
Mayan Prophecies: <http://www.knowledge.co.uk/xxx/cat/mayan>
Medicine in Ancient Egypt: <http://www.indiana.edu>
Mule: <http://www.imh.org>
Mystic Places: Stonehenge:
<http://exn.ca/mysticplaces/Construction.ctm>
Oared Warship Terms: <http://www.bright.net>
Phoenicia: <http://www.sarisssa.org>
Phoenician Ships, Navigation and Commerce: <http://phoenicia.org>
Primitive Forms: <http://www.wischik.com>
Process of Domestication: <http://www.mc.maricopa.edu>
Ptolemaic System: <http://es.rice.edu>
The Silk Road: <http://www.allaboutturkey.com>
To What Extent Is Modern Medical Theory and Practice Influenced by Its More Primitive Forms? <http://www.wischik.com/marcus/essay/med2.html>
Women in Medicine: <http://med.virginia.edu>

Name Index

Subject Index

About the Authors

ROBERT E. KREBS is retired Associate Dean for Research at the University of Illinois Health Sciences. He is also a former science teacher, science specialist for the U.S. Government, and university research administrator.

CAROLYN A. KREBS began her literary career in the textbook division of the McGraw-Hill Book Company in New York City. Before retiring to South Padre Island to assist her husband, Robert, in writing science books, she was the Executive Director of the Fort Wayne, Indiana Medical Society.

BOCA RATON PUBLIC LIBRARY, FLORIDA
3 3656 0536710 4